Advances in
PARASITOLOGY

VOLUME 43

Editorial Board

C. Bryant Division of Biochemistry and Molecular Biology, The Australian National University, Canberra, ACT 0200, Australia

M. Coluzzi Director, Istituto di Parassitologia, Università Degli Studi di Roma 'La Sapienza', P. le A. Moro 5, 00185 Roma, Italy

C. Combes Laboratoire de Biologie Animale, Université de Perpignan, Centre de Biologie et d'Ecologie Tropicale et Méditerranéenne, Avenue de Villeneuve, 66860 Perpignan Cedex, France

W.H.R. Lumsden 16A Merchiston Crescent, Edinburgh, EH10 5AX, UK

J.J. Shaw Instituto de Ciências Biomédicas, Universidade de São Paulo, av. Prof. Lineu Prestes, 1374, 05508-900, Cidade Universitária, São Paulo, SP, Brazil

Lord Soulsby of Swaffham Prior Department of Clinical Veterinary Medicine, University of Cambridge, Madingley Road, Cambridge, CB3 0ES, UK

K. Tanabe Laboratory of Biology, Osaka Institute of Technology, 5-16-1 Ohmiya, Asahi-Ku, Osaka 535, Japan

P. Wenk Falkenweg 69, D-72076 Tübingen, Germany

Advances in
PARASITOLOGY

Edited by

J.R. BAKER

*Royal Society of Tropical Medicine and Hygiene,
London, England*

R. MULLER

*International Institute of Parasitology,
St Albans, England*

and

D. ROLLINSON

*The Natural History Museum,
London, England*

VOLUME 43

ACADEMIC PRESS

San Diego   London   Boston
New York   Sydney   Tokyo   Toronto

ACADEMIC PRESS
a division of Harcourt Brace & Company
525 B Street, Suite 1900, San Diego,
California 92101–4495, USA
http://www.apnet.com

ACADEMIC PRESS
24–28 Oval Road
LONDON NW1 7DX
http://www.hbuk.co.uk/ap/

Copyright © 1999, by
ACADEMIC PRESS

This book is printed on acid-free paper

All Rights Reserved
No part of this publication may be reproduced or transmitted in any form
or by any means, electronic or mechanical, including photocopy,
recording, or any information storage and retrieval system,
without permission in writing from the publisher

A catalogue record for this book is available from the British Library

ISBN 0-12-031743-5

Typeset by Mathematical Composition Setters Ltd, Salisbury, Wiltshire
Printed in Great Britain by MPG, Bodmin, Cornwall

99 00 01 02 03 04 MP 9 8 7 6 5 4 3 2 1

CONTRIBUTORS TO VOLUME 43

P.J. BRINDLEY, *Molecular Parasitology Unit, and Australian Centre for International & Tropical Health & Nutrition, Queensland Institute of Medical Research, Post Office, Royal Brisbane Hospital, Queensland 4029, Australia*

J.P. DALTON, *School of Biological Sciences, Dublin City University, Dublin 9, Republic of Ireland*

W. GIBSON, *School of Biological Sciences, University of Bristol, Woodland Road, Bristol, BS8 1UG, UK*

P. GÖTZ, *Institute of Zoology, Free University of Berlin, Königin-Luise-Str. 1-3, 14195 Berlin, Germany*

A. HEMPHILL, *Institute of Parasitology, Faculties of Veterinary Medicine and Medicine, University of Bern, Langass-Strasse 122, CH-3012 Bern, Switzerland*

D. KNOX, *Moredun Research Institute, 408 Gilmerton Road, Edinburgh, EH17 7JH, UK*

P. ROSENTHAL, *Department of Medicine, Box 0811, San Francisco General Hospital, University of California, San Francisco, CA 94143-0811, USA*

J. STEVENS, *Department of Biological Sciences, University of Exeter, Exeter, EX4 14PS, UK*

J. TORT, *School of Biological Sciences, Dublin City University, Dublin 9, Republic of Ireland*

A. VILCINSKAS, *Institute of Zoology, Free University of Berlin, Königin-Luise-Str. 1-3, 14195 Berlin, Germany*

K.H. WOLFE, *Department of Genetics, University of Dublin, Trinity College, Dublin 2, Republic of Ireland*

PREFACE

This volume opens with an account by Wendy Gibson and Jamie Stevens, of the University of Bristol (UK), of what is known about sexual reproduction in the family Trypanosomatidae. The occurrence of sex in this group of parasitic flagellates is a relatively new discovery. When the late Cecil Hoare was writing his monograph on the trypanosomes of mammals about 25 years ago, he stated that 'All the available evidence indicates that the reproduction of trypanosomes throughout their life cycle ... is asexual' (Hoare, C.A. (1972): *The Trypanosomes of Mammals*. Oxford and Edinburgh: Blackwell Scientific Publications, pp. 48–49), although from very early in the history of 'trypanosomatology' various accounts of alleged sexual processes were reported in the literature. Hoare was rightly sceptical of these early accounts, referring to the authors' 'obsession with sexuality', 'fantastic views' and 'highly fanciful' descriptions. However, Gibson and Stevens discuss the careful work, in which the senior author played a significant part, beginning with the crossing experiments reported by Leo Jenni and his co-workers in 1986, which eventually led to general acceptance of the occurrence of genetic exchange — at least in the species *Trypanosoma brucei* and very probably in other species and genera as well, by a process which bore no relationship to the 'fanciful' processes described by the earlier authors. It remains true, however, that no one has yet succeeded in visualizing the process involved, so its mechanism remains unclear.

Andrew Hemphill, of the University of Bern (Switzerland), reviews the protozoan parasite, *Neospora caninum*. This genus was erected as recently as 1988 to describe an apicomplexan parasite, antigenically distinct from *Toxoplasma*, causing neuromuscular disease in dogs. The same parasite was then found to be responsible for abortion in cattle, and it has now been reported from other herbivores in many parts of the world. The life cycle has very recently been elucidated. The author describes new sensitive and specific molecular tools for the detection of the parasite which should help to determine its importance as a cause of disease in animals, and possibly in humans.

The volume continues with two linked reviews of the proteolytic enzymes of parasites. First, the proteases of almost all the medically important protozoan parasites of humans are discussed by Philip Rosenthal of the University of California, San Francisco (USA), who finds that almost all express multiple proteases. Cysteine proteases, which act either intracellularly or extracellularly in lysosome-like organelles, are present in all, and the importance of these and of other proteases in the life cycle of the parasites suggests that they may present a powerful new target for antiparasitic chemotherapy.

We are also extremely fortunate to have a chapter dedicated to the proteinases of parasitic helminths written by key workers in this field: Jose Tort, Kenneth Wolfe and John Dalton from Dublin, Republic of Ireland, Paul Brindley from Brisbane, Australia and Dave Knox from Edinburgh, Scotland.

The proteinases of helminths, and their associated genes, have been under close scrutiny in recent years because of their recognized importance in many aspects of the parasitic way of life. Proteinases are involved in tissue penetration, digestion and evasion of host immune responses. In this review, particular attention is given to digenean trematodes of medical and economic importance, cestodes and nematodes parasitizing animals and plants. Interesting points emerge from the wealth of literature reviewed, and of the four major groups of peptidases — serine, aspartic, cysteine and metalloproteinases — it is the papain superfamily of cysteine proteinases which has been most characterized from parasitic worms. The final section of the review is devoted to an in-depth phylogenetic analysis of this group of proteinases, which includes a comparison with similar proteinases from protozoan parasites.

The volume ends with an account by Andreas Vilcinskas and Peter Götz, from the Free University of Berlin (Germany), of parasitic fungi which infect insects. The authors examine the ways in which these two groups of disparate organisms interact at the molecular level, concentrating on fungal molecules involved in virulence and on insect molecules involved in the humoral immune response. Study of this rather unusual model system has shed new light on mechanisms of fungal pathogenesis, and the authors hope that their studies may be extended to the molecular interactions between insects which act as vectors and the parasites which they transmit to humans and other animals.

J.R. Baker
R. Muller
D. Rollinson

CONTENTS

CONTRIBUTORS TO VOLUME 43 v
PREFACE vii

Genetic Exchange in the Trypanosomatidae

W. Gibson and J. Stevens

Abstract	2
1. Introduction	2
2. Laboratory Experiments on Genetic Exchange	3
3. Evidence from Population Genetics and Evolutionary Studies .	20
4. The Biological Context	31
Acknowledgements	36
References	36

The Host–Parasite Relationship in Neosporosis

A. Hemphill

Abstract	48
1. Introduction	48
2. Historical Background and Phylogenetic Status of *Neospora caninum*	49
3. The Biology of *Neospora caninum*	51
4. Neosporosis	57
5. Host–Parasitic Interactions	67
6. Concluding Remarks	89
Acknowledgements	89
References	89

Proteases of Protozoan Parasites

P.J. Rosenthal

Abstract	106
1. Introduction	106
2. Classification of Proteases	106
3. Survey of Identified Protozoan Proteases	108
4. Protease Inhibitors as Antiprotozoan Drugs	110
5. *Leishmania*	110
6. African Trypanosomes	117
7. *Trypanosoma cruzi*	119
8. Malaria Parasites	122
9. *Entamoeba histolytica*	131
10. *Giardia lamblia*	135
11. *Cryptosporidium parvum*	136
12. *Trichomonas vaginalis*	137
13. *Toxoplasma gondii*	138
14. Summary	139
Acknowledgements	139
References	139

Proteinases and Associated Genes of Parasitic Helminths

J. Tort, P.J. Brindley, D. Knox, K.H. Wolfe and J.P. Dalton

Abstract	162
1. Introduction	163
2. Digenean Trematodes	164
3. Proteinases of Schistosomes	165
4. Proteinases of *Fasciola hepatica* and Other Fasciolidae	190
5. Proteinases of Other Flukes	194
6. Proteinases of Cestodes	196
7. Proteinases of Parasitic Nematodes	200
8. Phylogenetic Analysis of Cysteine Proteinases of the Papain Superfamily	225
9. Concluding Remarks	244
References	247

Parasitic Fungi and their Interactions with the Insect Immune System

A. Vilcinskas and P. Götz

Abstract 268
1. Insect Immunity 268
2. Entomopathogenic Fungi 270
3. Fungal Factors Determining Virulence 280
4. Humoral Immune Reactions of Insects Against Fungi 290
5. Comparisons with Other Pathogens and Parasites of Insects . . 301
References 303

INDEX 315

CONTENTS OF VOLUMES IN THIS SERIES 325

Genetic Exchange in the Trypanosomatidae

Wendy Gibson[1] and Jamie Stevens[2]

[1]*School of Biological Sciences, University of Bristol, Woodland Road, Bristol, BS8 1UG, UK, and*
[2]*Department of Biological Sciences, University of Exeter, Exeter, EX4 14PS, UK*

Abstract	2
1. Introduction	2
2. Laboratory Experiments on Genetic Exchange	3
2.1. Experimental crosses of *Trypanosoma brucei*	3
2.2. Other trypanosomatids	19
3. Evidence from Population Genetics and Evolutionary Studies	20
3.1. Introduction	20
3.2. Methods	21
3.3. Evidence for genetic exchange from population genetics analysis	25
4. The Biological Context	31
4.1. Implications for epidemiology and control of disease	31
4.2. Experimental versus population genetics analysis of *T. brucei*	32
4.3. Phylogenetic perspective	34
Acknowledgements	36
References	36

ABSTRACT

The only trypanosomatid so far proved to undergo genetic exchange is *Trypanosoma brucei*, for which hybrid production after co-transmission of different parental strains through the tsetse fly vector has been demonstrated experimentally. Analogous mating experiments have been attempted with other *Trypanosoma* and *Leishmania* species, so far without success. However, natural *Leishmania* hybrids, with a combination of the molecular characters of two sympatric species, have been described amongst both New and Old World isolates. Typical homozygotic and heterozygotic banding patterns for isoenzyme and deoxyribonucleic acid markers have also been demonstrated amongst naturally-occurring *T. cruzi* isolates.

The mechanism of genetic exchange in *T. brucei* remains unclear, although it appears to be a true sexual process involving meiosis. However, no haploid stage has been observed, and intermediates in the process are still a matter for conjecture. The frequency of sex in trypanosomes in nature is also a matter for speculation and controversy, with conflicting results arising from population genetics analysis.

Experimental findings for *T. brucei* are discussed in the first section of this review, together with laboratory evidence of genetic exchange in other species. The second section covers population genetics analysis of the large body of data from field isolates of *Leishmania* and *Trypanosoma* species. The final discussion attempts to put the evidence from experimental and population genetics into its biological context.

1. INTRODUCTION

Trypanosomatids are unicellular, obligate parasites found in a wide variety of invertebrates and vertebrates in nature. Some of these flagellates have monogenetic life cycles, but the notorious members of the family are the digenetic species responsible for the vector-borne diseases leishmaniasis and trypanosomiasis, which affect humans and domestic livestock in the warmer regions of the world.

Genetic exchange in trypanosomatids has been unequivocally demonstrated in only one species, *Trypanosoma brucei*, and these laboratory findings will be discussed in the first section of this review, together with experimental evidence of genetic exchange in other species. The second section covers population genetics analysis of the large body of data from field isolates of *Leishmania* and *Trypanosoma* species. Our final discussion attempts to put the evidence from experimental and population genetics into its biological context.

2. LABORATORY EXPERIMENTS ON GENETIC EXCHANGE

2.1. Experimental Crosses of *Trypanosoma brucei*

2.1.1. *Experimental Design*

The first successful laboratory cross was carried out by Jenni *et al.* (1986), who co-transmitted two clones of *Trypanosoma brucei* ssp. through tsetse flies (*Glossina* spp.) and demonstrated hybrid progeny among the metacyclic forms from the salivary glands. The two parent clones were distinguishable by isoenzymes and restriction fragment length polymorphisms (RFLPs), and the cloned hybrid progeny had inherited a mixture of markers from both parents. Further crosses followed this general scheme (Figure 1). Briefly, equal numbers of the two parental trypanosome clones were mixed and fed to groups of teneral tsetse flies. Not all flies became infected after the infected feed, and only some infected flies produced hybrids. Thus large numbers of flies and trypanosome populations needed to be screened to identify those containing hybrids. Mating is non-obligatory, so the metacyclic population from the salivary glands may contain a mixture of parents and hybrids. Before the use of selectable markers, hybrid-containing populations were identified by the presence of either heterodimeric isoenzyme bands unique to hybrids (Jenni *et al.*, 1986) or non-parental karyotypes after pulsed field gel electrophoresis (PFGE; Gibson, 1989).

The development of methods for the stable transformation of trypanosomes with exogenous deoxyribonucleic acid (DNA) (Lee and Van der Ploeg, 1990; Ten Asbroek *et al.*, 1990; Eid and Sollner-Webb, 1991), led to a second approach using selectable markers. In the cross described by Gibson and Whittington (1993), each of the parental clones was transformed with a different construct designed to integrate a gene for drug resistance into the tubulin locus by homologous recombination. In this way, parental clones resistant to the antibiotics hygromycin or G418 were created. After co-transmission through the fly, hybrid progeny were selected by resistance to both drugs. This strategy has obvious advantages over the previous 'finding a needle in a haystack' approach. However, since only one of the allelic tubulin arrays carried a drug resistance marker, and assuming Mendelian inheritance, only a quarter of the hybrid progeny would be expected to be doubly resistant and distinguishable from the parents by selection.

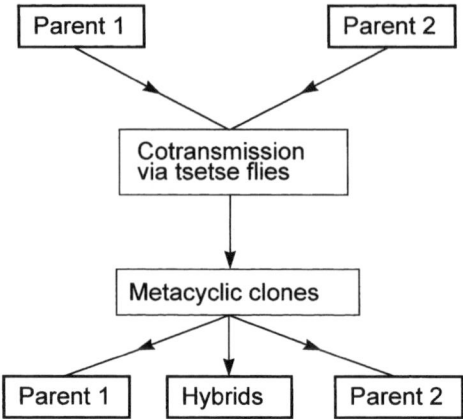

Figure 1 General scheme of an experimental cross of *Trypanosoma brucei*. Genetic exchange is non-obligatory and so both hybrids and parents are found among the metacyclics.

Little success with generation of hybrids *in vitro* has been reported, despite the use of selectable markers (Gibson and Whittington, 1993). Schweizer *et al.* (1991) observed a heterodimeric isoenzyme band after long-term cocultivation of procyclics of each homozygotic type, but did not succeed in isolating the putative hybrid trypanosomes. Later work (see Section 2.1.3) has indicated that genetic exchange probably takes place in the salivary glands of the fly, and thus the failure to reproduce the process *in vitro* probably results from the difficulty of growing these life-cycle stages outside the fly.

2.1.2. *Mating Compatibility*

A number of different crosses followed the pioneering experiment of Jenni and colleagues (1986), using all combinations of subspecies except Group 1 *T. b. gambiense* (as defined by Gibson, 1986); this is the classical avirulent type of *T. b. gambiense*, which is very difficult to transmit through tsetse flies in the laboratory. Most of the parental isolates used in these crosses originated from the two sides of Africa, an experimental design chosen to maximize allelic differences between the parents. However, Schweizer *et al.* (1994) crossed two parental isolates from East Africa (Tanzania and Uganda) and Degen *et al.* (1995) used two sympatric *T. b. brucei* isolates from Uganda.

From Table 1, which lists the laboratory crosses carried out to date, it appears that there is no barrier to genetic exchange between different

Table 1 Published experimental crosses of *T. brucei* spp.

	Parents[a]			References
A	*T. b. brucei* STIB 247	×	*T. b. gambiense* Group 2[b] STIB 386	Jenni et al., 1986; Paindavoine et al., 1986a; Wells et al., 1987; Sternberg et al., 1988, 1989
B	*T. b. brucei* STIB 247	×	*T. b. brucei* TREU 927/4	Turner et al., 1990
C	*T. b. brucei* TREU 927/4	×	*T. b. gambiense* Group 2[b] STIB 386	Turner et al., 1990
D	*T. b. brucei* STIB 247	×	*T. b. brucei* STIB 777	Schweizer et al., 1994
E	*T. b. brucei* TSW 196	×	*T. b. rhodesiense* 058	Gibson, 1989; Gibson & Garside, 1990; Gibson et al., 1992
F	*T. b. brucei* TSW 196	×	*T. b. brucei* J10	Gibson & Garside, 1991
G	*T. b. brucei* KP2N	×	*T. b. rhodesiense* 058H	Gibson & Whittington, 1993; Gibson & Bailey, 1994; Gibson et al., 1997a
H	*T. b. rhodesiense* 058H	×	*T. brucei* spp. P20[c]	Gibson et al., 1995
I	*T. b. brucei* STIB 826	×	*T. b. brucei* STIB 829	Degen et al., 1995
J	*T. b. rhodesiense* 058H	×	*T. b. gambiense* Group 2[b] TH2	Gibson et al., 1997a
K	*T. b. brucei* KP2N	×	*T. b. gambiense* Group 2[b] TH2	Gibson et al., 1997a

[a] Trypanosome origins: STIB 247, hartebeest, Tanzania, 1971; STIB 386, human, Côte d'Ivoire, 1978; TREU 927/4, tsetse, Kenya, 1970; STIB 777, tsetse, Uganda, 1971; TSW 196, pig, Côte d'Ivoire, 1978; 058, 058H, human, Zambia, 1974; J10, hyena, Zambia, 1973; KP2N, tsetse, Côte d'Ivoire, 1982; STIB 826, 829, tsetse, Uganda, 1990.
[b] Group 2 virulent *T. b. gambiense* as defined by Gibson (1986).
[c] F1 hybrid from KP2N × 058H cross (Gibson and Bailey, 1994).

subspecies or strains of *T. brucei*. Other lower eukaryotes have multiple mating types, designed to favour out-crossing, but it is as yet unclear what factors, if any, control compatibility in trypanosomes. The three-way cross carried out by Turner and colleagues (1990) to give all pairwise combinations of three stocks ruled out the possibility of a simple two-sex mating system. Evidence for intraclonal mating has been found in two independent crosses but, in both cases, it was in the presence of out-crossing trypanosomes. Tait *et al.* (1996) described five clones, which were identical to parent 1 for homozygous isoenzyme markers, but showed homozygosity at one or more loci where parent 1 was heterozygous. When heterozygous clones were singly transmitted through tsetse flies, similar reduction of heterozygosity to homozygosity was not found (Tait *et al.*, 1989). Gibson *et al.* (1997a) carried out a series of intraclonal and out-crosses, using parental trypanosomes with drug resistance markers to allow selection of hybrid progeny by double drug resistance. While three of the out-crosses successfully produced hybrids, none of the three intraclonal crosses gave rise to doubly drug-resistant progeny. However, further analysis of cloned progeny from one of the out-crosses using a battery of molecular markers revealed five trypanosome clones which were unequivocal products of intraclonal mating. These clones had no demonstrable input of genetic material from one parent and were identical to the other parent for most markers, except for five loci at which they were homozygous in contrast to the heterozygous parent. The authors concluded that intraclonal mating occurred only during out-crossing, and suggested that mating might be triggered by a diffusible factor.

2.1.3. *Location and Timing of Genetic Exchange in the Tsetse Fly*

Jenni originally showed that genetic exchange took place in the fly, since cloned metacyclic forms from the saliva of infected flies were hybrid (Jenni *et al.*, 1986); however, the exact location and life-cycle stage remained undefined. Schweizer and colleagues found a heterodimeric isoenzyme band suggestive of hybrid formation in mixed midgut procyclics both *in vitro* and *in vivo*, but did not succeed in cloning these hybrids (Schweizer and Jenni, 1991; Schweizer *et al.*, 1991). On the other hand, Gibson and Whittington (1993), using selectable markers, were able to demonstrate unequivocally that doubly drug-resistant hybrids could be retrieved from the salivary gland, but not midgut, trypanosome populations of single flies. In this experiment, batches of flies were dissected from weeks two to nine after the infected feed, and the drug resistance of trypanosome populations from the midguts and salivary glands was examined separately. No doubly drug-resistant hybrid was found in the midgut populations throughout the experiment, whereas, from week eight, doubly drug-resistant hybrids appeared in the salivary gland populations.

Although the evidence looks convincing that genetic exchange takes place only after trypanosomes reach the salivary glands, the results are also consistent with migration to the salivary glands of a cell cycle-arrested zygote produced in the midgut. This issue will be resolved only when intermediates in the process can be identified and visualized directly. In this context, it is appropriate to mention the electron microscope studies of Ellis et al. (1982), who found giant multinucleate cysts containing multiple organelles, including nuclei and kinetoplasts, in midgut stages of *T. brucei*; cysts were observed both in trypanosomes grown in culture *in vitro* and also inside tsetse midgut cells. With salivary gland forms, Nyindo et al. (1981) observed by phase contrast microscopy and electron microscopy close apposition between pairs of metatrypomastigotes and metacyclics *in vitro* in cultures initiated from infected tsetse flies. It remains to be seen whether further studies will prove these unusual forms to be intermediates in genetic exchange.

Hybrids are most likely to be found in flies infected for at least 28 days (Schweizer et al., 1988, 1994; Gibson and Whittington, 1993; Gibson et al., 1997a). Production of hybrids *in vivo* in infected flies was followed by Schweizer et al. (1988) from 12 to 52 days after infection by feeding individual flies on mice and characterizing the resultant trypanosome populations by isoenzyme electrophoresis. Nine of 23 flies with salivary gland infections secreted hybrids at some point during the course of the experiment, but hybrid secretion did not appear to be continuous, since pure populations of one or both parents were also found from time to time. Similar results were obtained by Schweizer et al. (1994) with a different cross. Turner et al. (1990) carried out three crosses, resulting in 12 hybrid populations in a total of 48 salivary gland infections. In Gibson and Whittington's experiment (1993) using selectable drug resistance markers, hybrids were detected only in flies infected for at least eight weeks, although salivary gland infections were present from week six onwards. Of 38 flies infected for at least eight weeks, 22 had salivary gland infections, of which five contained hybrids; the remaining 17 infections consisted of 12 single-parent infections and five mixtures without doubly drug-resistant hybrids. The compiled results for the three crosses described by Gibson et al. (1997a) gave a total of three flies with doubly drug-resistant hybrids among 26 with salivary gland infections at the end of the experiment (56 to 60 days after infection). Overall, only about one quarter of infected flies produced hybrids in these four sets of experiments (29/119 or 24%).

However, if we assume that genetic exchange takes place in the salivary glands, there is no chance of finding hybrids in a fly in which only a single parent has reached the salivary glands. Considering only flies with mixed salivary gland infections, the overall frequency of hybrid formation becomes 29/42 or 69%. In some experiments the two parental clones have clearly been unequal in their rate of development into salivary gland forms. In the cross performed by Schweizer *et al.* (1988), although parent 1 reached the metacyclic stage after 20 days, parent 2 took only 13 days. Similarly, in the cross reported by Gibson and Whittington (1993), salivary gland infections consisted only of parent 058 until eight weeks after infection, when the other parent, KP2N, also appeared. It is noteworthy that, in the three successful crosses carried out by Gibson *et al.* (1997a), the parents were well matched in developmental rate, while each pair of parents in the three crosses that failed differed by seven days in appearance of metacyclic forms. On balance, it seems reasonable to conclude that mating can be a frequent event in flies where both parental trypanosomes are present in the salivary glands, but this is often militated against by other circumstances.

2.1.4. *Clues to the Mechanism of Genetic Exchange*

(a) *Evidence for Mendelian inheritance of markers.* It is generally accepted that *T. brucei* is diploid, although this concept probably does not apply to the arrangement of single-copy, variant surface glycoprotein (VSG) genes and smaller chromosomes (Van der Ploeg *et al.*, 1984; Gibson *et al.*, 1985; Gottesdiener *et al.*, 1990). Therefore only the inheritance of certain markers, such as isoenzymes or RFLPs in housekeeping genes, would be expected to conform to the Mendelian pattern. Since most crosses have used parental clones with a high degree of homozygosity to maximize the levels of difference, hybrid progeny are, unsurprisingly, heterozygous at most loci. This is consistent with Mendelian inheritance, but also with other models such as fusion.

More informative evidence has come from the segregation of alleles at loci for which one or both parents are heterozygous, and this has been observed in several independent crosses (Sternberg *et al.*, 1988, 1989; Gibson, 1989; Turner *et al.*, 1990; Gibson and Garside, 1991; Gibson and Bailey, 1994; Schweizer *et al.*, 1994). However, two of these crosses also showed non-Mendelian inheritance of markers, as some hybrid progeny were homozygous instead of heterozygous as expected for markers where the parents were homozygous variants (Sternberg *et al.*, 1988; Gibson and Bailey, 1994). No satisfactory explanation has been put forward for these exceptions, other than gene conversion or multiple rounds of mating.

By far the greatest, and potentially the most significant, anomaly, considering straightforward Mendelian genetics, is the occurrence of a large number of hybrids found to have high DNA contents relative to the parents. This is dealt with in the next section.

(b) *Changes in ploidy.* The three original hybrid clones described by Jenni and colleagues were found to have much higher DNA contents than the parents (Paindavoine *et al.*, 1986a; Wells *et al.*, 1987). The DNA content of some of the hybrid clones was found to be unstable, which led to the hypothesis that genetic exchange involved fusion of parental trypanosomes followed by gradual loss of chromosomes to return to the diploid state (Paindavoine *et al.*, 1986a). However, several hybrid clones analysed subsequently were found to have DNA contents consistent with diploidy (Sternberg *et al.*, 1988, 1989; Turner *et al.*, 1990), and the original clones were relegated to the status of artifacts.

This was not the end of the story, as similar hybrids with high DNA contents reappeared in other crosses (Figure 2). Gibson *et al.* (1992) carried out detailed RFLP, isoenzyme and karyotype analysis of two such hybrid clones, showing them to have three alleles at several loci for housekeeping genes (phosphoglycerate kinase, aldolase, glucose phosphate isomerase, phospholipase C, superoxide dismutase, isocitrate dehydrogenase) and trisomy for two chromosomes carrying, respectively, the tubulin and phospholipase C genes. These results, coupled with DNA contents 1.5 times the parental values, strongly suggested that these clones were triploid. Both these clones originated from a single infected fly but, in a later cross employing drug-resistant parental trypanosomes (Gibson and Whittington, 1993), all five flies containing hybrids produced triploids, exclusively so in the case of three flies (Gibson and Bailey, 1994). A backcross between one of the diploid hybrid progeny from this cross and one parent also produced only triploid hybrid progeny (Gibson *et al.*, 1995), and similar triploid hybrid progeny were observed in a further cross (Gibson *et al.*, 1997a). Overall, five of six independent crosses, in which the DNA contents of hybrid progeny clones were measured, had hybrid clones with raised DNA contents consistent with triploidy (crosses A, E, G, H and J, Table 1; Paindavoine *et al.*, 1986a; Wells *et al.*, 1987; Gibson *et al.*, 1992, 1995, 1997a; Gibson and Bailey, 1994). Two of the five crosses which produced triploids used wild type parents, while the remaining three used genetically transformed parental trypanosomes.

Such a large proportion of supposedly aberrant hybrids demands an explanation in any proposed mechanism of genetic exchange. If subtetraploid hybrids were the product of simple fusion of parental cells followed by random chromosome loss, as suggested by Paindavoine *et al.* (1986a), a range of DNA contents between the 2n and 4n values would have been expected. Apart from the early report of Paindavoine *et al.* (1986a), triploids appear to be stable during growth and fly transmission (Wells *et al.*, 1987; Gibson *et al.*, 1992, 1997a). Figure 2 also shows the DNA contents of 12 progeny clones and parents from the cross described by Gibson *et al.* (1997a). There are slight but consistent differences in

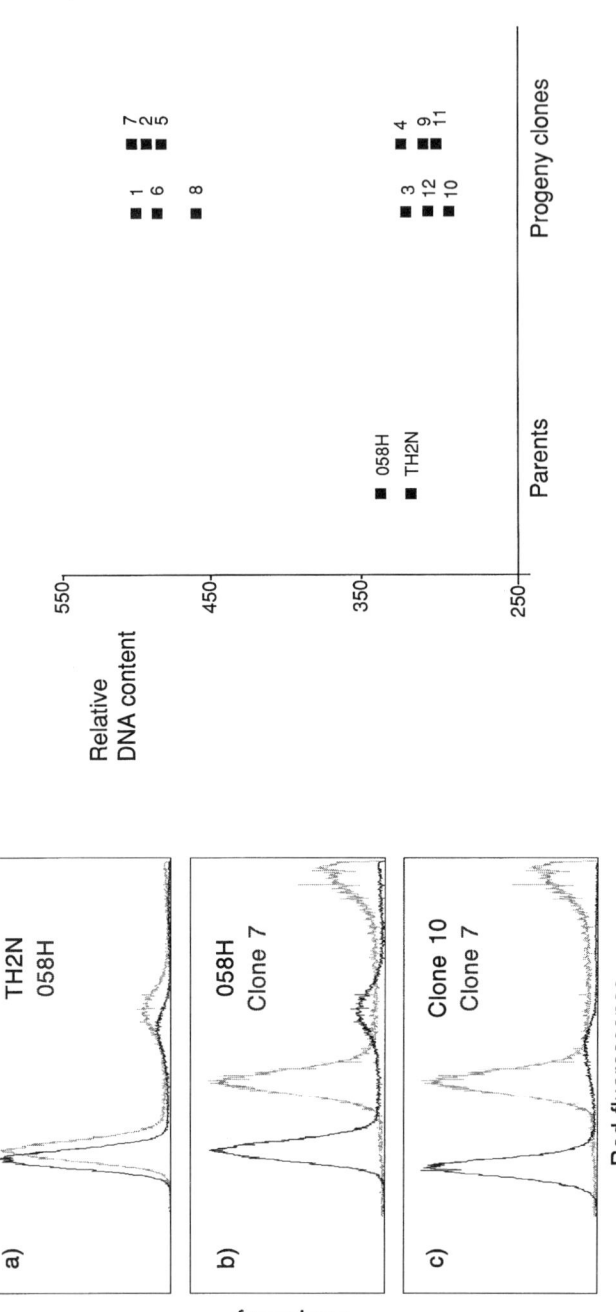

Figure 2 Left. Comparison of DNA contents of parents (panel a: TH2N, black trace; 058H, grey trace) and progeny clones (panels b and c: clone 7, grey trace; clone 10 black trace) from an experimental cross of *Trypanosoma brucei* by flow cytometry. The histograms plot, in arbitrary units against frequency, the intensity of linear red fluorescence resulting from laser excitation of trypanosomes stained with propidium iodide. Each histogram shows a large peak corresponding to cells in G1 and a smaller peak for G2. Although the DNA contents of the parental clones are approximately the same (panel a), progeny clone 7 has a DNA content midway between the G1 and G2 peaks of parent 058H (panel b). Given that the parental clones are diploid, the G1 peak corresponds to 2n; thus the DNA content of clone 7 corresponds to 3n. Progeny clone 10 has a DNA content similar to that of the parents (panel c).
Right. Flow cytometry data in graphical form from the same cross (Gibson *et al.*, 1997a). Relative DNA contents (mean G1 peak positions) in arbitrary units are plotted on a linear scale. There are clearly two clusters of progeny clones: the first cluster of clones (3, 4, 9–12) have DNA contents similar to the parental clones (058H and TH2N), i.e. 2n, while the other progeny clones (1, 2, 5–8) cluster at values corresponding to 3n.

DNA content between clones, perhaps reflecting differences in the inheritance of the smaller chromosomes; nevertheless, all progeny cluster at values consistent with either 2n or 3n. It seems plausible that triploids arise from the fusion of haploid and diploid nuclei, with the important corollary that the existence of haploid nuclei implies that meiosis has occurred. The initially anomalous observation of triploid hybrids has thus turned out to be evidence of meiosis.

(c) *Inheritance of kinetoplast DNA.* Kinetoplast DNA (kDNA) is the mitochondrial DNA of kinetoplastids (reviewed by Simpson, 1986; Stuart and Feagin, 1992; Shapiro and Englund, 1995). The structure is complex, consisting of two sets of interlocked DNA circles in a dense network. In *T. brucei* there are an estimated 50 maxicircles of 20–25 kb and 5000 minicircles of 1 kb per network. The maxicircles carry genes for mitochondrial function and are equivalent to the mitochondrial DNA of other eukaryotes. The minicircles, at first thought to be merely structural, are now known to encode the guide ribonucleic acids (RNAs) necessary for editing maxicircle transcripts (Sturm and Simpson, 1990).

So what happens to this complex structure during genetic exchange? Initial experiments, based on analysis of maxicircle polymorphisms, suggested that inheritance was uniparental and that either parent could contribute kDNA to the progeny (Sternberg *et al.*, 1988, 1989; Gibson, 1989). However, when the minicircle component of the kDNA was examined by restriction analysis and hybridization using cloned minicircle fragments from each parent, the surprising result was that the minicircle networks of the hybrid progeny were heteroplasmic and inheritance of kDNA must therefore be biparental (Gibson and Garside, 1990; Gibson *et al.*, 1997b). This result has been observed in three independent crosses (E, F, G; Table 1).

Two alternative mechanisms have been put forward to explain the different modes of inheritance of the mini- and maxicircles (Figure 3). The first mechanism is stochastic and assumes that a hybrid kDNA network initially contains a mixture of both mini- and maxicircles, but that subsequent mitotic divisions lead to the gradual loss of one maxicircle type by random segregation. The loss of heterogeneity by the non-stringent replication and partitioning of a relatively small number (<100) of molecules has been discussed by Birky (1983) with reference to the inheritance of yeast mitochondrial DNA. Such stochastic mechanisms would not have the same homogenizing effect on the minicircles, because of their much greater number and variety. In support of this model, Turner *et al.* (1995) found that four of 11 hybrid progeny analysed in the early stages of vegetative growth had maxicircles of both parental types and, furthermore, the results were in agreement with theoretical calculations based on a simple model of genetic drift.

However, it remains difficult to reconcile the initial formation of a hybrid kDNA network, which would involve decatenation, mixing and recatenation of the combined networks, with the elegant control of kDNA replication at the molecular level. A more appealing mechanism for the formation of hybrid networks was put forward by Shapiro and Englund (1995), based on some of their experimental observations. Shapiro (1993) showed, by selective removal

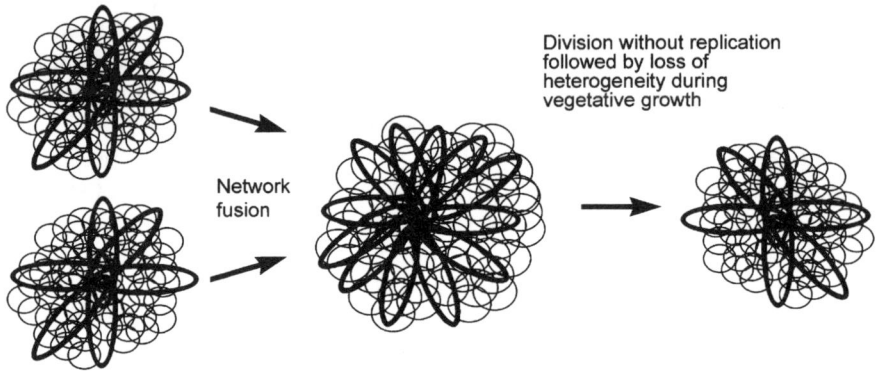

Model 1: Stochastic loss of maxicircle heterogeneity

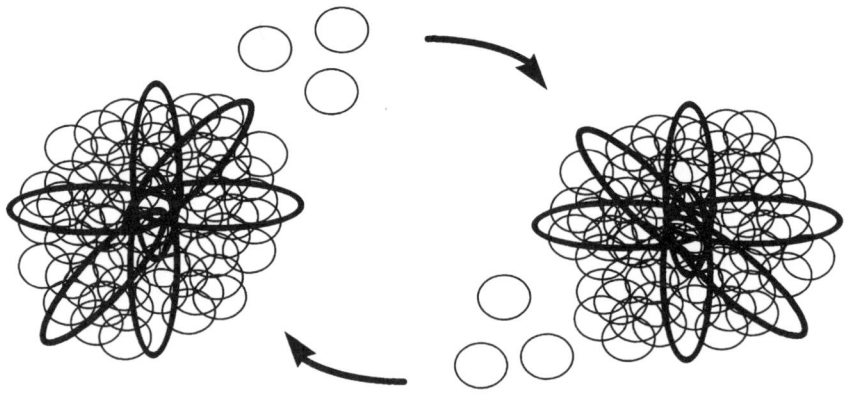

Model 2: Exchange of minicircles

Figure 3 Mechanism of formation of hybrid kinetoplast DNA networks in genetic crosses of *Trypanosoma brucei*. Model 1 illustrates the stochastic loss of maxicircle heterogeneity after network fusion, while model 2 illustrates minicircle exchange between adjacent parental networks (see Section 2.1.4.c).

of minicircles, that the maxicircles form an independent network interlocked to the minicircle network — a 'network within a network'. During division of the kinetoplast, individual minicircles detach from the network and replicate, and the daughter minicircles reattach to the network (Shapiro and Englund, 1995). If parental kDNA networks came into close proximity during genetic exchange, minicircles from one network might reattach to the other, leaving the maxicircle component of each network unchanged (Figure 3, model 2). Division of the heterokaryon would give rise to daughter trypanosomes with mixed minicircle networks and maxicircles of each parental type. In support of this model, hybrid clones have been observed which appear to be identical for all nuclear DNA characters analysed, but which have maxicircles of each parental type (Gibson, 1989; Gibson and Garside, 1990; Gibson et al., 1997b). Such clones are assumed to derive from a single heterokaryon, since the generation of identical sets of nuclear DNA polymorphisms following meiosis is extremely unlikely. The model predicts that minicircle repertoires of such hybrid clones should be more similar in clones which have the same maxicircle type than in those with a different maxicircle type. This was not the case for four clones analysed by Gibson et al. (1997b), which all had much the same minicircle network, and the authors concluded that minicircle mixing must occur on a gross scale. Much the same situation would result from network fusion and stochastic loss during replication, i.e. the stochastic mechanism described above. Gross disruption of the minicircle repertoire would be expected to have important consequences for hybrid viability, since correct maxicircle editing depends on a functionally complete set of minicircles. This might be one explanation for the high level of sequence redundancy observed among guide RNAs (Riley et al., 1994).

Both mechanisms assume the biparental inheritance of kDNA, which in turn implies fusion of parental mitochondria and hence fusion of parental cells. This has profound consequences for the copy number of other cell components, notably the flagellum and basal body, since the heterokaryon must initially have two flagella and two basal bodies. Robinson and Gull (1991) showed that the kinetoplast and basal body of the flagellum are intimately linked during cell division. Hence, if the heterokaryon proceeded through cell division with normal replication of these structures, the resultant trypanosomes would possess two flagella, etc., and a further doubling of these structures would occur after every subsequent mating. This is clearly not the case, and the simplest conclusion is that the trypanosome solves the problem by undergoing cell fission without replication after genetic exchange, perhaps triggered by the presence of two kinetoplast/basal body complexes in the heterokaryon.

Before leaving the kDNA, there is one further anomalous observation. In cross E (Table 1), one hybrid trypanosome clone was found that had exchanged only kDNA, its maxicircles being of type 196 in a hybrid kDNA network, while its nuclear genotype was identical to that of the *T. b. rhodesiense* parent (Gibson, 1989; Gibson and Garside, 1990). This anomalous clone may provide evidence that genetic exchange starts with trypanosome fusion and suggests that, in this case, the nuclear DNA component of the second parent failed to become incorporated.

(d) *Molecular karyotype analysis.* Initial karyotype analysis by PFGE showed that hybrids had new combinations of intermediate-sized (100–1000 kb) and minichromosome DNA bands (Paindavoine *et al.*, 1986a; Wells *et al.*, 1987; Sternberg *et al.*, 1988). Indeed, these striking karyotype differences were exploited to identify trypanosome populations containing hybrids by their non-parental karyotypes after PFGE (Gibson, 1989; Gibson and Garside, 1991). Comparison of the molecular karyotypes of parental and progeny clones using Southern analysis of intact and restriction-digested chromosomal DNAs showed that hybrid clones had new combinations of parental chromosomal bands throughout the resolvable range (1000 kb–2 Mb) and also sometimes had extra non-parental bands (Gibson, 1989; Gibson and Garside, 1991; Gibson *et al.*, 1992, 1995; Schweizer *et al.*, 1994; Degen *et al.*, 1995). The origins of some of these non-parental bands were determined by Southern analysis using various housekeeping gene probes, which revealed that substantial chromosomal length alterations, of the order of 10–20% of the original size of the chromosomal DNA molecule, had occurred (Gibson and Garside, 1991; Gibson *et al.*, 1992). The use of selectable markers in later crosses had the advantage of tagging individual chromosomes, which enabled chromosomal recombination to be demonstrated unequivocally (Gibson and Bailey, 1994). Chromosomal recombination occurred in hybrid progeny with some frequency, but was associated with both mitotic and meiotic division. The higher frequency of recombination in hybrid clones than in non-hybrid clones is, however, further evidence of meiotic division during genetic exchange (Gibson and Bailey, 1994).

(e) *Models of genetic exchange.* Various models of genetic exchange in *T. brucei* have been proposed (Figures 4 and 5). Some, though now recognized to have been based on erroneous or aberrant results, were the cause of some confusion in the earlier literature. The preliminary findings by Paindavoine *et al.* (1986a) of unstable subtetraploid hybrids led to the proposition that the mechanism of genetic exchange involved fusion followed by random elimination of DNA to return to a diploid state. However, at the same time, haploid metacyclics were reported in *T. brucei* by Zampetti-Bosseler *et al.* (1986), and the subtetraploid hybrid clone was also shown to reduce its DNA content by half following fly transmission (Paindavoine *et al.*, 1986a). These observations were combined into the somewhat complicated model shown in Figure 4 (Paindavoine *et al.*, 1986a). No further evidence to support this 'fusion followed by chromosome loss' model of Paindavoine et al. (1986a) has been obtained, and indeed all subsequent reports of subtetraploids described stable triploids (see Section 2.1.4.b). The idea of haploid metacyclics also failed to be substantiated by other workers (Kooy et al., 1989; Tait et al., 1989), although the fleeting occurrence of such a life-cycle stage cannot be ruled out. The technique of Feulgen microfluorometry for measuring DNA content seems to be somewhat prone to error, perhaps because so few cells (of the order of 100) can be measured compared to flow cytometry (of the order of 50 000); additionally, discrepancies may have arisen from different binding characteristics of the fluorescent dye to cell cycle-arrested metacyclics compared to actively dividing procyclic and bloodstream forms.

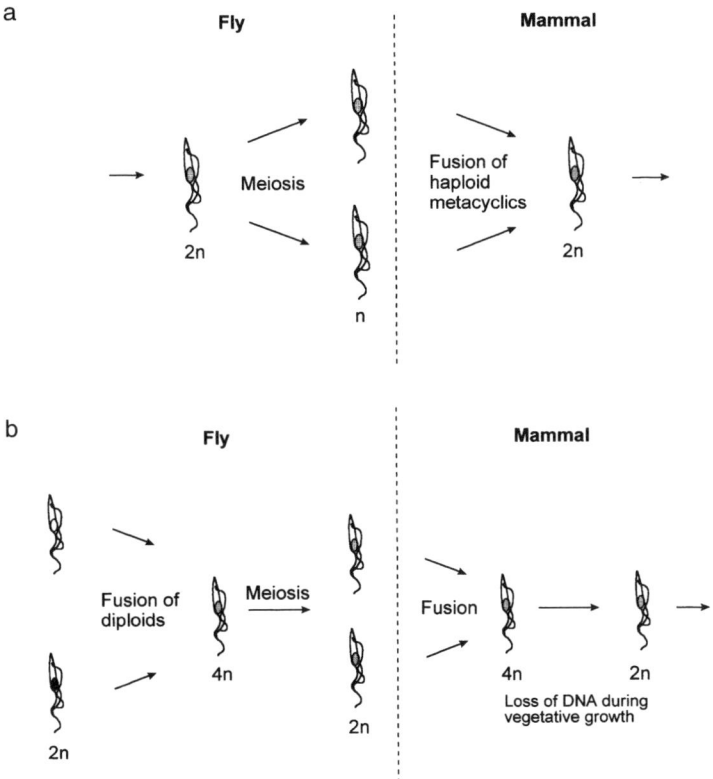

Figure 4 Models of genetic exchange in *Trypanosoma brucei* proposed by Paindavoine et al. (1986a). Scheme a shows the 'normal' sexual cycle with production of haploid metacyclics; scheme b shows the aberrant cycle deduced for the subtetraploid hybrids. No further evidence was found for either scheme (see Section 2.1.4.e).

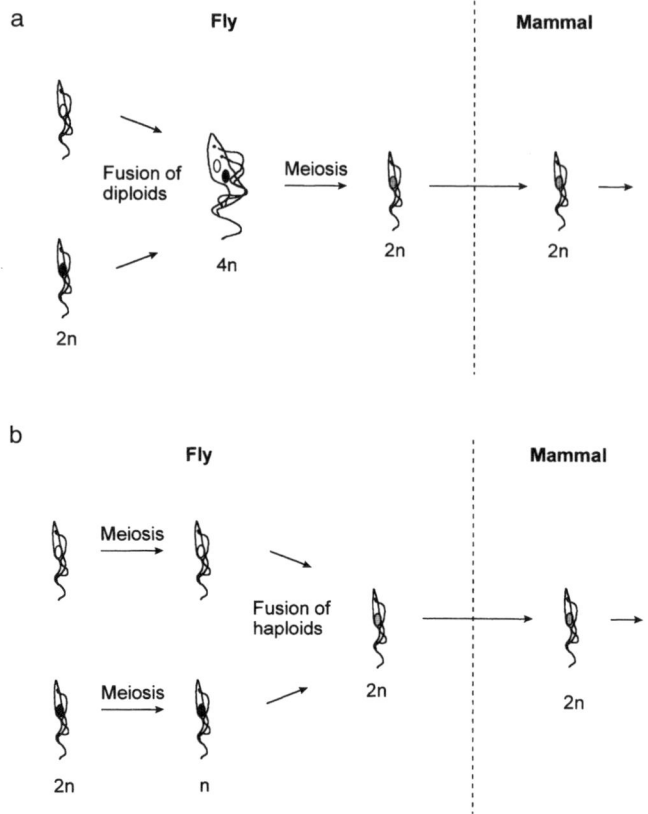

Figure 5 Models of genetic exchange in *Trypanosoma brucei* based on the proposals of Gibson (1995). Schemes a and b involve meiosis and fusion. In scheme a, fusion occurs first, and in scheme b, meiosis occurs first. Note that a haploid stage has not been substantiated during the life cycle of *T. brucei* (see Section 2.1.4.e).

By far the greatest weight of evidence indicates that normal genetic exchange in *T. brucei* involves meiosis and fusion (see Sections 2.1.4.a and c), but the order of these processes remains doubtful. Evidence for meiosis rests on (i) most markers being inherited in the Mendelian fashion, (ii) the high frequency of aberrant triploid hybrids, indicating that fusion of haploid and diploid nuclei is a common mistake during genetic exchange, and (iii) the higher frequency of recombination observed in hybrid than in non-hybrid progeny. The evidence for cell fusion rests on the observation of heteroplasmic inheritance of kDNA. Gibson (1995) attempted to combine all the experimental results into a single model, which in reality comprises the two scenarios illustrated in Figure 5. Each model includes fusion and meiosis, but in alternative orders. Figure 5a shows fusion of diploids to create a heterokaryon, which undergoes meiosis to produce diploid progeny. Figure 5b shows meiosis followed by fusion of haploids; note, however, that there is no evidence for a haploid stage in *T. brucei* (see above, this Section). Both models allow formation of triploids, as an error of either meiosis (Figure 5a) or fusion (Figure 5b). Currently there is no evidence, other than the lack of a haploid stage, to favour either model. One should also bear in mind that these are the simplest models consistent with the experimental observations; the real mechanism may be quite complex. As observed by Walker (1964), and considering the oddities of trypanosome biochemistry in general (e.g. discontinuous transcription, glycosomes, kinetoplast DNA, RNA editing), there is no reason to suppose that this ancient eukaryote may not also have evolved its own unique mode of sexual reproduction.

2.1.5. *Inheritance of Phenotypic Traits*

(a) *The trait for human-infectivity.* Several crosses have featured one parental trypanosome of known infectivity to humans, i.e. *T. b. rhodesiense* or Group 2 *T. b. gambiense*, mated to a non-human-infective *T. b. brucei* clone (see Table 1). This raises important questions, not only about the inheritance of the trait for human infectivity but also about the taxonomy of these subspecies, which is based on their ability to infect humans. The cross reported by Gibson (1989) between *T. b. rhodesiense* and *T. b. brucei* (cross E, Table 1) enabled analysis of the inheritance of the trait for human infectivity by the blood incubation infectivity test (BIIT; Rickman and Robson, 1970). In this test trypanosomes are incubated with human blood *in vitro* and their subsequent infectivity tested in mice; trypanosomes which are resistant to lysis are potentially human-infective. The test relies on differential killing of *T. b. brucei* by a trypanolytic factor in human blood (reviewed by Hajduk *et al.*, 1992). In the cross, the *T. b. rhodesiense* parent was found to be resistant to human serum, while the other parent was sensitive as expected. The BIIT response of two of the hybrid clones was identical to that of the *T. b. rhodesiense* parent, showing that the trait of human infectivity can be inherited undiminished. The other hybrid clones either were sensitive to human serum or gave an intermediate result, implying that human serum resistance is not inherited as a simple dominant trait.

Degen *et al.* (1995), using an *in vitro* version of the human serum resistance test whereby metacyclic or bloodstream form trypanosomes are cultured with human (test) or horse (control) serum (Jenni and Brun, 1982; Brun and Jenni, 1987), found that all parental and progeny clones from their cross of two Ugandan *T. brucei* ssp. stocks from wild-caught tsetse flies (cross I, Table 1) were sensitive to human serum and therefore most probably *T. b. brucei*. Gibson and Mizen (1997) tested metacyclic trypanosomes from flies infected with parental and progeny clones from cross G (Table 1) of *T. b. rhodesiense* and *T. b. brucei*. Again, the *T. b. rhodesiense* parent was found to be serum-resistant and the *T. b. brucei* parent serum-sensitive as expected, while five progeny clones were serum-resistant and five were serum-sensitive. However, two pairs of clones which were indistinguishable by DNA polymorphisms (Gibson and Bailey, 1994) had opposing responses to human serum. Although negative results from human serum resistance tests are always doubtful, positive results (i.e. serum resistance) are unequivocal. Thus, we can be certain that the trait for human serum resistance can be passed to hybrid progeny, with clear implications for the epidemiology of sleeping sickness and the taxonomy of *T. b. rhodesiense* and *T. b. brucei*.

(b) *Drug resistance*. Gibson and colleagues have used genetically engineered drug resistance markers extensively for selection of hybrid trypanosomes (Gibson and Whittington, 1993; Gibson and Bailey, 1994; Gibson *et al.*, 1995, 1997a). The advantage of rendering trypanosomes drug-resistant in this way is that it does not require prolonged passage with increasing levels of drug, during which fly transmissibility might be lost. On the other hand, inheritance of a drug resistance phenotype in experimental crosses allows the genetic basis of drug resistance to be determined directly. To this end, Scott *et al.* (1996) described clones of *T. brucei* with stable resistance to the two drugs cymelarsen (MelCy) and suramin.

(c) *Variant surface glycoproteins*. Genes for VSGs have been used as genetic markers to analyse progeny from most crosses. Interpretation of the results is not as straightforward as for housekeeping genes, however, because of the single-copy, non-diploid arrangement of VSG genes. Their organization into families of related genes (Frasch *et al.*, 1982) has also been exploited to provide 'VSG fingerprints' with which to compare parents and individual hybrids (Gibson, 1989).

Turner *et al.* (1991) examined expression of VSG genes in parents and progeny from the cross described by Sternberg *et al.* (1989) by using antisera raised against individual variant antigen types (VATs). As expected from the inheritance of VSG genes, each hybrid progeny clone contained some VATs from each parent, but not all. VAT expression in the hybrid progeny appeared to follow the same rules as in the parents, with parental metacyclic and bloodstream VATs being expressed in the appropriate hybrid stages, and individual progeny expressing only one VAT at a time. Importantly, the VAT repertoires of the hybrids appeared to be larger than those of the parents, since the hybrids had inherited more than 50% of the VATs of each parent. As emphasized by the authors, genetic recombination is a powerful mechanism for generating diversity in antigenic repertoires.

In addition to VSGs, Pearson and Jenni (1989) carried out a preliminary analysis of the parents and one hybrid clone from the cross reported by Jenni *et al.* (1986), using high-resolution two-dimensional gel electrophoresis. This technique, although demanding, extends phenotypic analysis to the hundreds of different proteins that can be labelled radioactively during culture. Surprisingly few protein differences were observed between parental and hybrid clones compared to the large number of RFLP and PFGE karyotype differences observed between the same clones previously (Paindavoine *et al.*, 1986a; Wells *et al.*, 1987). Nevertheless the hybrid clone's gel showed spots corresponding to both forms of a protein found separately in each parent, thus confirming its hybrid status, and the two-dimensional protein 'fingerprints' were stable after tsetse fly transmission and cloning.

2.2. Other Trypanosomatids

Compared to the wealth of experimental data available for *T. brucei*, evidence for genetic exchange in other trypanosomatids is scanty. The reviews by Walker (1964) and Tait (1983) cover the older literature on this subject.

Although several laboratories have doubtless attempted to set up experimental crosses of *T. cruzi*, *T. congolense* and other species of *Trypanosoma* and *Leishmania*, along similar lines to those carried out with *T. brucei*, these experiments have largely been unreported because of their lack of success. Some of these species have the advantage of being much more easily manipulated *in vitro* than *T. brucei*, which would have greatly facilitated analysis. Most of the published evidence is based instead on the observation of putative hybrid isoenzyme or RFLP patterns in field isolates (see Sections 3.3.4 and 3.3.5). However, there are also several intriguing studies which have described various unusual morphologies interpreted as sexual stages.

By electron microscopy, Deane and Milder (1972) demonstrated epimastigotes with cytoplasmic linkages to others among populations of *T. (Megatrypanum) conorhini*, a tropical rat trypanosome transmitted by triatomine bugs. These authors observed 'cyst-like bodies' with multiple nuclei, kinetoplasts and other organelles, on the edge of colonies of epimastigotes growing in rosettes, which they interpreted as products of the fusion of epimastigotes; the appearance of the cyst-like bodies was sudden and was not associated with depletion of nutrients in the culture medium. These cyst-like bodies were somewhat similar to the giant multinucleate forms observed by Ellis *et al.* (1982) in *T. brucei* (see Section 2.1.3) as well as in *T. rangeli* and another unidentified trypanosome of South American origin. Besides appearing in cultures, giant forms of *T. rangeli* also occurred in the gut cells of triatomine bugs and in the muscle layers around their salivary glands. These reports echo the description of intracellular stages of *T. lewisi*, which occasionally occur in infected fleas and have no satisfactory explanation in terms of the trypanosome's life cycle (Molyneux, 1969; Hoare, 1972).

Lanotte and Rioux (1990) directly observed apparent cell fusion between cultured promastigotes of *L. infantum* and *L. tropica* filmed by video microscopy, and Walters *et al.* (1993) described a similar phenomenon in *L. major* in the sandfly host. Possible evidence for cell fusion of amastigotes of *L. major* or *L. tropica* inside individual macrophages in cell culture was obtained by Kreutzer *et al.* (1994) using Feulgen microspectrophotometry. In these experiments, there were two populations of amastigotes with DNA contents of either 1x or 2x, with no intermediate value as would be expected for cells in vegetative growth; promastigotes of the same isolates formed a single population with a constant DNA content of 1x. This technique is limited by the small number of cells that can be analysed (see comment in Section 2.1.4.e).

An alternative approach to analysing genetic exchange in the medically important trypanosomatids with rather complicated life cycles is to use a monoxenous trypanosomatid as a model system. Glassberg *et al.* (1985) exploited the considerable technical advantages of using *Crithidia fasciculata*, a parasite of the gut of mosquitoes, which can be grown easily as colonies on agar plates. Stable, drug-resistant mutants were selected after chemical mutagenesis, mixed in culture and subsequently plated out to determine drug resistance phenotype. In two of six trials, doubly drug-resistant putative hybrid colonies developed. The problem of reproducibility led the authors to a rather cautious interpretation of the results of these promising experiments.

3. EVIDENCE FROM POPULATION GENETICS AND EVOLUTIONARY STUDIES

3.1. Introduction

Approaches to the investigation of genetic exchange in natural populations of trypanosomatids are now well established. Studies are generally based on molecular characterization data derived from a representative sample of isolates from the population under study, and data are then analysed with appropriate mathematical techniques using population genetics methods and/or phylogenetic methods. Although the former, as the name suggests, are used at the population level, the latter have been developed primarily for evolutionary studies and are therefore generally inappropriate for use with the rapidly evolving molecular markers employed in population genetics studies. Consequently, while certain phylogenetic methods have been shown to be useful in studies of recombination in micro-organisms, e.g. Trypanosomatidae (Cibulskis, 1986) and human immunodeficiency virus strains (Robertson *et al.*, 1995), their use is limited, and they will be discussed only briefly in this review.

The ability to study recombination and sexual processes at the population level in trypanosomatids was made possible by the application of molecular characterization methods, such as isoenzyme electrophoresis and RFLP analysis. The widespread introduction of such techniques, together with an increase in the use of computers, allowed the population genetics of trypanosomatids to be explored in depth. In 1990, Tibayrenc and colleagues proposed the theory of clonality of parasitic protozoa, which suggested that the predominant mode of reproduction in a number of genera was clonal. This opened a wide and continuing debate on the relative importance of sexual and clonal reproduction in the trypanosomatids, as well as other protozoa, and stimulated interest in this whole area of research (Gibson, 1990; Cibulskis, 1992; Stevens and Welburn, 1993; Hide *et al.*, 1994; Tibayrenc, 1995; Stevens and Tibayrenc, 1996; Gibson *et al.*, 1998).

3.2. Methods

The aim of population genetics methods is to provide a measure of gene flow within and between natural populations. In practice, this is achieved by testing for the presence of subdivisions within a given sample population, between which gene flow is either restricted or absent (see Tibayrenc, 1995); the resulting statistics thus equate to indirect measures of gene flow. The use of such measures to analyse field data can provide valuable information concerning the frequency and impact of recombination events in natural populations of trypanosomatids (Tait, 1980, 1983; Cibulskis, 1988; Tibayrenc *et al.*, 1990, 1991; Stevens and Welburn, 1993).

As a first step in determining the level of genetic exchange within a population, a null hypothesis (H_0) that the population is randomly mating, i.e. panmictic, is assumed. A significant variation from H_0 implies a non-panmictic population structure; statistics used to demonstrate departures from panmixia consider either the lack of segregation or the lack of recombination (see below). In turn, such departures may be interpreted as evidence that gene flow in the study population is restricted (Cibulskis, 1988; Tibayrenc *et al.*, 1990, 1991; Stevens and Welburn, 1993). Although such a result may be due to a genuine difference in reproductive strategy from that of a panmictic, sexual organism, the possibility of population substructuring, which can also affect gene flow, should also be considered (Maynard Smith *et al.*, 1993; Stevens and Tibayrenc, 1996). Non-significant variation from panmixia and the associated inability to reject H_0 represent a specialized case and are discussed below. Finally, as Tibayrenc (1996) indicated, although all tests rely on the same basic principle (departures from panmixia), levels of resolution will differ between tests. These differences should be borne in mind when drawing conclusions based on the results of different tests.

3.2.1. *Segregation Tests and Ploidy*

Segregation tests are based on the concept of Hardy–Weinberg equilibrium, in which there is random reassortment of different alleles at a given locus. Such tests require that alleles are identifiable and that the ploidy level of the organism being studied is known and greater than one.

In *T. brucei*, although the ploidy of the smaller chromosomes is problematic, the larger chromosomes which contain housekeeping genes are diploid (Gibson *et al.*, 1985; Gibson and Borst, 1986; Gottesdiener *et al.*, 1990). In *T. cruzi*, the situation is less clear. Although numerous isoenzyme-based characterization studies have detected heterozygous patterns which can be interpreted as diploid hybrids (e.g. see Tibayrenc *et al.*, 1981, 1985, 1986), direct studies of ploidy suggest that the situation is more complex. PFGE analysis by Gibson and Miles (1986) led them to conclude that chromosome size and organization is highly variable in *T. cruzi* and that, at the minimum, *T. cruzi* is diploid. Dvorak (1984) and McDaniel and Dvorak (1993) found substantial (30%–70%) differences in total DNA content between *T. cruzi* clones, leading them to propose that *T. cruzi* may be aneuploid. Diploidy can thus be taken only as a working hypothesis for population genetics studies of *T. cruzi*.

In *Leishmania* spp., a similar state of uncertainty is apparent (Bastien *et al.*, 1992). Although numerous characterization studies based on isoenzymes (e.g. by Maazoun *et al.*, 1981; Le Blancq *et al.*, 1986; Evans *et al.*, 1987), RFLPs (e.g. Kelly *et al.*, 1991) and molecular karyotype (e.g. Dujardin *et al.*, 1993, 1995) have identified band patterns indicative of diploid heterozygotes, direct determination of ploidy has proved difficult. Nevertheless, diploidy is again taken as a general working hypothesis for population genetics studies of *Leishmania* spp.

Segregation tests, assuming diploidy, have been applied to the analysis of trypanosomatid population data, predominantly isoenzyme patterns, in a number of highly informative studies. The seminal work of Tait (1980) used classical Hardy–Weinberg statistics to reveal genetic exchange in African trypanosomes isolated from the field, although a later study by Cibulskis (1988) underlined some of the pitfalls associated with single-locus Hardy–Weinberg analysis when sample sizes are small. The randomization approach developed by Cibulskis (1988) was extended by Stevens and Welburn (1993) to study genetic exchange in epidemic populations, together with a multilocus approach (Workman, 1969).

3.2.2. Recombination and Linkage Methods

Recombination tests are multilocus analyses which offer a powerful alternative to segregation methods. They have the advantage that they can be used whatever the ploidy level of the study organism. Indeed, such methods can even be used when the ploidy level is unknown (and perhaps indeterminate) and when individual alleles cannot be defined precisely. The only requirement is that loci are independent from one another (Tibayrenc, 1995, 1996).

Primarily, the tests rely on demonstrating departures from random assortment, in which the expected frequency of a given multilocus genotype is simply the product of the observed frequencies of the individual genotypes which make it up. Data which are randomly assorted conform to a random distribution; such a distribution is the only state for which statistical criteria can be readily defined, and is taken as the null hypothesis (H_0). Studies of organisms proven to be undergoing regular genetic exchange also indicate that disequilibrium between loci is rare and that departures from equilibrium are not generally observed (reviewed by Cibulskis, 1988). Thus a significant variation from H_0 can be interpreted as evidence of a non-panmictic population structure. Such variation can be measured by any one of a number of statistics based on randomization methods (e.g. see Tibayrenc *et al.*, 1990, 1991; Stevens and Tibayrenc, 1995), association indices (Brown and Feldman, 1981; Maynard Smith *et al.*, 1993) or a combination of the two (Souza *et al.*, 1992). All explore different aspects of the same variation: departures from panmixia or linkage disequilibrium (non-random association among loci, in which predictions of expected probabilities for multilocus genotypes are no longer satisfied).

Of course, although such tests facilitate the demonstration of departures from panmixia, the statistics say nothing about the underlying cause. Obstacles to gene flow can be classified under two main headings: physical (genetic isolation in either space or time) and biological (physical linkage of different genes on the same chromosome, clonality, cryptic speciation, natural selection). The relative importance of each will vary, depending on the population being considered.

Many of these methods have been employed in population studies of trypanosomatid parasites. Linkage analyses were used by Tibayrenc et al. (1990, 1991, 1993) in studies of a range of parasitic protozoa based on data derived from the literature, while an extended Mantel test (Mantel, 1967; Stevens and Tibayrenc, 1995) has been used to study a range of *T. brucei* populations (Stevens and Tibayrenc, 1995; Mathieu-Daudé et al., 1995; Kanmogne et al., 1996). Association indices have been employed for defining population structure in *T. brucei* (see Maynard Smith et al., 1993; Hide et al., 1994; Stevens and Tibayrenc, 1996) and *T. cruzi* (see Maynard Smith et al., 1993), and for investigating associations between parasite genotype and host or location (Cibulskis, 1992).

In order to interpret and appreciate the results of such tests, it is important to realize their somewhat unconventional nature, as tests for departures from equilibrium are effectively tests of the null hypothesis (Workman, 1969) and are heavily dependent upon the richness of the data under study. Accordingly, as overall variability within the data declines, evidence for accepting the alternative hypothesis, H_1, becomes insufficient; in such cases, failure to reject H_0 is sometimes mistakenly interpreted as acceptance of H_0. Such a conclusion is statistically invalid and is known as a Type II error. For segregation methods, the probability, b, of making a Type II error can, in certain cases, be calculated from the number of arrangements of genotype frequencies which conform to Hardy–Weinberg equilibrium by chance (Fairburn and Roff, 1980; Cibulskis, 1988). Once b is known, steps can be taken to reduce it to less than 5%, usually by increasing the sample size. If b cannot be reduced, then at least the probability of having avoided a Type II error can be estimated to provide some measure of confidence in the conclusions.

The very nature of randomization methods does not permit the calculation of such formal statistics, and their use remains dependent on the innate variability of the data being analysed and on the discriminative power of the technique employed. Stevens and Tibayrenc (1995), for example, used random amplified polymorphic DNA (RAPD) analysis to identify 13 genetically distinct populations, originating by cloning from two primary isolates of *T. brucei* from tsetse flies; isoenzyme characterization of the same stocks revealed only eight zymodemes. Correspondingly, all linkage analyses of the RAPD data demonstrated significant association, while only 70% of analyses of the isoenzyme data showed significant linkage; levels of significance obtained from analysis of the isoenzyme data were also much reduced, being at least one order of magnitude lower.

3.2.3. *Phylogenetic Methods*

Phylogenetic methods have classically been used to address higher levels of evolutionary divergence between taxa, i.e. above the species level, and so can be considered complementary to population genetics methods. Such techniques are now recognized as essential for exploring evolutionary questions and, in particular, for defining taxa in micro-organisms for which the biological species concept is often difficult or impossible to use (Tibayrenc, 1996).

Nevertheless, certain cladistic techniques have proved suitable for studies of genetic exchange in trypanosomatids. Cibulskis (1986, 1988) used Wagner networks (Farris, 1970) to explore sexual processes in a range of parasites, and showed that, although populations of *T. brucei* and *T. congolense* appear to undergo intra-specific genetic exchange, the results with populations of *T. cruzi* and *L. aethiopica* were atypical of organisms undergoing recombination (Cibulskis, 1986). Later, Hide *et al.* (1994) used the same approach to investigate gene flow in phylogenetic groups of *T. brucei* defined on the basis of isoenzyme patterns; their results suggested that the variation observed could be generated by mutation, and proposed, in conjunction with additional analyses, an 'epidemic' population structure (Maynard Smith *et al.*, 1993) for stocks within a clade corresponding to *T. b. rhodesiense*.

In future, it is to be expected that the use of phylogenetic methods in population genetics studies and recombination analyses will increase with the growing popularity of gene tree studies, gene genealogies and the application of coalescence theory (Hudson, 1990).

3.3. Evidence for Genetic Exchange from Population Genetics Analysis

3.3.1. *Trypanosoma brucei*

Among trypanosomatids, *T. brucei* is by far the most widely studied as a model of genetic recombination. The first tangible evidence that *T. brucei* might be a sexual organism was provided by isoenzyme data. Gibson *et al.* (1980) described isoenzyme patterns consistent with those expected from homo- and heterozygotes in an extensive analysis of *T. brucei* isolates from East and West Africa, while Tait (1980) showed that similar data from 17 Ugandan isolates of *T. brucei* conformed to Hardy–Weinberg equilibrium, indicating that the population was undergoing random mating. Tait (1983) analysed two more populations of *T. brucei* from Kenya, again using Hardy–Weinberg analysis, and found both to be in equilibrium for three of four enzyme systems; aspartate aminotransferase (ASAT, EC 2.6.1.1) showed fewer heterozygotes than expected. These studies preceded direct evidence of genetic exchange between trypanosome populations, and tacitly assumed meiosis. Jenni *et al.* (1986) subsequently demonstrated hybrid formation between two well-characterized clones of *T. brucei* after cyclical transmission through tsetse flies (see Section 2.1). Although the exact mechanism of genetic exchange and mode of inheritance of marker genes remain unknown, most evidence points to a meiotic division at some stage (see Section 2.1.4).

The simple idea that *T. brucei* formed a randomly mating population was first challenged by Cibulskis (1988), who demonstrated that, in analyses of genotype frequencies, agreement with Hardy–Weinberg predictions can be obtained even when genetic exchange has exerted no influence over such frequencies, especially if sample sizes are small; such a finding obviously had important implications for the results of Tait (1980, 1983). In the same study, however, cladistic analysis of three East African populations of *T. brucei* led Cibulskis (1988) to conclude that some diversity had arisen through a sexual process. Thus, the importance of genetic exchange in natural populations of *T. brucei* remained very much in debate, a situation not resolved by the inclusion of *T. brucei* in the clonal theory of Tibayrenc *et al.* (1990, 1991).

More recently, a growing body of evidence suggests that the degree of genetic exchange occurring, and hence the population structure observed for *T. brucei*, is highly dependent on epidemiological and ecological factors specific to a given population (Cibulskis, 1992; Stevens and Welburn, 1993; Hide *et al.*, 1994; Stevens and Tibayrenc, 1996). In particular, it appears that, during epidemics of human African trypanosomiasis, certain genotypes predominate and undergo rapid clonal reproduction; such populations can be classified as having an 'epidemic' population structure by association analysis (Maynard Smith *et al.*, 1993). Moreover, it appears that differences in the innate ability of certain *T. brucei* genotypes to reproduce epidemically may possibly be associated with their pathogenicity and phylogenetic position (Buyst, 1974; Godfrey *et al.*, 1987, 1990; Stevens and Tibayrenc, 1996). Certainly, index of association (I_A) analyses of the subspecies and various subtypes of *T. brucei* suggest that a range of population structures exist in nature, each presumably with a characteristic associated level of genetic exchange. I_A analysis of a population of *T. b. brucei* in Uganda (Hide *et al.*, 1994) suggested frequent recombination, while analysis of *T. b. rhodesiense* isolates from the same area indicated an 'epidemic' population structure (Hide *et al.*, 1994; Stevens and Tibayrenc, 1996). Thus, *T. b. rhodesiense* can be characterized epidemiologically by its ability to reproduce by explosive clonal reproduction in response to changes in environmental/ecological conditions, resulting in an outbreak of disease, whilst maintaining an important underlying level of genetic exchange. Analysis of *T. b. rhodesiense* from Zambia and *T. b. gambiense* from Côte d'Ivoire (Stevens and Tibayrenc, 1996) indicated clonal population structures for these populations and suggested that their reproduction was predominantly asexual. Indeed, while the taxonomic and phylogenetic status of *T. b. rhodesiense* and *T. b. brucei* remain the subject of considerable debate (e.g. Gibson *et al.*, 1980; Tait *et al.*, 1985; Paindavoine *et al.*, 1986b; Godfrey *et al.*, 1990; Hide *et al.*, 1990; Stevens and Godfrey, 1992; Mathieu-Daudé *et al.*, 1995), especially since genetic exchange has been demonstrated between the two subspecies (see Section 2.1.4.a), the monophyletic nature of Group 1 *T. b. gambiense* has been demonstrated clearly by a range of markers and techniques (e.g. Tait *et al.*, 1984; Gibson, 1986; Godfrey *et al.*, 1987; Paindavoine *et al.*, 1986b, 1989; Mathieu-Daudé *et al.*, 1994; Kanmogne *et al.*, 1996).

3.3.2. *T. congolense*

The second group of salivarian trypanosomes of which the number of characterized isolates is sufficient to allow analysis by a population genetics approach is the *T. congolense* group within the subgenus *Nannomonas* (see Young and Godfrey, 1983; Gashumba *et al.*, 1988; Tibayrenc *et al.*, 1990, 1991; Sidibé, 1997). Tibayrenc *et al.* (1990, 1991) analysed the extensive data set (114 stocks) studied by Gashumba *et al.* (1988), but concluded that, although there was some support for clonality, the limited number of markers used (seven isoenzyme loci) was not sufficient to equate the strains of *T. congolense* identified with actual clones. Consequently, no firm conclusion regarding the importance of sexual (or otherwise) processes within *T. congolense* has been reached on the basis of these data. Interestingly, six of the seven enzyme systems examined by Gashumba *et al.* (1988) showed one or more heterozygote patterns, with the appropriate accompanying homozygote patterns, which have been interpreted as circumstantial evidence of genetic exchange in a number of other parasites — e.g. *T. brucei* (see Gibson *et al.*, 1980; Tait, 1980), *T. cruzi* (see Tibayrenc *et al.*, 1981; Bogliolo *et al.*, 1996; Carrasco *et al.*, 1996) and *Leishmania* spp. (see Maazoun *et al.*, 1981; Evans *et al.*, 1987).

Analysis of *T. congolense* has been complicated by the discovery of several, previously unrecognized, subspecific groups. Isoenzyme analysis revealed the first of these groups, splitting *T. congolense* isolates into Savannah and Forest types (Young and Godfrey, 1983); a third type, Kilifi or Kenya coast, was then described on the basis of isoenzyme, DNA and karyotype markers (Gashumba *et al.*, 1988; Knowles *et al.*, 1988), followed by a fourth type, Tsavo (Majiwa *et al.*, 1993). The taxonomic status of these subgroups is uncertain, although Garside and Gibson (1995) have argued that Kilifi, at least, deserves species status on the basis of extensive genotype differences. Sidibé (1997; and see Tibayrenc, 1998) used a range of population genetics methods based on linkage disequilibrium statistics to analyse data from isoenzymes (18 loci) and RAPD analysis (23 primers), and demonstrated evidence of significant barriers to recombination between three of these groups (Forest, Kilifi, Savannah), confirming their distinct genetic identity. Moreover, separate analyses of isolates from each individual group also revealed significant linkage disequilibrium, suggesting that, even within a group, genetic exchange is rare, and that each group consists of non-recombining genotypes or clones. These findings imply that traits such as drug resistance are unlikely to spread through *T. congolense* by genetic recombination, a conclusion with important consequences for the long-term future control of animal trypanosomiasis in Africa.

3.3.3. *Trypanosoma vivax*, *T. evansi* and *T. equiperdum*

Difficulties in field isolation and laboratory cultivation of *T. vivax* in sufficient numbers for molecular characterization have severely limited population genetics studies of this trypanosome. As for *T. congolense*, characterization has revealed complexity within *T. vivax*, with at least two subspecific groups, broadly from East and West Africa, being recognized by DNA probes (Kukla *et al.*, 1987; Gibson and Dickin, 1989; Masake *et al.*, 1994). This complicates the interpretation of earlier studies on isoenzyme variation (Kilgour *et al.*, 1975; Murray, 1982; Allsopp and Newton, 1985). *T. vivax* relatively recently extended its range beyond the tsetse belt of Africa to South America, where, like *T. evansi*, it is non-cyclically transmitted by blood-sucking flies; isoenzyme and DNA analyses have shown a close relationship between the South American isolates and the West African form of *T. vivax* (see Murray, 1982; Gibson and Dickin, 1989; Dirie *et al.*, 1993), but it is doubtful whether the South American strains are still capable of cyclical development in the tsetse fly (Gardiner, 1989) and, by extrapolation, genetic exchange.

Similarly, *T. evansi* and *T. equiperdum*, despite their close relationship to *T. brucei*, have lost the ability to undergo cyclical transmission in tsetse flies, and presumably the ability to undergo genetic exchange as well. *T. evansi* is widely distributed throughout the tropics and subtropics, where it is non-cyclically transmitted by blood-sucking flies such as Tabanidae. Despite its immense geographic range, genetic diversity within *T. evansi*, as assessed by isoenzyme analysis and sequence variation in kDNA minicircles from African, South American and Asian isolates, is minimal compared to that of tsetse-transmitted *T. brucei* ssp. (Gibson *et al.*, 1983; Stevens *et al.*, 1989; Lun *et al.*, 1992). Far less is known about genetic variation in *T. equiperdum*, a venereally transmitted equine species, now rare, of which there are few laboratory isolates. Like *T. evansi*, *T. equiperdum* has defective kDNA and is thus unable to complete the full life cycle of the subgenus *Trypanozoon*, with the expectation that it too is asexual.

3.3.4. *Trypanosoma cruzi*

Of all trypanosomatids, *T. cruzi* is perhaps the species for which most population genetics evidence points in favour of asexual or clonal reproduction (Tibayrenc *et al.*, 1986, 1990, 1991), with an associated clonal population structure (Maynard Smith *et al.*, 1993). Population genetics analysis of data from isoenzyme-based characterization studies has repeatedly shown significant population substructuring, phylogenetic clustering and linkage disequilibrium characteristic of clonal populations in both domestic (Tibayrenc *et al.*, 1986) and sylvatic cycles (Lewicka *et al.*, 1995). More recently, evidence from a range of molecular analyses, e.g. RFLPs (Tibayrenc and Ayala, 1987), karyotype variability (Henriksson *et al.*, 1993; Sanchez *et al.*, 1993) and RAPD analysis (Tibayrenc *et al.*, 1993; Brisse, 1997; Tibayrenc, 1998), has corroborated the clonal status of *T. cruzi* indicated by earlier isoenzyme studies. As for *T. congolense* and *T. vivax*, detailed molecular characterization has revealed two major subspecific groups within *T. cruzi* (Souto *et al.*, 1996; Nunes *et al.*, 1997), suggesting a more complex population structure than first supposed.

Nevertheless, there is some evidence to suggest that genetic exchange may have, or may have had, an effect on the pattern of genetic variation in *T. cruzi*. Recent characterization studies using isoenzymes, RFLPs and RAPD analysis (Bogliolo *et al.*, 1996; Carrasco *et al.*, 1996) have identified typical, though rare, heterozygous profiles with corresponding homozygotes, suggesting the possibility of genetic exchange. Significantly, both studies detected these combinations of patterns in trypanosomes isolated from localized areas — a single Brazilian village (Bogliolo *et al.*, 1996) and triatomine bugs and sylvatic mammals from an area of Amazonian forest (Carrasco *et al.*, 1996) — raising the possibility that such sympatric genetic variants are the product of recent recombination within their respective epidemiological cycles. Carrasco *et al.* (1996) also presented a χ^2 analysis of their observed phosphoglucomutase (PGM) allele frequencies, showing agreement with Hardy–Weinberg equilibrium; the significance of agreement with Hardy–Weinberg expectations at a single locus (among the five systems analysed) requires further statistical support (see Cibulskis, 1988).

Although the contemporary existence of genetic exchange in *T. cruzi* seems likely to remain contentious, the possibility that some recombination could interfere with the evolutionary fate of clonal populations of *T. cruzi* has long been recognized (Tibayrenc *et al.*, 1986, 1990, 1991).

Using a phylogenetic approach to trace gene lineages, Brisse (1997) demonstrated genetic exchange between certain clonal populations of *T. cruzi*, suggesting that, although rare, the horizontal transfer of genetic material may have occurred at some time in the past within at least one of the two major *T. cruzi* lineages (Souto *et al.*, 1996; Nunes *et al.*, 1997; Tibayrenc, 1998).

Thus it appears that, although the prevailing population structure of *T. cruzi* is fundamentally clonal, there is good evidence to suggest that genetic exchange can occur and, indeed, has occurred in the evolutionary history of this parasite. However, the importance of such exchange in disrupting existing clonal lineages and thus affecting the epidemiology of the contemporary disease appears minimal.

3.3.5. *Leishmania*

As with other trypanosomatids, initial support for the idea of genetic exchange in *Leishmania* spp. relied on the observation of typical heterozygote patterns in isoenzyme characterization studies (e.g. Maazoun *et al.*, 1981; Le Blancq *et al.*, 1986). However, molecular characterization data together with circumstantial evidence in support of the natural occurrence of *Leishmania* hybrids is now extensive and convincing. The first putative hybrids showed biochemical characteristics of both *L. major* and *L. arabica*, and had been isolated from a feral dog and desert rat in an area of endemic leishmaniasis in the Eastern Province of Saudi Arabia, where both species were thought to be transmitted by a single sandfly species (Evans *et al.*, 1987; Kelly *et al.*, 1991). Similar evidence has been put forward for New World species. Belli and colleagues found that human cutaneous leishmaniasis in Nicaragua was caused by putative hybrids of *L. panamensis* and *L. braziliensis*, based on isoenzyme and DNA 'fingerprinting' results (Darce *et al.*, 1991; Belli *et al.*, 1994); the hybrid form was found in northern foci only, where the two species were sympatric. Hybrids have also been reported between *L. venezuelensis* and *L. braziliensis* in western Venezuela (Bonfante-Garrido *et al.*, 1992), between *L. braziliensis* and *L. peruviana* in Peru (Dujardin *et al.*, 1995), and in Ecuador between *L. braziliensis* and *L. panamensis/guyanensis* (Bañuls *et al.*, 1997). Blaineau *et al.* (1992), using PFGE, detected an association equilibrium between the different forms of chromosomes II and V in *L. infantum* strains isolated from humans and sandflies in southern France; such an association could be explained only by a high rate of recurrent mutations or genetic exchange.

Thus, while the ploidy of *Leishmania* remains debatable (Bastien *et al.*, 1992), evidence from several population studies lends support to the idea that some degree of genetic exchange exists within and between natural populations of *Leishmania* spp.

4. THE BIOLOGICAL CONTEXT

4.1. Implications for Epidemiology and Control of Disease

Genetic exchange in *T. brucei* has been demonstrated unequivocally in the laboratory, and crosses between human-infective and non-infective subspecies have shown that the trait for human infectivity is heritable (see Section 2.1.5). Thus, although the frequency of genetic exchange in natural populations may be variable or low, there can be no doubt that it results in the appearance of new genotypes of human-infective trypanosomes. The human-infective subspecies mated successfully with *T. b. brucei* are *T. b. rhodesiense* from East Africa and virulent or Group 2 *T. b. gambiense* from West Africa (Table 1). New human infective strains would thus be expected to occur from time to time in foci where either of these two trypanosomes is found. Indeed, there is abundant evidence of strain heterogeneity in *T. b. rhodesiense* from a number of foci—e.g. Busoga, Uganda (Gibson and Gashumba, 1983; Enyaru *et al.*, 1993, 1997; Hide *et al.*, 1994), Lambwe Valley, Kenya (Gibson and Wellde, 1985; Mihok *et al.*, 1990), north-western Tanzania (Komba *et al.*, 1997), and in Group 2 *T. b. gambiense* in Côte d'Ivoire (Mehlitz *et al.*, 1982; P. Truc, personal communication). The population structure of these trypanosomes certainly does not appear to be clonal.

One can speculate that Group 2 *T. b. gambiense* might itself be a hybrid, since it has the human infectivity of Group 1 *T. b. gambiense*, coupled with the virulence and fly transmissibility of *T. b. brucei*; all three are found together in Côte d'Ivoire (Mehlitz *et al.*, 1982). Although it is arguably the most important agent of sleeping sickness in Africa, having the widest distribution, Group 1 *T. b. gambiense* is difficult to transmit through tsetse flies in the laboratory and grows very slowly in experimental animals, and for these reasons has not been crossed successfully. This group is homogeneous in terms of biochemical markers (Gibson *et al.*, 1980; Pays *et al.*, 1983; Paindavoine *et al.*, 1986b, 1989; Godfrey *et al.*, 1987) and thus genetic exchange would not readily be detectable within the group. Unlike *T. b. rhodesiense*, Group 1 *T. b. gambiense* is always highly resistant to human serum *in vitro* (Mehlitz *et al.*, 1982) and appears to be better adapted to its human hosts in terms of prolonged infection and lack of dependence on animal reservoir hosts. The combination of this anthropophilic phenotype with the traits of high transmissibility and virulence seems potentially disastrous. Gibson *et al.* (1980) speculated that the highly virulent Busoga strain of *T. b. rhodesiense* in Uganda might have arisen in just this way.

Even if such hybridization events are relatively rare in the field, there are clearly extremely important implications for the epidemiology of sleeping sickness. Similarly, for leishmaniasis, where there is convincing evidence of natural hybrids (see Section 3.3.5), and possibly also for Chagas disease, the effects of genetic exchange in rapidly generating parasites with new combinations of determinants for virulence, pathogenicity and vector/host specificity cannot be disregarded.

4.2. Experimental Versus Population Genetics Analysis of *T. brucei*

Despite the high frequency of genetic exchange observable in laboratory crosses (see Section 2.1.3), the structure of natural populations appears to be largely clonal, as certain alleles and allelic combinations are not found. This results in substantial deviations from Hardy–Weinberg equilibrium, and there is often a disproportionately large number of heterozygotes (see Section 3.3.1). It is possible that some of these results stem from biased sampling; for example, isolates from wildlife are under-represented, while those from human patients are very much over-represented. However, there are several other reasons for the discrepancy between the results of field and laboratory analyses.

4.2.1. *Molecular Considerations*

A surprising number of loci for housekeeping genes in *T. brucei* show tandem repetition. For example, about half the loci for glycolytic enzymes which have been analysed contain repeated genes (phosphoglycerate kinase, PGK; aldolase; glycosomal glyceraldehyde phosphate dehydrogenase; pyruvate kinase); the known single genes include triose phosphate isomerase, glucose phosphate isomerase (GPI) and cytosolic glyceraldehyde phosphate dehydrogenase (Michels *et al.*, 1986, 1991; Michels, 1987; Allert *et al.*, 1991); the duplicate genes appeared to be identical except in the case of PGK (Osinga *et al.*, 1985). Other loci have also been found to have repeated genes, e.g. tubulin (Seebeck *et al.*, 1983; Thomashow *et al.*, 1983); calmodulin (Tschudi *et al.*, 1985) and procyclic acidic repetitive protein (PARP)/procyclin (König *et al.*, 1989). The trypanosome appears to use gene duplication as a means to increase gene expression (Michels *et al.*, 1986). Clearly, alleles in *cis* will dissociate far less frequently than those in *trans* during meiosis, giving rise to apparent fixed heterozygosity, a phenomenon frequently observed for isoenzyme data. The extent of gene duplication for the enzyme loci used in such analysis is unknown, but, by extrapolation, is predicted to be high.

With a few exceptions, most *T. brucei* isolates have been found to have tubulin, PGK and GPI genes on the same chromosome (Gibson and Borst, 1986). By chance, some of the crosses carried out used two West African *T. b. brucei* stocks in which the GPI genes were located on a different chromosome to the PGK and tubulin genes (TSW 196 and KP2 in crosses E, F and G, K respectively, Table 1). Mating parents with differences in gene linkage would potentially give rise to progeny with incorrect gene dosage, given that a meiotic division were involved in genetic exchange. If the progeny with incorrect gene dosage were metabolically inferior, such crosses would be partially infertile. Thus, incipient speciation might also underlie non-random mating.

4.2.2. *Biological Considerations*

In the laboratory, *T. brucei* isolates with widely separate geographic origins have been artificially crossed; the 'unnaturalness' of mating such distantly related trypanosomes may be one cause of the observed high frequency of triploid hybrids (see Section 2.1.4.b). In analysis of field isolates, too, there has been little regard for what might constitute a population of related trypanosomes. Yet defining the effective population is a fundamental problem for the population genetics analysis of any organism. Initial analyses considered *T. brucei* isolates from vast areas of Africa as one population. With a parasite such as *T. brucei*, which may exist in sympatric yet non-overlapping transmission cycles, depending on vector and host specificity, it seems likely that population structure will be multi-layered and complex. The assumption that isolates from one locality, however small, will constitute a single population is clearly simplistic.

For genetic exchange to occur, a tsetse fly must have a mixed infection. Yet tsetse are refractory to infection with *T. brucei*: field infection rates are generally less than 1%, and flies can readily be infected only at the first blood meal (Wijers, 1958). Flies can be superinfected with a second *T. brucei* strain at sequential blood meals, but the chance of a mixed infection is far higher when both trypanosomes are fed simultaneously at the first blood meal (Gibson and Ferris, 1992). The work of Maudlin and colleagues has revealed the basis for the susceptibility of teneral flies: lectin-mediated lysis of trypanosomes takes place in the midgut, and lectin secretion is stimulated by the first blood meal (Maudlin and Welburn, 1987; Welburn and Maudlin, 1992). So the chance of a mixed infection in the fly, and therefore of genetic exchange, is highest when the fly takes its first blood meal from a host with a mixed infection. This is unlikely to be a human host, since there are few reports of mixed infections among the many human isolates characterized by isoenzymes. Mixed infections of more than one trypanosome species have certainly been found in domestic animals such as pigs and cattle, and probably also occur in wild animals. These hosts also seem to be the likely source of mixed infections for the fly. Loss of these sources by increasing use of trypanocidal drugs and decline of wildlife may have disrupted natural transmission cycles to such an extent that genetic exchange in trypanosomes now occurs in only a few residual areas.

4.3. Phylogenetic Perspective

Parasitologists tend to regard the pathogens they work with as somehow special, and expect them to obey different rules from ordinary eukaryotes. For example, trypanosomatids have been assumed to be asexual, although there is no reason to believe they have avoided the general biological imperative for genetic recombination. On this basis it seems reasonable to assume that the group is primarily sexual, but has secondarily lost this capacity in some lineages. Alternatively, the branch leading to *T. brucei* may have evolved genetic exchange *de novo*; however, given that genetic exchange involves meiosis, this assumes the unlikely scenario that this complex process evolved more than once in eukaryotes. The existence of genetic exchange in *T. brucei* therefore implies it was the ancestral state.

Trypanosomatids form an anciently diverging branch on the eukaryote tree (Sogin *et al.*, 1986), and molecular phylogenetic analysis has further revealed that trypanosomes are monophyletic and that the *T. brucei* and *T. cruzi* lineages probably split about 100 million years ago (Figure 6; Alvarez *et al.*, 1996; Lukes *et al.*, 1997; Haag *et al.*, 1998; Stevens *et al.*, 1998). This suggests that, in addition to *T. brucei*, genetic exchange will most likely be found in related salivarian tsetse-transmitted species such as *T. congolense* and *T. vivax*. The evolutionary gulf between the salivarian trypanosomes and other lineages rules out assumptions about loss or retention of genetic exchange in other branches of the tree.

In phylogenetic trees with sufficient definition to assign meaningful time estimates to clades, the salivarian trypanosomes appear consistently as one of the most rapidly evolving groups. To what extent this might have been influenced by the capacity to undergo genetic exchange remains to be explored. This clade is also characterized by antigenic variation. Again, one can speculate whether genetic exchange was retained in order to enhance the generation of diversity in antigenic repertoires as part of the unceasing host–parasite arms race.

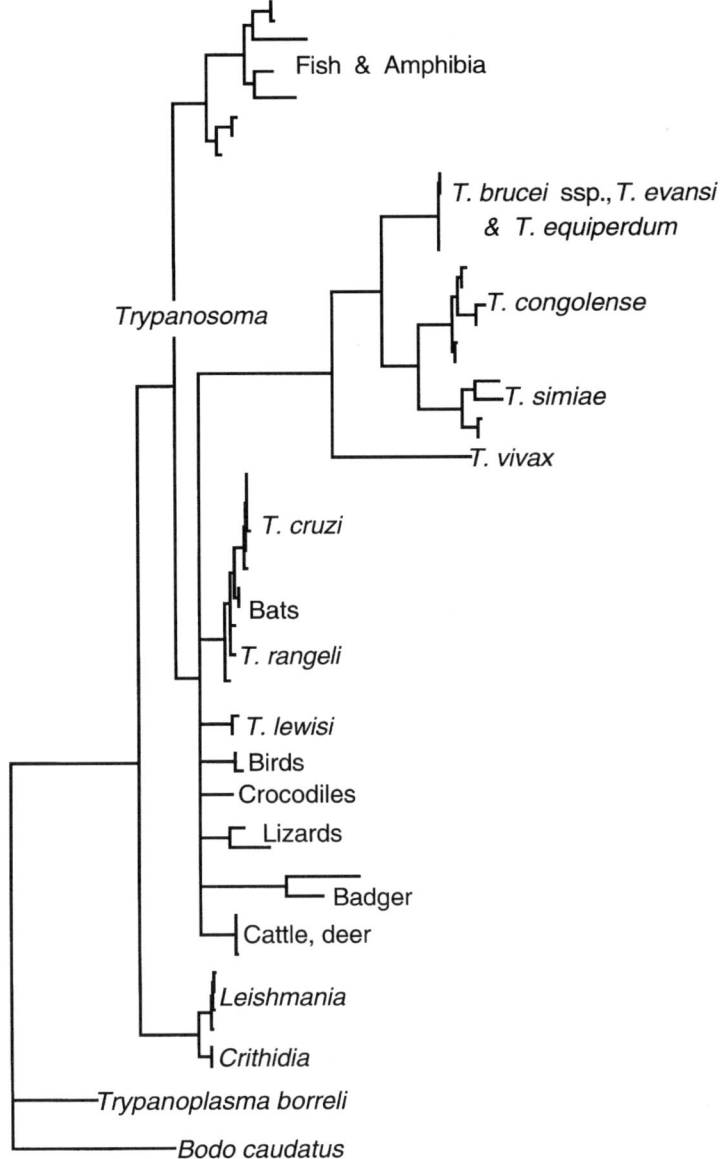

Figure 6 Phylogenetic tree of Trypanosomatidae based on maximum parsimony analysis of 18S ribosomal RNA sequences (after Stevens *et al.*, 1998). Salivarian trypanosomes are placed together in what appears to be the most rapidly evolving clade within the genus *Trypanosoma* (see Section 4.3).

ACKNOWLEDGEMENTS

W.G. and J.S. are currently supported by The Wellcome Trust. We thank S. Brisse, I. Sidibe, M. Tibayrenc, J. Haag and P. Overath for access to material before publication and M. Bailey for help with the Figures.

REFERENCES

Allert, S., Ernest, I., Poliszczak, A., Opperdoes, F.R. and Michels, P.A.M. (1991). Molecular cloning and analysis of two tandemly linked genes for pyruvate kinase of *Trypanosoma brucei*. *European Journal of Biochemistry* **200**, 19–27.

Allsopp, B.A. and Newton, S.D. (1985). Characterization of *Trypanosoma (Duttonella) vivax* by isoenzyme analysis. *International Journal for Parasitology* **15**, 265–270.

Alvarez, F., Cortinas, M.N. and Musto, H. (1996). The analysis of protein coding genes suggests monophyly of *Trypanosoma*. *Molecular Phylogenetics and Evolution* **5**, 333–343.

Bañuls, A.L., Guerrini, F., Le Pont, F., Barrera, C., Espinel, I., Guderian, R., Echeverria, R. and Tibayrenc, M. (1997). Evidence for hybridization by multilocus enzyme electrophoresis and random amplified polymorphic DNA between *Leishmania braziliensis* and *Leishmania panamensis/guyanensis* in Ecuador. *Journal of Eukaryotic Microbiology* **44**, 408–411.

Bastien, P., Blaineau, C. and Pagès, M. (1992). *Leishmania*: sex, lies and karyotype. *Parasitology Today* **8**, 174–177.

Belli, A.A., Miles, M.A. and Kelly, J.M. (1994). A putative *Leishmania panamensis/Leishmania braziliensis* hybrid is a causative agent of human cutaneous leishmaniasis in Nicaragua. *Parasitology* **109**, 435–442.

Birky, C.W. (1983). Relaxed cellular controls and organelle heredity. *Science* **222**, 468–475.

Blaineau, C., Bastien, P. and Pagès, M. (1992). Multiple forms of chromosome-I, chromosome-II and chromosome-V in a restricted population of *Leishmania infantum* contrasting with monomorphism in individual strains suggest haploidy or automixy. *Molecular and Biochemical Parasitology* **50**, 197–204.

Bogliolo, A.R., Lauria-Pires, L. and Gibson, W.C. (1996). Polymorphisms in *Trypanosoma cruzi*: evidence of genetic recombination. *Acta Tropica* **61**, 31–40.

Bonfante-Garrido, R., Meléndez, E., Barroeta, S., Mejía de Alejos, M.A., Momen, H., Cupolillo, E., McMahon-Pratt, D. and Grimaldi, G., jr (1992). Cutaneous leishmaniasis in western Venezuela caused by infection with *Leishmania venezuelensis* and *L. braziliensis* variants. *Transactions of the Royal Society of Tropical Medicine and Hygiene* **86**, 141–148.

Brisse, S. (1997). *Phylogénie moléculaire des clones naturels de* Trypanosoma cruzi, *agent de la maladie de Chagas: évolution clonale, recombinaison génétique, et relations phylogénétiques avec d'autres espèces du sous-genre* Schizotrypanum. PhD thesis, Université de Montpellier II, Montpellier, France.

Brown, A.H.D. and Feldman, M.W. (1981). Population structure of multilocus associations. *Proceedings of the National Academy of Sciences of the USA* **78**, 5913–5916.

Brun, R. and Jenni, L. (1987). Human serum resistance of metacyclic forms of *Trypanosoma brucei brucei*, *T. b. rhodesiense* and *T. b. gambiense*. *Parasitology Research* **73**, 218–223.

Buyst, H. (1974). The epidemiology, clinical features, treatment and history of sleeping sickness on the northern edge of the Luangwa fly belt. *Medical Journal of Zambia* **8**, 2–12.

Carrasco, H.J., Frame, I.A., Valente, S.A. and Miles, M.A. (1996). Genetic exchange as a possible source of genomic diversity in sylvatic populations of *Trypanosoma cruzi*. *American Journal of Tropical Medicine and Hygiene* **54**, 418–424.

Cibulskis, R.E. (1986). *Mutation and recombination in the Trypanosomatidae.* In: Leishmania. *Taxonomie et phylogenèse*, pp. 297–303. Montpellier: IMEEE, Colloque internationale.

Cibulskis, R.E. (1988). Origins and organization of genetic diversity in natural populations of *Trypanosoma brucei*. *Parasitology* **96**, 303–322.

Cibulskis R.E. (1992). Genetic variation in *Trypanosoma brucei* and the epidemiology of sleeping sickness in the Lambwe Valley, Kenya. *Parasitology* **104**, 99–109.

Darce, M., Moran, J., Palacios, X., Belli, A., Gomez-Urcuyo, F., Zamora, D., Valle, S., Gantier, J.C., Momen, H. and Grimaldi, G., jr (1991). Etiology of human cutaneous leishmaniasis in Nicaragua. *Transactions of the Royal Society of Tropical Medicine and Hygiene* **85**, 58–59.

Deane, M.P. and Milder, R. (1972). Ultrastructure of the cyst-like bodies of *Trypanosoma conorhini*. *Journal of Protozoology* **19**, 28–42.

Degen, R., Pospichal, H., Enyaru, J. and Jenni, L. (1995). Sexual compatibility among *Trypanosoma brucei* isolates from an epidemic area in southeastern Uganda. *Parasitology Research* **81**, 253–257.

Dickin, S.K. and Gibson, W.C. (1989). Hybridization with a repetitive DNA probe reveals the presence of small chromosomes in *Trypanosoma vivax*. *Molecular and Biochemical Parasitology* **33**, 135–142.

Dirie, M.F., Otte, M.J., Thatthi, R. and Gardiner, P.R. (1993). Comparative studies of *Trypanosoma (Duttonella) vivax* isolates from Colombia. *Parasitology* **106**, 21–29.

Dujardin, J.C., Llanos-Cuentas, A., Cacares, A., Arana, M., Dujardin, J.P., Guerrini, F., Gomez, J., Arroyo, J., De Doncker, S., Jacquet, D., Hamers, R., Guerra, H., Le Ray, D. and Arevalo, J. (1993). Molecular karyotype variation in *Leishmania (Viannia) peruviana*: indication of geographical populations in Peru distributed along a north–south cline. *Annals of Tropical Medicine and Parasitology* **87**, 335–347.

Dujardin, J.C., Bañuls, A.L., Llanos-Cuentas, A., Alvarez, E., DeDoncker, S., Jacquet, D., Le Ray, D., Arevalo, J. and Tibayrenc, M. (1995). Putative *Leishmania* hybrids in the eastern Andean Valley of Huanuco, Peru. *Acta Tropica* **59**, 293–307.

Dvorak, J.A. (1984). The natural heterogeneity of *Trypanosoma cruzi*: biological and medical implications. *Journal of Cellular Biochemistry* **24**, 357–371.

Eid, J. and Sollner-Webb, B. (1991). Stable integrative transformation of *Trypanosoma brucei* that occurs exclusively by homologous recombination. *Proceedings of the National Academy of Sciences of the USA* **88**, 2118–2121.

Ellis, D.S., Evans, D.A. and Stamford, S. (1982). Studies by electron microscopy of the giant forms of some African and South American trypanosomes found other than within their mammalian host. *Folia Parasitologia* (Praha) **29**, 5–11.

Enyaru, J.C.K., Stevens, J.R., Odiit, M., Okuna, N.M. and Carasco, J.F. (1993). Isoenzyme comparison of *Trypanozoon* isolates from two sleeping sickness areas of south-eastern Uganda. *Acta Tropica* **55**, 97–115.

Enyaru, J.C.K., Matovu, E., Odiit, M., Okedi, L.A., Rwendeire, A.J.J. and Stevens, J.R. (1997). Genetic diversity of *Trypanosoma brucei* from mainland and Lake Victoria island populations in south-east Uganda: epidemiological and control implications. *Annals of Tropical Medicine and Parasitology* **91**, 107–113.

Evans, D.A., Kennedy, W.P.K., Elbihari, S., Chapman, C.J., Smith, V. and Peters, W. (1987). Hybrid formation within the genus *Leishmania*? *Parassitologia* **29**, 165–173.

Fairburn, D.J. and Roff, D.A. (1980). Testing genetic models of isoenzyme variability without breeding data: can we depend on the X2? *Canadian Journal of Fisheries and Aquatic Science* **37**, 1149–1159.

Farris, J.S. (1970). Methods for computing Wagner trees. *Systematic Zoology* **19**, 83–92.

Frasch, A.C.C., Borst, P. and Van den Burg, J. (1982). Rapid evolution of genes coding for variant surface glycoproteins in trypanosomes. *Gene* **17**, 197–211.

Gardiner, P.R. (1989). Recent studies of the biology of *Trypanosoma vivax*. *Advances in Parasitology* **28**, 229–317.

Garside, L.H. and Gibson, W.C. (1995). Molecular characterisation of trypanosome species and subgroups within subgenus *Nannomonas*. *Parasitology* **111**, 301–312.

Gashumba, J.K., Baker, R.D. and Godfrey, D.G. (1988). *Trypanosoma congolense*: the distribution of enzymic variants in East and West Africa. *Parasitology* **96**, 475–486.

Gibson, W.C. (1986). Will the real *Trypanosoma brucei gambiense* please stand up? *Parasitology Today* **2**, 255–257.

Gibson, W.C. (1989). Analysis of a genetic cross between *Trypanosoma brucei rhodesiense* and *T. b. brucei*. *Parasitology* **99**, 391–402.

Gibson, W.C. (1990). Trypanosome diversity in Lambwe Valley, Kenya — sex or selection? *Parasitology Today* **6**, 342–343.

Gibson, W. (1995). The significance of genetic exchange in trypanosomes. *Parasitology Today* **11**, 465–468.

Gibson, W. and Bailey, M. (1994). Genetic exchange in *Trypanosoma brucei*: evidence for meiosis from analysis of a cross between drug resistant transformants. *Molecular and Biochemical Parasitology* **64**, 241–252.

Gibson, W.C. and Borst, P. (1986). Size-fractionation of the small chromosomes of *Trypanozoon* and *Nannomonas* trypanosomes by pulsed field gradient gel electrophoresis. *Molecular and Biochemical Parasitology* **18**, 127–140.

Gibson, W. and Ferris, V. (1992). Sequential infection of tsetse flies with *Trypanosoma congolense* and *T. brucei*. *Acta Tropica* **50**, 345–352.

Gibson, W. and Garside, L. (1990). Kinetoplast DNA minicircles are inherited from both parents in genetic hybrids of *Trypanosoma brucei*. *Molecular and Biochemical Parasitology* **42**, 45–54.

Gibson, W. and Garside, L. (1991). Genetic exchange in *Trypanosoma brucei brucei*: variable location of housekeeping genes in different trypanosome stocks. *Molecular and Biochemical Parasitology* **45**, 77–90.

Gibson, W.C. and Gashumba, J.K. (1983). Isoenzyme characterization of some *Trypanozoon* stocks from a recent trypanosomiasis epidemic in Uganda. *Transactions of the Royal Society of Tropical Medicine and Hygiene* **77**, 114–118.

Gibson, W.C. and Miles, M.A. (1986). The karyotype and ploidy of *Trypanosoma cruzi*. *EMBO Journal* **5**, 1299–1305.

Gibson, W.C. and Mizen, V.H. (1997). Heritability of the trait for human infectivity in genetic crosses of *Trypanosoma brucei* spp. *Transactions of the Royal Society of Tropical Medicine and Hygiene* **91**, 236–237.

Gibson, W.C. and Wellde, B.T. (1985). Characterization of *Trypanozoon* stocks from the South Nyanza sleeping sickness focus in Western Kenya. *Transactions of the Royal Society of Tropical Medicine and Hygiene* **79**, 671–676.

Gibson, W. and Whittington, H. (1993). Genetic exchange in *Trypanosoma brucei*: selection of hybrid trypanosomes by introduction of genes conferring drug resistance. *Molecular and Biochemical Parasitology* **60**, 19–26.

Gibson, W.C., Marshall, T.F de C. and Godfrey, D.G. (1980). Numerical analysis of enzyme polymorphism: a new approach to the epidemiology and taxonomy of trypanosomes of the subgenus *Trypanozoon*. *Advances in Parasitology* **18**, 175–246.

Gibson, W.C., Wilson, A.J. and Moloo, S.K. (1983). Characterisation of *Trypanosoma (Trypanozoon) evansi* from camels in Kenya using isoenzyme electrophoresis. *Research in Veterinary Science* **34**, 114–118.

Gibson, W.C., Osinga, K.A., Michels, P.A.M. and Borst, P. (1985). Trypanosomes of subgenus *Trypanozoon* are diploid for housekeeping genes. *Molecular and Biochemical Parasitology* **16**, 231–242.

Gibson, W., Garside, L. and Bailey, M. (1992). Trisomy and chromosome size changes in hybrid trypanosomes from a genetic cross between *Trypanosoma brucei rhodesiense* and *T. b. brucei*. *Molecular and Biochemical Parasitology* **52**, 189–200.

Gibson, W., Kanmogne, G. and Bailey, M. (1995). A successful backcross in *Trypanosoma brucei*. *Molecular and Biochemical Parasitology* **69**, 101–110.

Gibson, W., Winters, K., Mizen, G., Kearns, J. and Bailey, M. (1997a). Intraclonal mating in *Trypanosoma brucei* is associated with outcrossing. *Microbiology* **143**, 909–920.

Gibson, W., Crow, M. and Kearns, J. (1997b). Kinetoplast DNA minicircles are inherited from both parents in genetic crosses of *Trypanosoma brucei*. *Parasitology Research* **83**, 483–488.

Gibson, W., Stevens, J. and Truc, P. (1998). Identification of trypanosomes: from morphology to molecular biology. In: *Progress in Human African Trypanosomiasis Research* (B. Bouteille, M. Dumas and A. Buguet, eds). Paris: Springer-Verlag.

Glassberg, J., Miyazaki, L. and Rifkin, M.R. (1985). Isolation and partial characterization of mutants of the trypanosomatid *Crithidia fasciculata* and their use in detecting genetic recombination. *Journal of Protozoology* **32**, 118–125.

Godfrey, D.G., Scott, C.M., Gibson, W.C., Mehlitz, D. and Zillmann, U. (1987). Enzyme polymorphism and the identity of *Trypanosoma brucei gambiense*. *Parasitology* **94**, 337–347.

Godfrey, D.G., Baker, R.D., Rickman, L.R. and Mehlitz, D. (1990). The distribution, relationships and identification of enzymic variants within the subgenus *Trypanozoon*. *Advances in Parasitology* **29**, 1–74.

Gottesdiener, K., Garcia-Anoveros, J., Lee, G.-S.M. and Van der Ploeg, L.H.T. (1990). Chromosome organization of the protozoan *Trypanosoma brucei*. *Molecular and Cellular Biology* **10**, 6079–6083.

Haag, J., O'Huigin, C. and Overath, P. (1998). The molecular phylogeny of trypanosomes: evidence for an early divergence of the Salivaria. *Molecular and Biochemical Parasitology* **91**, 37–49.

Hajduk, S.L., Hager, K. and Esko, J.D. (1992). High density lipoprotein-mediated lysis of trypanosomes. *Parasitology Today* **8**, 95–98.

Henriksson, J., Pettersson, U. and Solari, A. (1993). *Trypanosoma cruzi*: correlation between karyotype variability and isoenzyme classification. *Experimental Parasitology* **77**, 334–348.

Hide, G., Cattand, P., Le Ray, D., Barry, D.J. and Tait, A. (1990). The identification of *Trypanosoma brucei* subspecies using repetitive DNA sequences. *Molecular and Biochemical Parasitology* **39**, 213–226.

Hide, G., Welburn, S.C., Tait, A. and Maudlin, I. (1994). Epidemiological relationships of *Trypanosoma brucei* stocks from south east Uganda: evidence for different population structures in human infective and non-infective isolates. *Parasitology* **109**, 95–111.

Hoare, C.A. (1972). *The Trypanosomes of Mammals*. Oxford: Blackwell Scientific Publications.

Hudson, R.R. (1990). Gene genealogies and the coalescent process. *Oxford Surveys in Evolutionary Biology* **7**, 1–44.

Jenni, L. and Brun, R. (1982). A new *in vitro* test for human serum resistance of *Trypanosoma (T.) brucei*. *Acta Tropica* **39**, 281–284.

Jenni, L., Marti, S., Schweizer, J., Betschart, B., Lepage, R.W.F., Wells, J.M., Tait, A., Paindavoine, P., Pays, E. and Steinert, M. (1986). Hybrid formation between African trypanosomes during cyclical transmission. *Nature* **322**, 173–175.

Kanmogne, G.D., Stevens, J.R., Asonganyi, T. and Gibson, W.C. (1996). Genetic heterogeneity in the *Trypanosoma brucei gambiense* genome analysed by random amplification of polymorphic DNA. *Parasitology Research* **82**, 535–541.

Kelly, J.M., Law, J.M., Chapman, C.J., Van Eyes, G.J.J.M. and Evans, D.A. (1991). Evidence of genetic recombination in *Leishmania*. *Molecular and Biochemical Parasitology* **46**, 253–264.

Kilgour, V., Godfrey, D.G. and Na'isa, B.K (1975). Isoenzymes of two aminotransferases among *Trypanosoma vivax* in Nigerian cattle. *Annals of Tropical Medicine and Parasitology* **69**, 329–335.

Knowles, G., Betschart, B., Kukla, B.A., Scott, J.R. and Majiwa, P.A.O. (1988). Genetically discrete populations of *Trypanosoma congolense* from livestock on the Kenya coast. *Parasitology* **96**, 461–474.

Komba, E.K., Kibona, S.N., Ambwene, A.K., Stevens, J.R. and Gibson, W.C. (1997). Genetic diversity among *Trypanosoma brucei rhodesiense* isolates from Tanzania. *Parasitology* **115**, 571–579.

König, E., Delius, H., Carrington, M., Williams, R.O. and Roditi, I. (1989). Duplication and transcription of procyclin genes in *Trypanosoma brucei*. *Nucleic Acids Research* **17**, 8727–8739.

Kooy, R.F., Hirumi, H., Moloo, S.K., Nantulya, V.M., Dukes, P., Van der Linden, P.M., Duijndam, W.A.L., Janse, C.J. and Overdulve, J.P. (1989). Evidence for diploidy in metacyclic forms of African trypanosomes. *Proceedings of the National Academy of Sciences of the USA* **86**, 5469–5472.

Kreutzer, R.D., Yemma, J.J., Grogl, M., Tesh, R.B. and Martin, T. I. (1994). Evidence of sexual reproduction in the protozoan parasite *Leishmania* (Kinetoplastida: Trypanosomatidae). *American Journal of Tropical Medicine and Hygiene* **51**, 301–307.

Kukla, B.A., Majiwa, P.A.O., Young, C.J., Moloo, S.K. and Ole-Moiyoi, O.K. (1987). Use of species-specific DNA probes for the detection and identification of trypanosome infections in tsetse flies. *Parasitology* **95**, 1–26.

Lanotte, G. and Rioux, J.A. (1990). Fusion cellulaire chez les *Leishmania* (Kinetoplastida, Trypanosomatidae). *Comptes Rendus de l'Académie des Sciences, Paris* **310**, 285–288.

Le Blancq, S.M., Cibulskis, R.E. and Peters, W. (1986). *Leishmania* in the Old

World: 5. Numerical analysis of isoenzyme data. *Transactions of the Royal Society of Tropical Medicine and Hygiene* **80**, 517–524.

Lee, G.-S. M. and Van der Ploeg, L.H.T. (1990). Homologous recombination and stable transfection in the parasitic protozoan *Trypanosoma brucei*. *Science* **250**, 1583–1587.

Lewicka, K., Brenière, S.F., Barnabé, C., Dedet, J.P. and Tibayrenc, M. (1995). An isoenzyme survey of *Trypanosoma cruzi* genetic variability in sylvatic cycles from French Guiana. *Experimental Parasitology* **81**, 20–28.

Lukes, J., Jirku, M., Dolezel, D., Kral'ova, I., Hollar, L. and Maslov, D.A. (1997). Analysis of ribosomal RNA genes suggests that trypanosomes are monophyletic. *Journal of Molecular Evolution* **44**, 521–527.

Lun, Z.R., Brun, R. and Gibson, W.C. (1992). Kinetoplast DNA and molecular karyotypes of *Trypanosoma evansi* and *T. equiperdum* from China. *Molecular and Biochemical Parasitology* **50**, 189–196.

Maazoun, R., Lanotte, G., Rioux, J.A., Pasteur, N., Killick-Kendrick, R. and Pratlong, F. (1981). Signification du polymorphisme enzymatique chez les leishmanies. *Annales de Parasitologie Humaine et Comparée* **56**, 467–475.

Majiwa, P.A.O., Maina, M., Waitumbi, J.N., Mihok, S. and Zweygarth, E. (1993). *Trypanosoma (Nannomonas) congolense*: molecular characterisation of a new genotype from Tsavo, Kenya. *Parasitology* **106**, 151–162.

Mantel, N. (1967). The detection of disease clustering and a generalized regression approach. *Cancer Research* **27**, 209–220.

Masake, R.A., Nantulya, V.M., Pelle, R., Makau, J.M., Gathuo, H. and ole-MoiYoi, O.K. (1994). A species-specific antigen of *Trypanosoma (Duttonella) vivax* detectable in the course of infection is encoded by a differentially expressed tandemly reiterated gene. *Molecular and Biochemical Parasitology* **64**, 207–218.

Mathieu-Daudé, F., Bicart-See, A., Bosseno, M.F., Brenière, S.F. and Tibayrenc, M. (1994). Identification of *Trypanosoma brucei gambiense* group 1 by a specific kinetoplast DNA probe. *American Journal of Tropical Medicine and Hygiene* **50**, 13–19.

Mathieu-Daudé, F., Stevens, J.R., Welsh, J., Tibayrenc, M. and McClelland, M. (1995). Genetic diversity and population structure of *Trypanosoma brucei*: clonality versus sexuality. *Molecular and Biochemical Parasitology* **72**, 89–101.

Maudlin, I. and Welburn, S.C. (1987). Lectin mediated establishment of midgut infections of *T. congolense* and *T. brucei* in *Glossina morsitans*. *Tropical Medicine and Parasitology* **38**, 167–170.

Maynard Smith, J., Smith, N.H., O'Rourke, M. and Spratt, B.G. (1993). How clonal are bacteria? *Proceedings of the National Academy of Sciences of the USA* **90**, 4384–4388.

McDaniel, J.P. and Dvorak, J.A. (1993). Identification, isolation and characterization of naturally-occurring *Trypanosoma cruzi* variants. *Molecular and Biochemical Parasitology* **57**, 213–222.

Mehlitz, D., Zillmann, U., Scott, C.M. and Godfrey, D.G. (1982). Epidemiological studies on the animal reservoir of gambiense sleeping sickness. III. Characterization of *Trypanozoon* stocks by isoenzymes and sensitivity to human serum. *Tropenmedizin und Parasitologie* **33**, 113–118.

Michels, P.A.M. (1987). Genomic organization and gene structure in African trypanosomes. In: *Gene Structure in Eukaryotic Microbes* (J.R. Kinghorn, ed.), pp. 243–262. Oxford: IRL Press.

Michels, P.A.M., Poliszczak, A., Osinga, K.A., Misset, O., Van Beeumen, J., Wierenga, R.K., Borst, P. and Opperdoes, F.R. (1986). Two tandemly-linked identical genes code for the glycosomal glyceraldehyde-phosphate dehydrogenase in *Trypanosoma brucei. EMBO Journal* **5**, 1049–1056.

Michels, P.A.M., Marchand, M., Kohl, L., Allert, S., Wierenga, R.K. and Opperdoes, F.R. (1991). The cytosolic and glycosomal isoenzymes of glyceraldehyde-3-phosphate dehydrogenase in *Trypanosoma brucei* have a distant evolutionary relationship. *European Journal of Biochemistry* **198**, 421–428.

Mihok, S., Otieno, L.H. and Darji, N. (1990). Population genetics of *Trypanosoma brucei* and the epidemiology of human sleeping sickness in the Lambwe Valley, Kenya. *Parasitology* **100**, 219–233.

Molyneux, D.H. (1969). Intracellular stages of *Trypanosoma lewisi* in fleas and attempts to find such stages in other trypanosomes. *Parasitology* **59**, 737–744.

Murray, A.K. (1982). Characterisation of stocks of *Trypanosoma vivax*. I. Isoenzyme studies. *Annals of Tropical Medicine and Parasitology* **76**, 275–282.

Nunes, L.R., de Carvalho, M.R.C. and Buck, G.A. (1997). *Trypanosoma cruzi* strains partition into two groups based on the structure and function of the spliced leader RNA and rRNA gene promoters. *Molecular and Biochemical Parasitology* **86**, 211–224.

Nyindo, M., Chimtawi, M. and Owor, J. (1981). *Trypanosoma brucei*: evidence suggesting existence of sexual forms of parasites cultured from the tsetse, *Glossina morsitans morsitans*. *Insect Science and its Application* **1**, 171–175.

Osinga, K.A., Swinkels, B.W., Gibson, W.C., Borst, P., Veeneman, G.H., Van Boom, J.H., Michels, P.A.M. and Opperdoes, F.R. (1985). Topogenesis of microbody enzymes: a sequence comparison of the genes for the glycosomal (microbody) and cytosolic phosphoglycerate kinases of *Trypanosoma brucei*. *EMBO Journal* **4**, 3811–3817.

Paindavoine, P., Zampetti-Bosseler, F., Pays, E., Schweizer, J., Guyaux, M., Jenni, L. and Steinert, M. (1986a). Trypanosome hybrids generated in tsetse flies by nuclear fusion. *EMBO Journal* **5**, 3631–3636.

Paindavoine, P., Pays, E., Laurent, M., Geltmeyer, Y., Le Ray, D., Mehlitz, D. and Steinert, M. (1986b). The use of DNA hybridisation and numerical taxonomy in determining relationships between *Trypanosoma brucei* stocks and subspecies. *Parasitology* **92**, 31–50.

Paindavoine, P., Zampetti-Bosseler, F., Coquelet, H., Pays, E. and Steinert, M. (1989). Different allele frequencies in *Trypanosoma brucei brucei* and *T. b. gambiense* populations. *Molecular and Biochemical Parasitology* **32**, 61–72.

Pays, E., Dekerck, P., Van Assel, S., Babiker, E.A., Le Ray, D., Van Meirvenne, N. and Steinert, M. (1983). Comparative analysis of a *Trypanosoma brucei gambiense* antigen gene family and its potential use in sleeping sickness epidemiology. *Molecular and Biochemical Parasitology* **7**, 63–74.

Pearson, T.W. and Jenni, L. (1989). Detection of hybrid phenotypes in African trypanosomes by high resolution two-dimensional gel electrophoresis. *Parasitology* **76**, 63–67.

Rickman, L.R. and Robson, J. (1970). The testing of proven *Trypanosoma brucei* and *T. rhodesiense* strains by the blood incubation infectivity test. *Bulletin of the World Health Organization* **42**, 911–916.

Riley, G.R., Corell, R.A. and Stuart, K. (1994). Multiple guide RNAs for identical editing of *Trypanosoma brucei* apocytochrome *b* mRNA have an unusual minicircle location and are developmentally regulated. *Journal of Biological Chemistry* **269**, 6101–6108.

Robertson, D.L., Sharp, P.M., McCutchan, F.E. and Hahn, B.H. (1995). Recombination in HIV-1. *Nature* **374**, 124–126.

Robinson, D.R. and Gull, K. (1991). Basal body movements as a mechanism for mitochondrial genome segregation in the trypanosome cell cycle. *Nature* **352**, 731–733.
Sanchez, G., Wallace, A., Munoz, S., Venegas, J., Ortiz, S. and Solari, A. (1993). Characterization of *Trypanosoma cruzi* populations by several molecular markers supports a clonal mode of reproduction. *Biological Research* **26**, 167–176.
Schweizer, J. and Jenni, L. (1991). Hybrid formation in the lifecycle of *Trypanosoma (T.) brucei*: detection of hybrid trypanosomes in a midgut-derived isolate. *Acta Tropica* **48**, 319–321.
Schweizer, J., Tait, A. and Jenni, L. (1988). The timing and frequency of hybrid formation in African trypanosomes during cyclical transmission. *Parasitology* **75**, 98–101.
Schweizer, J., Pospichal, H. and Jenni, L. (1991). Hybrid formation between African trypanosomes *in vitro*. *Acta Tropica* **49**, 237–240.
Schweizer, J., Pospichal, H., Hide, G., Buchanan, N., Tait, A. and Jenni, L. (1994). Analysis of a new genetic cross between 2 East African *Trypanosoma brucei* clones. *Parasitology* **109**, 83–93.
Scott, A.G., Tait, A. and Turner, C.M.R. (1996). Characterization of cloned lines of *Trypanosoma brucei* expressing stable resistance to MelCy and suramin. *Acta Tropica* **60**, 251–262.
Seebeck, T., Whittaker, P.A., Imboden, M.A., Hardman, N. and Braun, R. (1983). Tubulin genes of *Trypanosoma brucei*: a tightly clustered family of alternating genes. *Proceedings of the National Academy of Sciences of the USA* **80**, 4634–4638.
Shapiro, T.A. (1993). Kinetoplast DNA maxicircles: networks within networks. *Proceedings of the National Academy of Sciences of the USA* **90**, 7809–7813.
Shapiro, T.A. and Englund, P. (1995). The structure and replication of kinetoplast DNA. *Annual Review of Microbiology* **49**, 117–143.
Sidibé I. (1997). *Variabilité génétique de* Trypanosoma congolense, *agent de la trypanosomose animale: implications taxonomiques et épidémiologiques*. PhD thesis, Université de Montpellier II, Montpellier, France.
Simpson, L. (1986). The mitochondrial genome of kinetoplastid protozoa: genomic organisation, transcription, replication and evolution. *Annual Review of Microbiology* **41**, 363–382.
Sogin, M.L., Elwood, H.J. and Gunderson, J.H. (1986). Evolutionary diversity of eukaryotic small subunit rRNA genes. *Proceedings of the National Academy of Sciences of the USA* **83**, 1383–1387.
Souto, R.P., Fernandes, O., Macedo, A.M., Campbell, D.A. and Zingales, B. (1996). DNA markers define two major phylogenetic lineages of *Trypanosoma cruzi*. *Molecular and Biochemical Parasitology* **83**, 141–152.
Souza, V., Nguyen, T.T., Hudson, R.R., Piuero, D. and Lenski, R.E. (1992). Hierarchichal analysis of linkage disequilibrium in *Rhizobium* populations: evidence for sex? *Proceedings of the National Academy of Sciences of the USA* **89**, 8389–8393.
Sternberg, J., Tait, A., Haley, S., Wells, J.M., Lepage, R.W.F., Schweizer, J. and Jenni, L. (1988). Gene exchange in African trypanosomes: characterization of a new hybrid genotype. *Molecular and Biochemical Parasitology* **27**, 191–200.
Sternberg, J., Turner, C.M.R., Wells, J.M., Ranford-Cartwright, L.C., Lepage, R.W.F. and Tait, A. (1989). Gene exchange in African trypanosomes: frequency and allelic segregation. *Molecular and Biochemical Parasitology* **34**, 269–280.
Stevens, J.R. and Godfrey, D.G. (1992). Numerical taxonomy of *Trypanozoon* based on polymorphisms in a reduced range of enzymes. *Parasitology* **104**, 75–86.

Stevens, J.R. and Tibayrenc, M. (1995). Detection of linkage disequilibrium in *Trypanosoma brucei* isolated from tsetse flies and characterised by RAPD analysis and isoenzymes. *Parasitology* **110**, 181–186.
Stevens, J.R. and Tibayrenc, M. (1996). *Trypanosoma brucei* s.l.: evolution, linkage and the clonality debate. *Parasitology* **112**, 481–488.
Stevens, J.R. and Welburn, S.C. (1993). Genetic processes within an epidemic of sleeping sickness in Uganda. *Parasitology Research* **79**, 421–427.
Stevens, J.R., Nunes, V.L.B., Lanham, S.M. and Oshiro, E.T. (1989). Isoenzyme characterization of *Trypanosoma evansi* from capybaras and dogs in Brazil. *Acta Tropica* **46**, 213–222.
Stevens J.R., Noyes, H.A., Dover, G.A. and Gibson, W.C. (1998). The ancient and divergent origins of the human pathogenic trypanosomes *Trypanosoma brucei* and *T. cruzi*. *Parasitology* **118**, in press.
Stuart, K. and Feagin, J.E. (1992). Mitochondrial DNA of kinetoplastids. *International Review of Cytology* **141**, 65–88.
Sturm, N.R. and Simpson, L. (1990). Kinetoplast DNA minicircles encode guide RNAs for editing of cytochrome oxidase subunit II mRNA. *Cell* **61**, 879–884.
Tait, A. (1980). Evidence for diploidy and mating in trypanosomes. *Nature* **287**, 536–538.
Tait, A. (1983). Sexual processes in the Kinetoplastida. *Parasitology* **86**, 29–57.
Tait, A., Babiker, E.A. and Le Ray, D. (1984). Enzyme variation in *Trypanosoma brucei* spp. I. Evidence for the subspeciation of *T. b. gambiense*. *Parasitology* **89**, 311–326.
Tait, A., Barry, J.D., Wink, R., Sanderson, A.D. and Crowe, J.S. (1985). Enzyme variation in *Trypanosoma brucei* spp. II. Evidence for *T. b. rhodesiense* being a set of variants of *T. b. brucei*. *Parasitology* **90**, 89–100.
Tait, A., Turner, C.M.R., Le Page, R.F.W. and Wells, J.M. (1989). Genetic evidence that metacyclic forms of *Trypanosoma brucei* are diploid. *Molecular and Biochemical Parasitology* **37**, 247–256.
Tait, A., Buchanan, N., Hide, G. and Turner, M. (1996). Self-fertilization in *Trypanosoma brucei*. *Molecular and Biochemical Parasitology* **76**, 31–42.
Ten Asbroek, A.L.M.A., Ouellette, M. and Borst, P. (1990). Targeted insertion of the neomycin phosphotransferase gene into the tubulin gene cluster of *Trypanosoma brucei*. *Nature* **348**, 174–175.
Thomashow, L.S., Milhausen, M., Rutter, M. and Agabian, N. (1983). Tubulin genes are tandemly linked and clustered in the genome of *Trypanosoma brucei*. *Cell* **32**, 35–43.
Tibayrenc, M. (1995). Population genetics of parasitic protozoa and other microorganisms. *Advances in Parasitology* **36**, 47–115.
Tibayrenc, M. (1996). Towards a unified evolutionary genetics of microorganisms. *Annual Review of Microbiology* **50**, 401–429.
Tibayrenc, M. (1998). Genetic epidemiology of parasitic protozoa and other infectious agents: the need for an integrated approach. *International Journal for Parasitology* **28**, 85–104.
Tibayrenc, M. and Ayala, F. (1987). Forte corrélation entre classification isoenzymatique et variabilité de l'ADN kinétoplastique chez *Trypanosoma cruzi*. *Comptes Rendus de l'Académie des Sciences, Paris* **304**, 89–93.
Tibayrenc, M., Cariou, M.L. and Solignac, M. (1981). Interprétation génétique des zymogrammes de flagellés des genres *Trypanosoma* et *Leishmania*. *Comptes Rendus de l'Académie des Sciences, Paris* **292**, 623–625.
Tibayrenc, M., Cariou, M.L., Solignac, M., Dedet, J.-P., Poch, O. and Desjeux, P.

(1985). New electrophoretic evidence of genetic variation and diploidy in *Trypanosoma cruzi*, the causative agent of Chagas' disease. *Genetica* **67**, 223–230.

Tibayrenc, M., Ward, P., Moya, A. and Ayala, F.J. (1986). Natural populations of *Trypanosoma cruzi*, the agent of Chagas disease, have a complex multiclonal structure. *Proceedings of the National Academy of Sciences of the USA* **83**, 115–119.

Tibayrenc, M., Kjellberg, F. and Ayala, F.J. (1990). A clonal theory of parasitic protozoa: the population structures of *Entamoeba*, *Giardia*, *Leishmania*, *Naegleria*, *Plasmodium*, *Trichomonas* and *Trypanosoma* and their medical and taxonomical consequences. *Proceedings of the National Academy of Sciences of the USA* **87**, 2414–2418.

Tibayrenc, M., Kjellberg, F., Arnaud, J., Oury, B., Brenière, S. F., Darda, M. L. and Ayala, F.J. (1991). Are eucaryotic microorganisms clonal or sexual? A population genetics vantage. *Proceedings of the National Academy of Sciences of the USA* **88**, 5129–5133.

Tibayrenc, M., Neubauer, K., Barnabé, C., Guerrini, F., Skarecky, D. and Ayala, F.J. (1993). Genetic characterization of six parasitic protozoa: parity between random-primer DNA typing and multilocus enzyme electrophoresis. *Proceedings of the National Academy of Sciences of the USA* **90**, 1335–1339.

Tschudi, C., Young, A.S., Ruben, L., Patton, C.L. and Richards, F.F. (1985). Calmodulin genes in trypanosomes are tandemly repeated and produce multiple mRNAs with a common 5' leader sequence. *Proceedings of the National Academy of Sciences of the USA* **82**, 3998–4002.

Turner, C.M.R., Sternberg, J., Buchanan, N., Smith, E., Hide, G. and Tait, A. (1990). Evidence that the mechanism of gene exchange in *Trypanosoma brucei* involves meiosis and syngamy. *Parasitology* **101**, 377–386.

Turner, C. M., Aslam, N., Smith, E., Buchanan, N. and Tait, A. (1991). The effects of genetic exchange on variable antigen expression in *Trypanosoma brucei*. *Parasitology* **103**, 379–386.

Turner, C.M.R., Hide, G., Buchanan, N. and Tait, A. (1995). *Trypanosoma brucei*: inheritance of kinetoplast DNA maxicircles in a genetic cross and their segregation during vegetative growth. *Experimental Parasitology* **80**, 234–241.

Van der Ploeg, L.H.T., Cornelissen, A.W.C.A., Michels, P.A.M. and Borst, P. (1984). Chromosome rearrangements in *Trypanosoma brucei*. *Cell* **39**, 213–221.

Walker, P.J. (1964). Reproduction and heredity in trypanosomes. A critical review dealing mainly with the African species in the mammalian host. *International Review of Cytology* **17**, 51–98.

Walters, L.L., Irons, K.P., Chaplin, G. and Tesh, R.B. (1993). The lifecycle of *Leishmania major* (Kinetoplastida: Trypanosomatidae) in the neotropical sandfly *Lutzomyia longipalpis* (Diptera: Psychodidae). *Journal of Medical Entomology* **30**, 179–198.

Welburn, S.C. and Maudlin, I. (1992). The nature of the teneral state in *Glossina* and its role in the acquisition of trypanosome infection in tsetse. *Annals of Tropical Medicine and Parasitology* **86**, 529–536.

Wells, J.M., Prospero, T.D., Jenni, L. and Le Page, R.W.F. (1987). DNA contents and molecular karyotypes of hybrid *Trypanosoma brucei*. *Molecular and Biochemical Parasitology* **24**, 103–116.

Wijers, D.J.B. (1958). Factors that may influence the infection rate of *Glossina palpalis* with *Trypanosoma gambiense*. I. The age of the fly at the time of the infected feed. *Annals of Tropical Medicine and Parasitology* **52**, 385–390.

Workman, P.L. (1969). The analysis of simple genetic polymorphisms. *Human Biology* **41**, 97–114.

Young, C.J. and Godfrey, D.G. (1983). Enzyme polymorphism and the distribution of *Trypanosoma congolense* isolates. *Annals of Tropical Medicine and Parasitology* **77**, 467–481.

Zampetti-Bosseler, F., Schweizer, J., Pays, E., Jenni, L. and Steinert, M. (1986). Evidence for haploidy in metacyclic forms of *Trypanosoma brucei*. *Proceedings of the National Academy of Sciences of the USA* **83**, 6063–6064.

The Host–Parasite Relationship in Neosporosis

Andrew Hemphill

Institute of Parasitology, Faculties of Veterinary Medicine and Medicine, University of Bern, Länggass-Strasse 122, CH-3012 Bern, Switzerland

Abstract 48
1. Introduction 48
2. Historical Background and Phylogenetical Status of *Neospora caninum* 49
3. The Biology of *Neospora caninum* 51
   3.1. The life cycle 51
   3.2. Morphology and ultrastructure 52
   3.3. Natural infections and experimental animal models 54
   3.4. *In vitro* cultivation 56
4. Neosporosis 57
   4.1. Clinical signs 57
   4.2. Diagnostic tools 59
5. Host–Parasite Interactions 67
   5.1. Adhesion and invasion of host cells by apicomplexan parasites 69
   5.2. The physical interaction of *N. caninum* tachyzoites with host cells .. 73
   5.3. Identification and characterization of intracellular and cell surface-associated *N. caninum* tachyzoite proteins 77
   5.4. Immunology of *N. caninum* infections 83
6. Concluding Remarks 89
Acknowledgements 89
References 89

ABSTRACT

Neospora caninum is an apicomplexan parasite which invades many different cell types and tissues. It causes neosporosis, namely stillbirth and abortion in cattle and neuromuscular disease in dogs, and has been found in several other animal species. *N. caninum* is closely related to *Toxoplasma gondii*, and controversial opinions exist with respect to its phylogenetical status. Initially, two stages of *N. caninum* had been identified, namely asexually proliferating tachyzoites and bradyzoites. The sexually produced stage of this parasite, oocysts containing sporozoites, has been found only recently. In order to answer the many open questions regarding its basic biology and its relationship with the host, a number of diagnostic tools have been developed. These techniques are based on the detection of antibodies against parasites in body fluids, the direct visualization of the parasite within tissue samples by immunohistochemistry, or the specific amplification of parasite DNA by PCR. Other studies have been aiming at the identification of specific antigenic components of *N. caninum*, and the molecular and functional characterization of these antigens with respect to the cell biology of the parasite. Clearly, molecular approaches will also be used increasingly to elucidate the immunological and pathogenetic events during infection, but also to prepare potential new immunotherapeutic tools for future vaccination against *N. caninum* infection.

1. INTRODUCTION

Neospora caninum is an apicomplexan parasite which is structurally very similar to, but antigenically distinct from, the closely related *Toxoplasma gondii*. *N. caninum* causes neosporosis, namely neuromuscular disorders, paralysis and death in dogs, and abortion and neonatal morbidity in cattle, sheep, goats, horses and deer. Because of its high prevalence in cattle, *N. caninum* has now emerged as an important cause of bovine abortion worldwide, and neosporosis has been recognized as an economically important disease affecting the livestock industry (Dubey and Lindsay, 1993, 1996).

The pioneering work of Dubey and co-workers has established *N. caninum* as an independent species (Dubey *et al.*, 1988a,b; Dubey and Lindsay, 1993). However, during the last ten years other researchers have also contributed largely to the characterization of this parasite, especially with regard to its phylogenetical status, its epidemiology and significance as a pathogenic agent, and with respect to the development of molecular diagnostic tools. In addition, research has started to tackle questions dealing with the host–parasite relationship, such as how *N. caninum* triggers and influences the host's immune system and how the parasite is capable of physically interacting with its host cell on the molecular level.

N. caninum tachyzoites can be cultured easily in the laboratory using similar techniques previously developed for *in vitro* cultivation of *T. gondii* (Dubey and Beattie, 1988). Thus, like *T. gondii*, *N. caninum* represents an excellent model to study host cell invasion mechanisms, survival strategies, and the respective parasite and host molecules which are required for these interactions. Although much could be learned from the research previously carried out on the closely related *T. gondii*, several studies have shown considerable differences between these two, and other intracellular parasites (for reviews see Kasper and Mineo, 1994; Dubremetz and McKerrow, 1995; Smith, 1995; Mauël, 1996; Sam-Yellowe, 1996; Dubremetz, 1998). The aims of this article are to provide an overview of the research carried out on *N. caninum* since its isolation and characterization as an independent species in 1988, and to review the more recent achievements in the field of neosporosis, with special emphasis directed towards the molecular mechanisms taking place at the host–parasite interface. Many aspects will be discussed in relation to findings obtained in studies of other intracellular protozoan parasites, especially the members of the phylum *Apicomplexa*.

2. HISTORICAL BACKGROUND AND PHYLOGENETICAL STATUS OF *NEOSPORA CANINUM*

Neosporosis was first reported in Norway by Bjerkås *et al.* in 1984, as encephalomyelitis and myositis in dogs caused by an unidentified cyst-forming sporozoon. In 1988, Dubey and co-workers identified a similar parasite in dogs in the USA causing the same symptoms, and thus proposed a new genus, *Neospora*, and species, *N. caninum* (Dubey *et al.*, 1988a). Isolation of *N. caninum* in cell culture was achieved, and experimental infection of dogs with parasites isolated from cell cultures showed that *N. caninum* was indeed the causative agent of neurological disorders, paresis, paralysis, and even death, in infected dogs (Dubey *et al.*, 1988b). Also in 1988, an indirect antibody fluorescent test (IFAT) was developed for the detection of anti-*N. caninum* antibodies. Polyclonal antisera against *N. caninum* tachyzoites were raised in rabbits, and these sera were used for immunohistochemical detection of the parasite in paraffin-embedded tissue sections (Lindsay and Dubey, 1989b). Using these diagnostic tools, Bjerkås and Dubey (1991) demonstrated that the parasites identified in 1984 in Norway were indeed *N. caninum*.

The first experimental animal models for investigating the biology, immunology, pathogenesis and chemotherapy of *N. caninum* infections were developed in mice and rats (Lindsay and Dubey, 1989c, 1990). It was also shown that transplacental transmission of the parasite could be induced experimentally in dogs, cats and sheep (Dubey and Lindsay, 1989a,b, 1990b). Thilstedt and Dubey were the first to identify neosporosis as a cause of abortion in dairy cattle (Thilstedt and Dubey, 1989), and in 1991 *N. caninum* was demonstrated to be a major cause of abortion of cattle in California (Anderson *et al.*, 1991; Barr *et al.*, 1991). Dubey *et al.* (1992) then showed that, upon experimental infection of cattle, transplacental transmission of the parasite could be induced. Isolation of *N. caninum* from aborted bovine foetuses was first achieved by Conrad *et al.* (1993a), and, using this first bovine isolate, bovine foetal infection and death were reproduced experimentally (Barr *et al.*, 1994b).

Specific enzyme-linked immunosorbent assays (ELISAs) were developed for the serological diagnosis of *N. caninum* infection in dogs and cattle (Björkman *et al.*, 1994a; Paré *et al.*, 1995; Dubey *et al.*, 1996a), and a direct agglutination test for serological diagnosis of *Neospora* infection has recently been established (Romand *et al.*, 1998). Lally *et al.* (1996a) produced the first bacterially expressed recombinant antigens for diagnosis. For the molecular detection of the parasite DNA within infected tissues, a number of *Neospora*-specific polymerase chain reactions (PCRs) were developed (Ho *et al.*, 1996; Holmdahl and Mattson, 1996; Lally *et al.*, 1996b; Müller *et al.*, 1996; Payne and Ellis, 1996; Yamage *et al.*, 1996).

Investigations dealing with the basic biology and pathology of neosporosis have now started to benefit from the development of molecular genetic tools for *N. caninum*. Using the existing DNA vectors originally developed for *T. gondii*, it has recently been shown that *T. gondii* proteins are faithfully expressed and correctly targeted in *N. caninum* tachyzoites (Beckers *et al.*, 1997; Howe *et al.*, 1997). This points out the suitability of *N. caninum* as a heterologous expression system for studying the functional significance of *T. gondii* proteins, and vice versa.

The phylogenetical status of *N. caninum* in relation to the genera *Toxoplasma, Sarcocystis* and other members of the tissue cyst-forming coccidia is still controversial (Tenter and Johnson, 1997). Most information on the taxonomic position of this parasite is derived from molecular data. Analysis of the small subunit ribosomal RNA (ssrRNA) sequences of several isolates of *N. caninum* and *T. gondii* revealed a consistent four-nucleotide difference between these two species (Marsh *et al.*, 1995). Several authors suggested that *N. caninum* and *T. gondii* should be placed into the same genus, namely *Toxoplasma*, since their 16s-like rRNA genes exhibit striking homologies (Ellis *et al.*, 1994; Holmdahl *et al.*, 1994). In contrast, Guo and Johnson (1995) investigated the genomes of *Neospora, Sarcocystis* and *Toxoplasma* by random amplified polymorphic DNA (RAPD)-PCR, and did not find a significantly close relationship between *Neospora* and *Toxoplasma*. In fact, this study suggested that an even closer genetic relationship exists between *Sarcocystis muris* and *T. gondii* than between *Neospora* and *Toxoplasma*. In addition it was demonstrated that homologues of the three dominant *T. gondii* genes B1, p22 and p30 are absent in *N. caninum* (Brindley et al., 1993; Müller et al., 1996), and considerable variability between the ITS 1 regions of *N. caninum* and *T. gondii* was also reported (Holmdahl and Mattsson, 1996; Homan et al., 1997). *N. caninum* is currently placed into the family *Sarcocystidae* and is established as a sister group to *T. gondii* in the phylum *Apicomplexa* (Ellis et al., 1994). However, the taxonomic position of *N. caninum* remains uncertain because the life cycle of this parasite has not yet been elucidated, so the definitive host and a sexual stage of the parasite have not been found.

3. THE BIOLOGY OF *NEOSPORA CANINUM*

3.1. The Life Cycle

Two asexually produced stages have been identified for *N. caninum*. These are the tachyzoite stage, represented by rapidly dividing parasites which can be transmitted transplacentally from mother to offspring during pregnancy, and the bradyzoite stage, which represents a slowly dividing stage, present within intracellular tissue cysts, and surrounded by a cyst wall which protects the parasites from immunological and physiological reactions on the part of the host.

N. caninum is an obligate intracellular parasite, and tachyzoites have been detected in a variety of tissues, such as the brain, spinal cord, heart, lung, liver, foetal membrane, muscle, placenta, and skin (Dubey and Lindsay, 1993, 1996). Many different cell types, including neural cells, fibroblasts, vascular endothelial cells, myocytes, renal tubular epithelial cells, hepatocytes and macrophages, were shown to harbour the parasite. This suggests that *N. caninum*, as is the case for *T. gondii*, is capable of invading a wide range of, if not all, nucleated cells. The tachyzoites, enclosed in a parasitophorous vacuole, proliferate by endodyogeny, producing several hundred new parasites in a few days p.i. Proliferating tachyzoites form a pseudocyst which is lacking a cyst wall, and as this pseudocyst has reached a critical mass, host cell lysis occurs and newly formed tachyzoites infect neighbouring cells.

As is the case for *T. gondii*, *N. caninum* bradyzoites are capable of forming intracellular tissue cysts, but they are surrounded by a solid cyst wall. These tissue cysts can persist within an infected host for several years without causing significant clinical manifestations (Dubey and Lindsay, 1996). *N. caninum* tissue cysts containing bradyzoites were detected exclusively in the central nervous system (CNS, Lindsay et al., 1993), with the one exception of a single cyst found in the ocular muscles of a foal (Lindsay et al., 1996c).

T. gondii exhibits three infective stages within its life cycle, namely tachyzoites, bradyzoites, and the sporozoites which are the product of a sexual process taking place within the intestine of its feline final host (Dubey and Beattie, 1988). Despite numerous attempts to elucidate the life cycle of *N. caninum*, a sexually produced sporozoite stage of the parasite was not found for a long time. However, a carnivorous definitive host is suspected to be involved (Dubey and Lindsay, 1996). Transplacental infection is, as in the case of congenital toxoplasmosis, a recognized mode of transmission of neosporosis. Experimentally, this process has been induced in several species, e.g. dogs (Cole et al., 1995b), cats (Dubey and Lindsay, 1989a), sheep (McAllister et al., 1996; Buxton et al., 1997a), cattle (Barr et al., 1994b), goats (Lindsay et al., 1995b), mice (Cole et al., 1995a; Long and Baszler, 1996), and even in non-human primates (Barr et al., 1994a).

Although *T. gondii* has long been recognized as an important pathogen in man, and acute toxoplasmosis caused by the activation of tissue cysts and bradyzoites can occur within immunocompromised individuals (Dubey and Beattie, 1988; Darcy and Santoro, 1994), no human cases of neosporosis have been reported to date. However, experimental infection of pregnant rhesus macaques with a bovine *Neospora* isolate resulted in transplacental transmission and foetal infections comparable to those seen in human fetuses infected with *T. gondii* (Barr et al., 1994a; Ho et al., 1997a).

3.2. Morphology and Ultrastructure

N. caninum tachyzoites and bradyzoites are morphologically and ultrastructurally very similar to the corresponding stages of *T. gondii*. Tachyzoites are approximately 5–7 µm in length and 1–2 µm wide. They are ovoid, lunate or globular, depending on the stage of division (Dubey and Lindsay, 1996). Ultrastructurally, tachyzoites derived from cell culture are virtually identical to those found within *in vivo* infected cells (Speer and Dubey, 1989; Lindsay et al., 1993). They possess a three-layered plasma membrane and an apical complex composed of microtubules, apical rings, conoid, and a polar ring.

Neospora and *Toxoplasma* contain a set of secretory organelles which is found in most members of the apicomplexa: micronemes, rhoptries and dense granules. Micronemes are small vesicular structures, and the number of these organelles is highly variable, although up to 150 micronemes per tachyzoite can be found in one cell (Dubey and Lindsay, 1996). The 8-18 rhoptries of *N. caninum* are arranged along the longitudinal axis of the cell, and they are filled with amorphous electron-dense material, while the contents of *T. gondii* tachyzoite rhoptries exhibit a distinct and defined structured morphology. This provides one means of distinguishing these two parasites by electron microscopy. Dense granules are globular organelles which are located at the anterior and posterior ends of the parasites, and they have been shown to contain molecules which are secreted shortly after the invasion of host cells (Hemphill *et al.*, 1998). In addition, *N. caninum* tachyzoites contain a Golgi complex, rough and smooth endoplasmic reticulum, a nucleus and a nucleolus, and mitochondria (Speer and Dubey, 1989; Lindsay *et al.*, 1993).

Within their host cells tachyzoites reside within a parasitophorous vacuole, a specialized compartment which is separated from the host cell cytoplasm by a parasitophorous vacuole membrane (PVM). The PVM is originally derived from the surface membrane of the host cell, but is modified by the parasite shortly after the host cell is invaded (Hemphill *et al.*, 1996, 1998). As the number of parasites within this vacuole increases, the size of the PVM is enhanced as well. A single host cell can be infected by several parasites, and therefore could contain several parasitophorous vacuoles. The intercellular space within these vacuoles is comprised of the intravacuolar tubular network, which is at least partially derived from the secretory products originating from the tachyzoites (Hemphill, 1996; Hemphill *et al.*, 1998).

N. caninum tissue cysts are up to approximately 100 µm in diameter, round to oval in shape, and contain several hundred bradyzoites. The bradyzoites are surrounded by a cyst wall (up to 4 µm in thickness) which stains variably using periodic acid Schiff (PAS), and which is argyrophilic (Dubey, 1993). This cyst wall ensures that the parasites are enclosed in a chemically and physiologically stable entity. It contains two distinct components: an outer, electron-dense, single plasma membrane, and an inner, thick, granular layer which harbours branched tubule-like structures (Bjerkås and Dubey, 1991). In *T. gondii* cyst walls, which are usually much thinner, chitin represents a significant part of the overall composition, rendering these cysts resistant to acid pepsin digestion. This property allows them to survive through the stomach and release bradyzoites in the gut (Boothroyd *et al.*, 1998). However, in order to discriminate reliably between the two parasites, the thickness of the cyst wall is not an unambiguous criterion, since its size is most likely also dependent on how long the infection has persisted, and small cysts exhibit a thinner wall than older ones (Jardine, 1996). Bradyzoites are approximately 6-8 µm long and 1-2 µm wide, and the same organelles found in tachyzoites are also present in bradyzoites. In addition, bradyzoites harbour vesiculo-membranous organelles containing short flat membranous segments and smaller vesicles (Jardine, 1996). There are fewer rhoptries than in tachyzoites, and the bradyzoites contain more amylopectin granules. The ground substance within the cysts contains tortuous and branched vesicles, small, irregular, electron-dense bodies, and lobulated lipid-like inclusions (Bjerkås and Dubey, 1991; Jardine, 1996).

3.3. Natural Infections and Experimental Animal Models

Natural infections of *N. caninum* were originally identified in dogs, and later in cattle, sheep, goats, horses and deer. The parasite has been reported in many different geographical areas worldwide (reviewed by Dubey and Lindsay, 1996). As the parasite infects its host, dissemination of the pathogen into many different tissues can take place because of the infection of, and proliferation within, cells of the reticuloendothelial system such as macrophages and lymphocytes. The predilection site for primary parasite proliferation and for the establishment of the hypobiotic cyst stage is the CNS. Tachyzoites can multiply rapidly, and repeated processes of host cell invasion, proliferation, host cell lysis, and subsequent infection of neighbouring cells produce significant necrotic lesions within affected tissues. As a consequence, severe neuromuscular disease occurs because of the destruction of neural cells in the brain and within cranial and spinal nerves, affecting the conductivity of the neural tissue (Mayhew *et al.*, 1991; Dubey and De Lahunta, 1993). In contrast, *N. caninum* tissue cysts containing the slowly dividing, hypobiotic bradyzoite stage of the parasite do not cause any significant host reaction, although formation of granulomas around degenerating tissue cysts or bradyzoites has been observed (Dubey *et al.*, 1990a, 1992; Mayhew *et al.*, 1991). Cyst rupture most likely occurs now and then, and can cause foci of inflammation (Dubey and Lindsay, 1996).

Infection probably occurs in two ways: first, oral infection can take place through the ingestion of tissue harbouring *N. caninum* tissue cysts. It has been shown that bradyzoites within tissue cysts were resistant to HCl-pepsin solution, and experimental infections of cats and dogs confirmed that *N. caninum* could be transmitted from an animal by the ingestion of tissues containing viable tissue cysts (Dubey and Lindsay, 1989b; Dubey *et al.*, 1990b). Transplacental infection of the foetus during pregnancy is the second way infection can occur. This process can occur repeatedly in the same animal (Bjerkås *et al.*, 1984; Dubey *et al.*, 1988b; Barr *et al.*, 1993). However, even if prenatal infection of the foetus takes place it does not always lead to disease, and the parasite might reside silently within tissues in clinically normal offspring.

Experimental infections of *N. caninum* have been induced in several species, including mice, rats, rabbits, gerbils, dogs, foxes, cats, sheep, goats, cattle and pigs (reviewed by Dubey and Lindsay, 1996). This was done to gain information on the life cycle and possible definitive host, to study the dissemination of the parasite and the pathogenesis of neosporosis, and to search for a suitable animal model which enables the host–parasite relationship to be studied in more detail. Experimental infection can be carried out by inoculating the parasites orally, subcutaneously, intramuscularly or by the intraperitoneal route (Dubey and Lindsay, 1993).

Several attempts were made to identify the definitive host by searching for oocysts in the faeces of experimentally infected animals. Many of those experiments focused on cats, since the cat is the definitive host of *T. gondii* (Dubey and Beattie, 1988). Cuddon *et al.* (1992) used naturally infected dog brain tissue to infect cats orally. Oral infection of cats was also carried out with tissues of naturally infected cows, experimentally infected cat brain and experimentally infected mice. In addition, subcutaneous, intramuscular and oral infection of cats using tachyzoites of the NC-1 isolate were also performed (reviewed by Dubey and Lindsay, 1996). Experimental infection of cats showed for the first time that *N. caninum* could be transmitted congenitally from an animal infected before pregnancy, and that the parasite could be transmitted orally by ingesting infected tissues containing tissue cysts (Dubey *et al.*, 1990b).

Other experimental infections of dogs (Dubey and Lindsay, 1989b), raccoons (Dubey *et al.*, 1993), coyotes (Lindsay *et al.*, 1996b), the red-tailed hawk, turkey vulture, barn owl and American crow (Baker *et al.*, 1995) were also performed. No oocysts were found in the faeces of any of these animals. However, these results are not all definitive, since in some of these experiments the presence of parasites was not always ascertained (Dubey and Lindsay, 1996). Thus an efficient method of obtaining tissue material infected with *N. caninum* tissue cysts containing bradyzoites was urgently needed.

McGuire *et al.* (1997a) developed a protocol for the production of *N. caninum* tissue cysts in mice. They tested different mouse strains, *N. caninum* isolates, tachyzoite inoculum doses and treatments for immunosuppression. They concluded that male ICR mice, treated with methylprednisolone acetate (MPA) and infected with the *N. caninum* Liverpool isolate, were most efficient with respect to production of tissue cysts and survival of mice (McGuire *et al.*, 1997a). In addition, a method for the separation and cryopreservation of *N. caninum* tissue cysts from the murine brain was worked out, and this makes it possible to store viable cysts for oral infectivity trials and other studies (McGuire *et al.*, 1997b).

Other mouse models have been developed for studying experimental *N. caninum* infection. Adult outbred Swiss Webster mice did not develop clinical signs, but encysted parasites could be detected in the CNS (Dubey et al., 1988b). Congenital infection of litters was also achieved (Cole et al., 1995a). By administering corticosteroids, severe signs of clinical neosporosis in mice could be induced. In addition, the severity of the disease was dependent on the isolate used for infection (Dubey and Lindsay, 1993). The use of immunodeficient mice such as nude mice (Yamage et al., 1996; Sawada et al., 1997), mice deficient in gamma interferon (Dubey and Lindsay, 1996) and μMT (antibody knock-out) mice (Éperon et al., 1998), has shown that they are highly susceptible to infection and fatal disease caused by *N. caninum*. Inbred BALB/c mice were also susceptible to neosporosis, although the effect was dependent on the isolate used for infection. BALB/c mice developed encephalomyelitis (Lindsay et al., 1995a), and the parasite was transmitted congenitally in these mice, resulting either in foetal death and resorption (Long and Baszler, 1996), or mice born alive but infected with parasites (Liddell et al., 1997).

Other experimental models for studying *N. caninum* infections include sheep, pygmy goats and cats. Although natural neosporosis is rarely observed in sheep (Dubey et al., 1990a), experimental infection can lead to abortion, birth of weak lambs, and birth of infected lambs without clinical signs (McAllister et al., 1996b; Buxton et al., 1997a,b). Accordingly, the sheep model provides a good alternative to study the typical features of bovine neosporosis at a reduced cost (Dubey and Lindsay, 1996). Basically the same applies to pygmy goats (Lindsay et al., 1995b).

3.4. *In vitro* Cultivation

In vitro cultivation of *N. caninum* tachyzoites has been achieved in many cell types, both primary cells and established cell lines. Essentially the same techniques for cultivation and cryopreservation as previously described for *T. gondii* tachyzoites can be used (Dubey and Beattie, 1988). Care should be taken in the selection of foetal calf serum (FCS). Many batches of commercially available batches of FCS contain antibodies directed against *N. caninum*. This can lead to agglutination and the rapid death of parasites once they are released into the medium following host cell lysis. It is not known whether this is due to the high prevalence of subclinical fetal neosporosis (Dubey and Lindsay, 1996), or whether other reasons account for this. In order to avoid complications, *in vitro* cultivation of *N. caninum* tachyzoites can be carried out using IgG-free horse serum, or FCS can be omitted towards the end of the cultivation. Tachyzoites cultured through serial passages for eight years have still been able to retain their infectivity in mice (Dubey and Lindsay, 1996).

The proliferation rate seems to vary between different isolates, and is also dependent on which host cells are used. Tachyzoites of the NC-1 isolate cultured in bovine aorta endothelial cells were going through endodyogeny within six hours p.i., and host cell lysis occurs around 72 hours p.i. (Hemphill et al., 1996). Extracellular maintenance of *N. caninum* tachyzoites in the presence of growth medium for longer than four hours results in a rapid loss of infectivity. This is in contrast to what has been reported for *T. gondii* tachyzoites, which remain infective upon extracellular maintenance for up to 72 hours (De Braganca et al., 1996).

Weiss and Ma (1997) recently reported that they succeeded in generating *N. caninum* tissue cysts containing bradyzoites *in vitro*. This was achieved through cultivation in human foreskin fibroblast, employing the protocol previously described for *in vitro* tissue cyst formation of *T. gondii* (Boothroyd et al., 1997). With *in vitro* cultivation of *N. caninum* tachyzoites, the tools for immunodiagnosis, immunohistochemical and molecular detection of the parasite by PCR could be established (see Section 4.2). More recently, the tissue culture system has also been used for investigating in more detail the processes which occur during adhesion and invasion of host cells (Hemphill et al., 1996), and has also enabled genetic manipulation of the parasite (Howe et al., 1997; Howe and Sibley, 1997; Beckers et al., 1997). *In vitro* cultures were also employed for assaying the susceptibility of *N. caninum* tachyzoites to more than 40 chemotherapeutic agents (Lindsay and Dubey, 1989a; Lindsay et al., 1994; Lindsay et al., 1996a, 1997). Two mutants resistant towards pyrimethamine have been generated. In contrast, only a few compounds have been evaluated for their efficacy against *N. caninum in vivo* (Lindsay and Dubey, 1990; Dubey et al., 1995).

4. NEOSPOROSIS

4.1. Clinical Signs

Natural neosporosis was first diagnosed in dogs; clinical symptoms include hind limb paresis and hyperextension, progressive hind limb paralysis (Cuddon et al., 1992; Barber and Trees, 1996), difficulty in swallowing, paralysis of the jaw (Hay et al., 1990), muscle flaccidity, muscle atrophy and even heart failure (Odin and Dubey, 1993). Most clinical cases have been described in Labrador retrievers, boxers, greyhounds, golden retrievers and basset hounds (Dubey and Lindsay, 1996). Parasites could be identified in most organs. *N. caninum* can also cause severe dermatitis (Dubey et al., 1988a, 1995). Most affected individuals are young congenitally infected dogs, although fatal neosporosis has also been documented in older animals (Dubey and Lindsay, 1993).

Results of experimental infections indicate that subclinical neosporosis can be reactivated by suppressing the immune system. Congenital neosporosis has been experimentally reproduced in dogs (Dubey and Lindsay, 1989b; Cole *et al.*, 1995b). Following the administration of large doses of corticosteroids, clinically normal, congenitally infected offspring developed hepatic lipidosis, pneumonia and myositis caused by tachyzoites (Dubey and Lindsay, 1990a). Although infection with *N. caninum* early in pregnancy may be associated with early foetal death and the birth of weak offspring, there is no report of abortion caused by neosporosis in dogs (Dubey and Lindsay, 1996). Postnatally infected pups can also develop clinical neosporosis (Cole *et al.*, 1995b).

The successful treatment of neosporosis in dogs is dependent on the stage of disease at the time the treatment is initiated. Clindamycin, pyrimethamine and sulfadiazine, either alone or in combination, have shown some effect in a restricted number of cases (Barber and Trees, 1996). However, there is at present no treatment to prevent diaplacentar transmission of *N. caninum*.

In cows, abortion is the most relevant clinical feature of *N. caninum* infection. Both dairy and beef cattle are affected, although most reports are from dairy cattle (Dubey and Lindsay, 1993, 1996). Cows have been reported to abort up to eight years of age. Foetuses may die *in utero*, be resorbed, mummified, autolyzed, stillborn, born alive but weak and/or diseased, or born clinically normal but chronically infected. Foetal death can occur throughout gestation although abortions of foetuses younger than three months of age have not been observed (Dubey and Lindsay, 1996). Repeated abortions have been shown to occur in the same cow (Anderson *et al.*, 1995; Dannatt *et al.*, 1995; Obendorf *et al.*, 1995). Cows may abort sporadically or in groups within a few weeks, or abortions may persist within a herd. There are several reports of abortion storms (Thornton *et al.*, 1994; Yaeger *et al.*, 1994; Moen *et al.*, 1995; McAllister *et al.*, 1996a), but diagnosis was based only on histologic examination of a few foetuses, so it is not certain whether all these abortions were really caused exclusively by *N. caninum* or whether other pathogens were also involved. A recent study has shown that *N. caninum* infection has an economic impact, in that milk production in first-lactation dairy cows is significantly reduced (Thurmond and Hietala, 1997).

Subclinical congenital neosporosis is a common phenomenon. It is likely that repeat congenital infection is more common than repeat abortion, and there is evidence that only a small percentage of congenitally infected calves develop clinical neosporosis. The clinical signs of congenitally infected calves born alive are mostly neurological. They exhibit hind limb and/or forelimb hyperextension, ataxia, decreased patellar reflexes, loss of proprioception, paralysis, asymmetrical appearance of eyes, exophthalmia, or deformities associated with embryonic neural cells (Parish *et al.*, 1987;

Barr *et al.*, 1993; Dubey and De Lahunta, 1993; Bryan *et al.*, 1994). No chemotherapeutical treatment for neosporosis in cattle has so far been developed.

Other species affected by natural neosporosis include pygmy goats (Barr *et al.*, 1992; Dubey *et al.*, 1992), dairy goats (Dubey *et al.*, 1996b) and horses (Dubey and Porterfield, 1990; Marsh *et al.*, 1996) and there has been one report of neosporosis in sheep (Dubey *et al.*, 1990a). In these cases, clinical signs resembled those seen in cattle, with infection predominantly localized within the foetal CNS. In horses, unusual cases of *N. caninum* infection have been reported, such as the observation of tissue cysts outside the CNS (Lindsay *et al.*, 1996c) and an unusual case of visceral neosporosis (Gray *et al.*, 1996). Two reports exist on the occurence of neosporosis in deer, one of which was identified in full-term stillborn deer from the Paris Zoo, France (Dubey *et al.*, 1996c).

4.2. Diagnostic Tools

There are a number of diagnostic tools available, especially for discriminating neosporosis from infections with the closely related *Toxoplasma* and *Sarcocystis*. These include indirect methods such as the demonstration of the presence of antibodies in blood, or direct methods such as electronmicroscopical, immunohistopathological and histopathological visualization of the parasite or its immunopathological effects, *in vitro* isolation of parasites or the detection of parasite DNA by PCR.

4.2.1. *Indirect Detection Techniques*

Indirect diagnosis of *N. caninum* infections relies on the detection of antibodies specifically directed against *N. caninum*. Although these techniques can demonstrate that an individual animal has been exposed to the parasite, they do not prove that *N. caninum* is in fact the causative agent of the corresponding symptoms. This can be done only by demonstrating its physical presence within affected tissue (see Section 4.2.2). However, serological assays are an important measure for demonstrating exposure to the parasite and assessing the risks of acquiring neosporosis.

(a) Indirect fluorescence antibody technique (IFAT). In the IFAT, cell culture-derived *N. caninum* tachyzoites are placed on to glass slides and are incubated with the serum to be tested. If the serum contains antibodies against *N. caninum* tachyzoites, these will bind to the parasites, and can be visualized by fluorescent reagents. The IFAT for neosporosis was first introduced to detect parasite-specific antibodies in dogs (Dubey *et al.*,

1988a). This test is the most widely used serological *N. caninum* assay and is considered the most specific (Dubey and Lindsay, 1996). No or only a little cross-reaction with antibodies to *T. gondii* tachyzoites has been observed, none with *Sarcocystis* sp., *Cryptosporidium parvum* or *Eimeria bovis*. The IFAT has been used to detect antibodies in peripheral blood and the cerebrospinal fluid of infected dogs, and in maternal and foetal sera of infected cattle. The IFAT has also been shown to represent a valuable tool for the diagnosis of bovine foetal *Neospora* infection (Barr et al., 1995). The cut-off IFAT-titres which determine the specificity of a reaction vary between different laboratories, mainly because of differences in buffers, incubation conditions and conjugates. The type and quality of a fluorescence microscope can also influence detection limits. The titres usually vary between 1:160 and 1:640 for a positive reaction in maternal serum. For the detection of antibodies in foetal serum, the titres are significantly lower (B. Hentrich, personal communication). A major setback of the IFAT is that the result relies on a personal evaluation rather than on objective assessment. However, it is generally agreed that in case of a positive signal the whole surface of the tachyzoite should fluoresce (Dubey and Lindsay, 1996). Apical fluorescence occurs in many sera of uninfected individuals, and is considered to be due to unspecific binding of serum components.

Since many batches of FCS contain antibodies against *N. caninum*, only tachyzoites cultured in the absence of FCS should be used as antigen (see Section 3.4). Alternatively, parasites can be grown in the presence of IgG-free horse serum. It has been shown that IFAT titres in *N. caninum*-infected cows can decrease after abortion, and perhaps also vary during pregnancy (Conrad et al., 1993b). Therefore individual cows might erroneously be considered free of the parasite if blood samples collected later than two months after abortion or early in pregnancy are tested. Further serological assays with increased sensitivity, but also with at least the same specificity, are therefore required.

(b) Enzyme-linked immunsorbent assay (ELISA). Major efforts were directed towards the development of ELISAs, which offer the advantages of automation and provide objective results. The use of ELISA for demonstration of antibodies to *N. caninum* can be problematic, since high background absorbances and non-specific reactions, most likely caused by cross-reactivity with related coccidian parasites, may be a common phenomenon dependent upon the type of antigen tested (Dubey and Lindsay, 1996). It has been suggested that assays based on surface membrane proteins as antigens could be more specific for the serological detection of protozoan infections (Huldt, 1981). The fact that cross-reactivity between *T. gondii* and *N. caninum* in canine sera is low in the IFAT (where mainly surface proteins are accessible to the antibodies)

points out the suitability of parasite surface molecules as antigens (Dubey and Lindsay, 1993). Different antigen preparation methods have therefore been developed in order to increase the specificity and sensitivity of ELISA assays for the detection of anti-*N. caninum* antibodies. These include the application of membrane proteins incorporated into iscoms (immunostimulating complexes, Björkman *et al*., 1994a,b), the preparation of water-soluble extracts to capture antibodies (Paré *et al*., 1995), generation of a monoclonal antibody directed against an *N. caninum* surface protein for the serological diagnosis of bovine neosporosis in a competitive ELISA (Baszler *et al*., 1996), the use of recombinant antigens (Lally *et al*., 1996a; Jenkins *et al*., 1997) and of chemically fixed parasites (Williams *et al*., 1997).

The integration of parasite proteins into iscoms for application in immunodiagnosis was first described by Loevgren *et al*. (1987). Björkman *et al*. were the first to introduce the iscom ELISA approach for *N. caninum* serology (Björkman and Lunden, 1998). At first it was used for the analysis of dog serum (Björkman *et al*., 1994a,b; Koudela *et al*., 1998), and was later applied to the study of bovine sera (Holmdahl *et al*., 1995; Björkman *et al*., 1996; Stenlund *et al*., 1997a,b). The iscom ELISA is also a valuable tool for demonstrating the presence of antibodies to *N. caninum* in the milk of cattle (Björkman *et al*., 1997).

Another ELISA system was developed in which water-soluble antigens are used to capture the antibodies to be detected. Such soluble *N. caninum* tachyzoite extracts have been used for serological investigations with sera from cattle (Paré *et al*., 1995; Gottstein *et al*., 1998), and gave reliable results with no cross-reactivity with sera from cattle infected with *T. gondii. C. parvum* and *E. bovis*. However, some cross-reaction with sera from cattle infected with *S. cruzi* was observed (Paré *et al*., 1995). As most cattle have been exposed to *Sarcocystis* sp. (especially in the USA), the absence of cross-reactivity with these species should be documented for any ELISA (Dubey and Lindsay, 1996).

The use of recombinant antigens for the immunodiagnosis of bovine *N. caninum* infections was first reported by Lally *et al*. (1996a). By immunoscreening of an *N. caninum* tachyzoite cDNA expression library with sera from naturally and experimentally infected cattle, two cDNA clones were selected and were subcloned into an expression plasmid (pTrcHisB) and expressed in *E. coli*. On Western blots of *N. caninum* tachyzoite extracts, antibodies directed against these two recombinant antigens bind to two proteins of 33 and 36 kDa respectively, and both localize to the *N. caninum* tachyzoite dense granules (Lally *et al*., 1997). The purified recombinant antigens were evaluated by ELISA. Both antigens could discriminate between sera from *Neospora*-infected cows and sera from uninfected control animals. No evidence for cross-reactivity with *T. gondii*,

S. cruzi, *S. hominis* and *S. hirsuta* was found, indicating that these two recombinant proteins represent promising antigens for further diagnostic applications. Indeed, Jenkins *et al.* (1997) showed that antibodies directed against these two recombinant antigens could be detected in cows which aborted *N. caninum*-infected foetuses and in calves born to seropositive mothers. However, the recombinant antigen ELISA was not useful for the detection of foetal antibodies to *N. caninum* (Wouda *et al.*, 1997).

A similar approach to identifying recombinant antigens for bovine serology was used by Louie *et al.* (1997). They obtained two clones. Antibodies directed against the corresponding recombinant proteins recognized several bands on Western blots of *N. caninum* extracts, with molecular weights of between 64–97 kDa and 28–34 kDa, respectively. When a defined 'gold standard' panel of bovine sera from confirmed cases and uninfected control animals was tested by ELISA, both recombinant protein-based ELISAs exhibited higher sensitivities and higher or same specificities compared to the whole tachyzoite lysate ELISA (Louie *et al.*, 1997).

A recent study of antibody responses of cows during an outbreak of neosporosis applying the different serological assays: IFAT, whole tachyzoite lysate ELISA, iscom ELISA, recombinant antigen ELISA (Lally *et al.*, 1996a) and competitive inhibition ELISA, showed that in all tests the antibody levels of aborting cows as a group were higher than in non-aborting cows. However, serological tests did not allow individual determination of neosporosis as a cause of abortion in an individual cow (Dubey *et al.*, 1997). The recombinant antigens described by Louie *et al.* (1997) have not yet been used in this kind of study.

4.2.2. Direct Detection Techniques

(a) Transmission electron microscopy (TEM). As visualized by TEM, *Toxoplasma* and *Neospora* exhibit distinct differences, such as the appearance and number of rhoptries, and sometimes the positioning of micronemes (see Figure 1). However, these morphological features can vary a great deal, depending on the prefixation of a given sample. Furthermore, if fixation and processing of the specimen is not standardized, comparative studies are difficult to perform. In addition, processing and viewing of samples is time-consuming, and the low chances of sectioning through a tissue area harbouring parasite material makes TEM an unsuitable approach for routine diagnosis.

(b) In vitro *isolation*. Isolation of *N. caninum* in cell cultures was achieved using samples from infected neural bovine tissue. However, the success rate was very low when investigating aborted foetuses. For example, Conrad

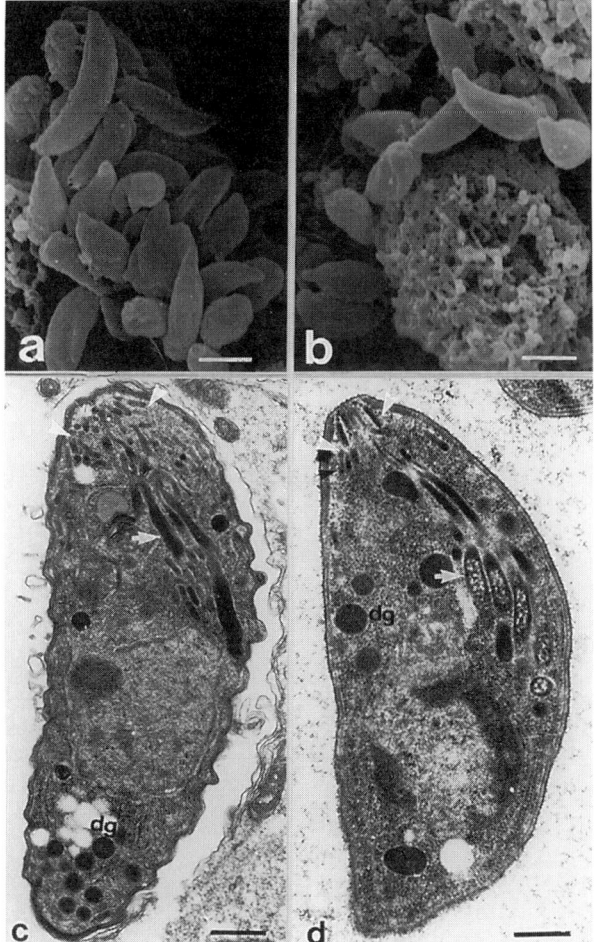

Figure 1 SEM and TEM of cell culture-derived *Neospora caninum* (a, c) and *Toxoplasma gondii* (b, d) tachyzoites. Note the very few differences such as ultrastructure of rhoptry contents (arrow) and arrangement of micronemes (arrowheads) at the apical tip. dg = dense granules. Scale bars in a and b = 1.5 μm, in c and d = 250 nm.

et al. (1993a) undertook the monumental effort of isolating the parasite out of 100 bovine foetuses, and succeeded in only two cases. It is probable that autolysis of the tissue before inoculation is critical, and success is dependent on the presence of intact tissue cysts containing viable bradyzoites rather than tachyzoites, which, because of their proliferative behaviour, cause massive tissue destruction leading to abortion. There is also the problem of opportunistic microbial contamination of cultures because of the condition of the aborted foetal tissue at the time of the attempted isolation. More consistent success has been achieved using neural tissue from stillborn or prenatally infected newborn calves, or from congenitally infected dogs (Dubey and Lindsay, 1996). In these cases, more tissue cysts containing viable bradyzoites are present. These are more resistant to the harsh treatment of the tissues before inoculation into cell culture. In any case, cultures have to be observed for two months before any statement can be made (Dubey and Lindsay, 1993).

Ten recognized isolates of *N. caninum* have been reported. Seven of these isolates originate from the USA, three are of canine origin (Dubey *et al.*, 1988b; Hay *et al.*, 1990; Cuddon *et al.*, 1992), and four are bovine isolates (Barr *et al.*, 1993; Conrad *et al.*, 1993a; Marsh *et al.*, 1995). One isolate from a dog was obtained from England (Barber *et al.*, 1993; 1995), and two further bovine isolates were obtained in Sweden (Stenlund *et al.*, 1997a) and Japan (Yamane *et al.*, 1997). Marsh *et al.* (1996) reported the isolation of *N. caninum* in the brain and spinal cord of a horse with neurological signs.

Several lines of evidence suggest that all canine and bovine isolates obtained so far are identical: no differences were found when sequences of their 16S-like-rRNA were compared, while differences were noted in relation to *T. gondii* 16S-like-rRNA (Barber *et al.*, 1995; Marsh *et al.*, 1995; Stenlund *et al.*, 1997a). Sequencing of the more variable ITS 1 region also revealed no differences between the different *N. caninum* isolates, suggesting strongly that they all belong to the same species. Within this sequence, a large number of differences exist when compared with *T. gondii*. Protein analysis and electron microscopy also suggest that all *N. caninum* isolates are identical.

Apart from the isolation of *N. caninum* in cell cultures, there is also the possibility of obtaining isolation of the parasite via the inoculation of infected tissue into mice. Although this method works well for the detection of *T. gondii* (Dubey and Beattie, 1988), it is more difficult for *N. caninum*, since mice in general are not very suitable hosts for this parasite.

(c) Histology and immunohistochemistry. By histological investigation, well-developed tissue cysts of *N. caninum* can be distinguished from *T. gondii* tissue cysts. *N. caninum* cysts have been found almost exclusively in neural tissues, and in dogs the cyst wall is up to 4 µm thick, while *T. gondii* cysts containing bradyzoites may be found in many organs, and the cyst wall is always less than 1 µm thick (Dubey and Lindsay, 1996). However, since the thickness of the cyst wall probably also depends on how long the infection has been persisting, this should not be the sole diagnostic criterion.

In chronic latent infections, most tissue cysts are not surrounded by any structures which would indicate an immunological host reaction. In contrast, in cases of acute neosporosis histopathological lesions induced by *N. caninum* tachyzoites can be found in many organs. In the CNS they are mostly characterized by a central necrotic focus surrounded by infiltrating inflammatory cells. Immunocytochemical staining of histological sections is necessary in order to confirm the diagnosis. For this purpose several reagents have been developed. Polyclonal antisera directed against cell culture-derived tachyzoites can be used to detect the parasites and to distinguish *N. caninum* from *T. gondii*. (Dubey and Lindsay, 1996). However, the problems mainly encountered using such polyclonal antisera are possible cross-reaction with antigenic components from other related coccidian parasites such as *T. gondii* or *Sarcocystis* sp., the fact that the specificity of a given serum can vary depending on which tissue is used and how it is processed before antibody labelling; and that general unspecific binding can occur. For example, Sundermann *et al.* (1997) recently reported that commercially available antibodies directed against *T. gondii* used for immunohistochemistry also reacted with *N. caninum*. McAllister *et al.* (1996c) showed that *T. gondii* and *N. caninum* bradyzoites stain both with antibodies directed against a *Toxoplasma*-bradyzoite specific antigen (BAG 5), which has been expressed as a recombinant protein in *E. coli*. In order to overcome the problem of cross-reactivity between *N. caninum* and *T. gondii* in formalin-fixed paraffin-embedded tissue sections, a murine monoclonal antibody which reacted specifically with *N. caninum* tachyzoites and tissue cysts was generated (Cole *et al.*, 1993). It recognizes *N. caninum* tachyzoites and bradyzoites in dogs, cattle, mice, rats, sheep and goats. In addition, a whole set of monoclonal antibodies directed against antigens incorporated into *Neospora* iscoms (Björkman *et al.*, 1994a) has recently been characterized. These antibodies react with antigens located on the surface and at the apical complex of *N. caninum* tachyzoites, and no cross-reaction has been observed with *T. gondii*, either on Western blots or by immunofluorescence (Björkman and Hemphill, 1998). These mAbs stained *N. caninum* tachyzoites on paraffin sections upon pretreatment of the slides with pronase (N. Fuchs, A. Hemphill, and C. Björkman, unpublished), and the absence of any labelling on tissue sections harbouring *T. gondii* tachyzoites suggests that they could be applied in immunohistochemical studies. Polyclonal antibodies directed against the *N. caninum* surface proteins Nc-p43 (Hemphill and Gottstein, 1996) and Nc-p36 (Hemphill *et al.*, 1997a) are being evaluated for a diagnostic application. Nc-p43-specific antibodies stain both *N. caninum* tachyzoites and bradyzoites, while anti-Nc-p36 antibodies label tachyzoites only. These antibodies do not cross-react with *T. gondii* (Fuchs *et al.*, 1998). Use of these antibodies in differential staining of tachyzoites and bradyzoites could also provide information on the state of the disease (chronic versus acute infection), and the application of these antibodies might be useful in investigations dealing with the differentiation of the parasite.

(d) The polymerase chain reaction (PCR). As immunohistochemistry is often not sensitive enough, more sensitive techniques for direct demonstration of parasites within tissue samples and body fluids are urgently needed. Accordingly, many laboratories focused on the development of *Neospora*-specific PCR assays for the detection of parasite DNA. The efficiency of the amplification by PCR permits the use of very small amounts of starting material. Furthermore, the procedure can be rendered even more efficient by the use of nested primers (Lally *et al.*, 1996b; Ellis *et al.*, 1997). However, extreme care must be taken to prevent contamination, from other samples, from the environment, or from PCR products previously amplified.

In many diagnostic labs, PCR diagnosis of *T. gondii* relies on methods based on the amplification of either the p30 (SAG1) gene (Savva *et al.*, 1990) or the B1 gene (Burg *et al.*, 1989). The B1 gene is present in a high copy number, and is highly conserved between different strains of *T. gondii*, but is not present in *N. caninum. Sarcocystis, Plasmodium* and others. The PCR protocol for amplification of the B1 gene was modified by introducing a nested primer system, which resulted in a significant increase in sensitivity. Using the B1-PCR, the parasite was successfully demonstrated in amniotic fluid, blood samples (Ho-Yen *et al.*, 1992), cerebrospinal fluid (Parmley *et al.*, 1992), aqueous humour (Brezin *et al.*, 1991) and in blood, lymph and tissues of *T. gondii*-infected sheep (Wastling *et al.*, 1993).

Current PCR tests used in the diagnosis of neosporosis are based on specific sequences, such as the internal transcribed spacer 1 (ITS 1) between 5.8S and 16S-like rRNA genes (Holmdahl and Mattson, 1996), the ITS 1 between 5.8S and 18S rRNA genes (Payne and Ellis, 1996), nuclear small-subunit rRNA gene sequences and hybridization with a *Neospora*-specific oligonucleotide probe (Ho *et al.*, 1996) a fragment of the 14-3-3 gene in *N. caninum* (Lally *et al.*, 1996b), and a *Neospora*-specific genomic DNA sequence named Nc5 (Kaufman *et al.*, 1996; Müller *et al.*, 1996; Yamage *et al.*, 1996).

The ITS 1 as a target sequence offers the advantage that it is present in a high copy number, being conserved within species but variable between species (Holmdahl and Mattsson, 1996; Payne and Ellis, 1996; Homan *et al.*, 1997). The ITS 1 PCR was tested successfully on tissue samples from experimentally infected mice and on the cerebrospinal fluid, buffy coat cells, amniotic fluid, placenta, spinal cord, heart and brain of experimentally infected ewes and foetuses. Ellis *et al.* (1997) recently introduced new primers for use in nested PCR in order to amplify as short a fragment as possible, since DNA obtained from clinical material such as autolysed foetuses is often degraded. This test has also been succesfully applied using formalin-fixed, paraffin-embedded tissues.

Ho *et al.* (1996) developed a PCR assay which amplifies a conserved region of the ssrRNA gene, and the amplification product can be specifically identified by hybridization with specific probes which exhibit only a single base-pair difference from a similar *Toxoplasma* specific probe and hybridized to *Neospora* PCR products. This system was able to detect *N. caninum* DNA in the bovine brain, spinal cord, heart, lung, kidney, diaphragm, skeletal muscle and placenta, as well as in amniotic fluid samples of *Neospora*-infected cattle (Ho et al., 1997b). In addition, PCR products were amplified from DNAs of different foetal tissues from experimentally infected rhesus macaques, including brain, lung, heart, liver, spleen, skeletal muscle, skin and placenta (Ho et al., 1997a).

The 14-3-3 PCR is based on a sequence of a 14-3-3 homologue of *N. caninum* which was identified using serum from an experimentally infected cow. Although the 14-3-3 proteins are highly conserved, primers annealing with less conserved regions could be designed and, using a nested PCR system, a 614 bp fragment was amplified from *N. caninum* DNA (Lally et al., 1996b). The PCR test did not amplify any fragment from *T. gondii, S. cruzi, S. tenella* and *S. muris*, and amplification of DNA from tissue samples taken from experimentally infected mice was possible, as well as amplification of DNA from an infected mouse brain that was extensively autolysed.

The Nc5-PCR originated from differential hybridization of genomic DNA libraries of *N. caninum* and *T. gondii* (Kaufmann et al., 1996). This approach resulted in the identification of clone pNc5 which contained a *N. caninum*-specific DNA fragment. Oligonucleotide primers were designed which amplified specifically a 944 base pair fragment from *N. caninum* tachyzoite DNA. Using a further set of primers, a single tachyzoite could be detected from a 2 mg sample of bovine brain tissue (Yamage et al., 1996). Müller et al. (1996) further optimized this test for application in routine diagnosis. They introduced the uracil DNA glycosidase system, which eliminated potential contamination of amplified target DNA from previous reactions, and the PCR reaction was followed by a DNA hybridization immunoassay (DIA) which unambiguously identified the *Neospora*-specific amplification products. These protocols were also introduced for optimizing the B1-PCR for the detection of *T. gondii*. In addition, Müller et al. (1997) reported on a protocol for the detection of both *N. caninum* and *T. gondii* bradyzoites from formalin-fixed, paraffin-embedded sections containing brain tissue of experimentally infected mice. Thus differential diagnosis of *N. caninum* and *T. gondii* within the same sample is now possible.

5. HOST–PARASITE INTERACTIONS

The histopathological effects of *N. caninum* infections have been quite extensively characterized, and were reviewed recently by Dubey and Lindsay

(1996). In dogs, lesions induced by *N. caninum* tachyzoites were found predominantly in the CNS, but also in the liver and in muscle biopsies. *N. caninum* can cause gross lesions such as granulomas in visceral tissues, cerebrellar atrophy and ulcerative dermatitis. The parasite was also identified in lung aspirates and in dermal pustular exudate of dogs. In cattle, lesions are most common in the CNS, heart, skeletal muscle and liver (Anderson *et al.*, 1991; Barr *et al.*, 1991; Wouda *et al.*, 1997; Gottstein *et al.*, 1998). Lesions within the CNS consist of nonsuppurative encephalomyelitis, characterized by multifocal nonsuppurative infiltration, with or without multifocal necrosis and multifocal to diffuse nonsuppurative leukocytic infiltration of the meninges (Dubey and Lindsay, 1996). Characteristically, lesions in the brain exhibit a focus of mononuclear cell infiltration around a central region of necrosis, with occasional glial proliferation. Lesions in the heart are severe, but often masked by autolysis, and hepatic lesions consist of periportal infiltrations of mononuclear cells and variable foci of hepatocellular necrosis (Dubey and Lindsay, 1996; Wouda *et al.*, 1997). The destructive effects described above, imposed upon tissues of individuals suffering from acute neosporosis, are consequences of the ways *N. caninum* interacts with its host or host cells.

Generally, investigations on the host–parasite relationship include two main aspects. First, the host–parasite relationship is strongly dependent on the host immune reaction, which will obviously be targeted mainly towards those parasites, or parasite components, which are directly accessible to the actions of the host immune system. The pattern and type of parasite antigens recognized by the host immune system are probably important for the further outcome of the infection, elimination or survival of the parasite. However, the actual responsible effectors causing formation of the necrotic lesions observed in *N. caninum*-infected tissues are partially mediated by the immunopathological processes on part of the host. Thus, the immunobiology of *N. caninum* infections needs to be investigated further. Second, host–parasite interactions take place on the cellular level, where the parasite and the target cell establish direct physical contact, probably through one (or several) host cell surface membrane receptor(s) which will bind to one (or several) parasite ligand(s). This level of interaction is most important for intracellular parasites, since recognition of suitable receptors on the host cell surface could be crucially involved in triggering subsequent host cell invasion. It is known that *N. caninum* tachyzoites cause cell destruction because of their extensive proliferation which eventually leads to host cell lysis and the infection of neighbouring cells. Thus the process of host cell invasion is essential and merits further investigation. *In vitro* models (see Section 5.1.1) represent ideal experimental systems for dissecting the processes involved, as parasites and host cells can interact with each other under defined conditions. They have also been used for studies of adhesion and invasion mechanisms of a number of other protozoan parasites, including several members of the phylum *Apicomplexa*. Investigations into the physical interaction between *N. caninum* and host cells have been strongly influenced by the work carried out on *Plasmodium* and especially *T. gondii*. Therefore some of these studies which have provided most important information, relevant also with respect to *Neospora*–host cell interactions, are presented briefly in Section 5.1.

5.1. Adhesion and Invasion of Host Cells by Apicomplexan Parasites

5.1.1. *The Role of* in vitro *Models for Studies on Adhesion and Invasion of Host Cells by Apicomplexan Parasites*

The application of *in vitro* systems for studying adhesion and invasion of host cells by apicomplexan parasites has been widely used, and has led to the identification of molecules, either within the parasite or the host cell, which mediate these interactions. Detailed studies of the mechanisms by which parasites adhere to, and invade, their host cells were carried out on *Plasmodium* (reviewed by Bannister and Dluzewski, 1990; Pasvol *et al.*, 1992; Holder, 1994; Sam-Yellowe, 1996), *Eimeria* (Augustine, 1989; Chobotar *et al.*, 1993; Bauer *et al.*, 1995; Werner-Meier and Entzeroth, 1997), *T. gondii* (reviewed by Werk, 1985; Bonhomme *et al.*, 1992; Joiner and Dubremetz, 1993; Kasper and Mineo, 1994; Dubremetz and McKerrow, 1995; Mauël, 1996; Dobrowolski and Sibley, 1997; Dubremetz, 1998), and *Theileria* (reviewed by Shaw, 1997). These studies showed that the invasive procedures employed by most members of the *Apicomplexa* are distinct from phagocytic processes. The invasive processes applied by apicomplexan parasites can be divided into three successive steps, namely recognition of the host cell membrane and attachment to host cell receptors via parasite ligands, active invasion of the host cell and intracellular development of the parasite.

In vitro models mostly consist of host cell primary cultures, or cell lines, which grow adherent to coverslips, within ELISA wells or other matrices. They are then allowed to interact with parasites under defined conditions. By altering parameters such as the temperature, and the composition of the incubation medium, and by biochemical and enzymatic pretreatments of either parasites or host cells prior and during the incubations, it has been possible to gain information on the metabolic requirements and the reactive groups necessary for the adhesion and invasion process to take place. The fact that *T. gondii* and *N. caninum* exhibit virtually no host cell specificity has permitted investigations on a wide range of cultured cells and established cell lines. Recently, reproducing the adhesion and invasion of host cells by *T. gondii in vitro* has made it possible for the first time to apply electrophysiological methods to the study of host cell invasion during parasitic infection (Suss-Toby *et al.*, 1996). In contrast to *T. gondii* and *N. caninum*, other apicomplexan parasites such as the members of the genus *Plasmodium* are much more limited in their host range (Braun-Breton and Perreira Da Silva, 1993; Galinski and Barnwell, 1996).

Many investigations employing *in vitro* models not only extracted information about the mechanisms by which parasites adhere to and invade their host cells, but also on the different strategies by which survival and proliferation of intracellular parasites is achieved (reviewed by Mauël, 1996). Studies employing *in vitro* infection experiments have been used to demonstrate the mechanism employed by *T. gondii* to escape the normal endocytic pathway after being taken up by nonactivated macrophages or other phagocytic cells (Joiner *et al.*, 1990; Sibley *et al.*, 1993). After entry into its host cell, *T. gondii* tachyzoites reside in an intracellular vacuole that is completely unable to fuse with other endocytic or biosynthetic organelles. This fusion block requires the entry of viable organisms, but is irreversible even if the parasites are killed after entry. However, the fusion block is dependent on the route of entry, and does not take place when parasites are taken up through the phagocytic pathway, indicating that the fusion inhibition reflects active invasion. Thus a specific modification of the vacuole membrane at the time of entry into the host cell has to take place (Sibley *et al.*, 1993; Sibley, 1995). Secretion of a soluble inhibitor by *T. gondii* appears not to play an important role with respect to fusion inhibition, although maturation by the vacuole in many cell types is accompanied by exocytosis from rhoptries (Saffer *et al.*, 1992; Dubremetz and Schwartzman, 1993), and later by dense granules from the parasite (Charif *et al.*, 1990; Achbarou *et al.*, 1991b; Cesbron-Delauw, 1994). More recent experiments performed by Mordue and Sibley (1997) showed that *T. gondii* parasitophorous vacuoles formed through active invasion of macrophages evade endocytic processing because of the absence of host regulatory proteins essential for endocytic fusion. In contrast to this successful survival strategy applied by *T. gondii* within normal macrophages, it was shown that in activated macrophages the vacuoles containing live tachyzoites fuse with lysosomes, and that this is followed by destruction of the parasite (Sibley *et al.*, 1993).

Another valuable advantage of *in vitro* systems is that detailed electron microscopical investigations on the ultrastructure of adhesion and invasion processes and of intracellular development can be carried out. Most parasites actively invading a host cell have been shown to employ a very similar, but not necessarily identical, 'self-zippering mechanism' between the parasite and host cell surface membranes. The host cell invasion process has been investigated in detail for *Theileria parva* (Shaw, 1997), *Plasmodium* spp.

(Tait and Sacks, 1988; Holder, 1994), and *T. gondii* (Werk, 1985; Bonhomme *et al.*, 1992, Sibley, 1995; Dubremetz, 1998). Additional immunocytochemical studies on invading apicomplexan parasites demonstrated the roles of host and parasite organelles and molecules during invasive processes (Dubremetz and Schwartzman, 1993; Saffer *et al.*, 1992; Grimwood and Smith, 1995; Sibley *et al.*, 1995; Morisaki *et al.*, 1995; Grimwood *et al.*, 1996; Carruthers and Sibley, 1997).

5.1.2. *The Roles of Secretory Organelles in Adhesion, Invasion and Intracellular Development of Apicomplexan Parasites*

All apicomplexan parasites share certain features and specialized organelles which are currently considered to be involved in host cell invasion and intracellular development. These include the conoid and secretory organelles such as rhoptries, micronemes and dense granules. Most infective stages of apicomplexan parasites actively invade their host cells apical end first. Thus the function most often attributed to the conoid is to aid in the penetration of host cells. This idea was confirmed by ultrastructural observations carried out on the interaction between *Plasmodium* and red blood cells (Bannister and Dluzewski, 1990), where a reorientation step was observed after attachment. This reorientation step may result from a gradient of receptor distribution on the merozoite surface, or from the presence of apically located higher affinity receptors (Galinski and Barnwell, 1996). In other sporozoa, such a reorientation step and apical contact with the host cell surface could be a result of gliding motility or conoid flexing (Werk, 1985; Bonhomme *et al.*, 1992; Chobotar *et al.*, 1993; Dobrowolsky and Sibley, 1996). It was demonstrated that conoid extrusion which accompanies host cell invasion by *T. gondii* tachyzoites is a Ca^{2+} dependent process (Mondragon and Frixione, 1996), and Dobrowolsky and Sibley (1996, 1997) have shown that an actin-myosin-based system plays a crucial role in host cell invasion, although actin filaments themselves have not been demonstrated (Dobrowolski *et al.*, 1997).

Micronemes are organelles which, before invasion, exocytose molecules involved in recognition and binding with the host cell on to the parasite surface. In *Plasmodium* the first of these micronemal proteins to be identified was the 135 kD *P. knowlesi* Duffy-binding protein, which is known to bind to an erythrocyte membrane glycoprotein bearing the Duffy blood group determinants (Adams *et al.*, 1990). This protein was subsequently found to be closely related to the *P. falciparum* erythrocyte binding antigen-175 (EBA-175), which is also localized to the micronemes, but bound to sialic acid residues specifically on erythrocyte glycophorin A (Sim *et al.*, 1990, 1992). Thus these two micronemal proteins, although related, are responsible for the differential receptor–ligand interactions of different malaria species. Three micronemal proteins (MIC1–3) were identified in *T. gondii* (Achbarou *et al.*, 1991a; Dubremetz and Schwartzman, 1993). The cDNA coding for MIC1 has been cloned and sequenced, and it contains a duplicated receptor-like domain with distant homology to the *Plasmodium* micronemal protein TRAP-SSP2 (Fourmaux *et al.*, 1996). MIC1 from tachyzoite lysates and a recombinant peptide of the N-terminal duplicated domain of the protein bound to the surface of putative host cells, suggesting a role for MIC1 in host cell adhesion or recognition (Fourmaux *et al.*, 1996). Using the expressed sequence tag (EST) dataset of *T. gondii*, the gene encoding MIC2 was identified and characterized (Wan *et al.*, 1997). It contains five copies of the conserved thrombospondin-like motif present in a number of molecules with adhesive properties, as well as a conserved region implicated with the adhesive characteristics of several integrins (Wan *et al.*, 1997).

Rhoptries are large club-shaped anterior organelles with a slender duct through which organellar contents are discharged at the time of invasion of host cells. It was shown that this process occurs upon cell–cell contact, implying that it would aid invasion by altering the plasma membrane and the cortical actin skeleton of the target cells. Many rhoptry proteins were identified in *Plasmodium* (reviewed in Perkins, 1992). One of these, the apical membrane antigen-1 (AMA-1) was proposed to act as a receptor (Cheng and Saul, 1994). In *T. gondii*, up to ten rhoptry proteins with molecular weights ranging from 40–220 kD were identified by monoclonal antibodies (Leriche and Dubremetz, 1991). The best characterized of these proteins, ROP1, was previously known as penetration enhancing factor (PEF) originally identified by Lycke and Norrby (1966). DNA sequence analysis revealed that this 60 kD protein is a hydrophilic polypeptide which may bind to other molecules due to its charge asymmetry (Ossorio *et al.*, 1992). ROP1 was shown to associate with the PVM after parasite entry (Saffer *et al.*, 1992). It does not contain a transmembrane domain, and thus must interact directly with the phospholipid bilayer or with other transmembranous particles. During further maturation of the parasitophorous vacuole, ROP1 disappears. However, analysis *in vitro* and *in vivo* of *T. gondii* deletion mutants which do not express ROP1 showed no significant alterations in growth rate, host specificity, invasiveness or virulence, suggesting that the ROP1 gene product is not essential under the conditions tested (Soldati *et al.*, 1995). Other rhoptry proteins such as ROP2, 3, 4, 5 were shown to exhibit a basic pI, which would be in agreement with the idea that polycationic polypeptides play a role during invasion (Werk, 1985). The lipid content of *T. gondii* rhoptries is enriched in cholesterol and lysolipids (Foussard *et al.*, 1991), suggesting that these components were probably involved in driving the inward expansion of the PVM. This was also supported by the finding that phospholipase A2 could be involved in invasion (Saffer and Schwartzman, 1991; Gomez et al., 1996).

Dense granules resemble the secretory vesicles of mammalian cells, and are probably formed by budding from the Golgi apparatus. Seven dense granule proteins have been identified in *T. gondii* to date, and sequencing has shown that six of them contain typical hydrophobic signal sequences which target them in the secretory pathway (Cesbron-Delauw, 1994; Lecordier *et al.*, 1995; Fischer *et al.*, 1998; Jacobs *et al.*, 1998). It is not yet clear whether they are involved in the invasion process itself, such as enclosure of the tachyzoite in the vacuole. Immunogold electronmicroscopy showed that the proteins of dense granules are targeted either to the vacuolar space, the vacuolar membranous network and/or the vacuole membrane (reviewed by Cesbron-Delauw *et al.*, 1996) or, in the case of erythrocytes infected by *P. falciparum*, the inner side of the host cell membrane (Aikawa *et al.*, 1990).

Carruthers and Sibley (1997) demonstrated the sequential secretion of proteins from micronemes, rhoptries and dense granules during invasion of human fibroblasts, using immunofluorescence, immunoelectronmicroscopy and quantitative immunoassays. They showed that binding to the host cell triggered apical release of MIC2 at the attachment zone, that subsequent invagination of the host cell membrane was triggered through the release of ROP1, forming a nascent parasitophorous vacuole, and that release of the dense granule proteins GRA1 and NTPase (Sibley *et al.*, 1994) occurred only after the parasite was completely enclosed by the parasitophorous vacuole membrane (about 20 minutes after invasion). Furthermore, as *T. gondii* tachyzoites were treated with cytochalasin D, a drug which depolymerizes actin filaments, the invasion process of cytochalasin D-resistant host cells was interrupted at the point where a nascent parasitophorous vacuole is formed (Dobrowolski and Sibley, 1996). Thus, each organelle or its secretory products appear to be involved in distinct steps during the invasion process, and the actin skeleton of the parasite is responsible for exerting the driving force for invading the host cell cytoplasm.

5.2. The Physical Interaction of *N. caninum* Tachyzoites with Host Cells

Hemphill *et al.* (1996) investigated the interaction between *N. caninum* and bovine aorta endothelial (BAE) cells *in vitro*. *N. caninum* adhesion, invasion, and intracellular proliferation was investigated using transmission electronmicroscopy. In addition, a fluorescence-based assay modified from Schenkman *et al.* (1991) was employed in order to quantitate the effect of various treatments of parasites and host cells prior to their interaction on adhesion and invasion.

Both adhesion and invasion took place within a relatively short time (five minutes) after incubation of the tachyzoites with host cells, and occurred in a non-homogeneous manner, similar to that previously described for *T. gondii* (Kasper and Mineo, 1994). After 45–60 minutes tachyzoite invasion had reached a plateau, although there were still additional parasites adhering to the endothelial cell surface, and there were still a considerable number of uninfected host cells present. This indicates that not every parasite which adheres to the host cell surface will automatically penetrate the host cell surface membrane, and that the tachyzoites had lost their potential to infect target cells upon prolonged extracellular maintenance. It is also very likely that, as has been demonstrated for *T. gondii* tachyzoites, both adhesion- and infection-efficiencies are dependent on the host cell cycle (Youn *et al.*, 1991; Grimwood *et al.*, 1996). Thus it has been suggested that certain host cell surface proteins which are expressed at different times on the host cell surface would serve as parasite receptors (Grimwood *et al.*, 1996).

Transmission electronmicroscopy indicated that *N. caninum* tachyzoites initially attach to BAE cells by any part of their surface. No preference could be seen for any specialized host cell surface domain. Invasion of BAE cells by *N. caninum* tachyzoites was a multi-step process. As the parasite encountered a suitable location for penetration of the endothelial cell surface, the host and the parasite plasma membranes bound closely to each other at the site of initial contact. A recess developed on the endothelial cell surface, and the parasite moved into the host cell, the membrane of which curved around the invading parasite, which eventually became intracellular, completely encircled by the parasitophorous vacuole membrane. The tachyzoites within their parasitophorous vacuole moved towards the interior of the cell, preferentially close to the nucleus, and were soon surrounded by host cell mitochondria. Massive secretion into the lumen of the parasitophorous vacuolar space takes place, indicating that the parasite modifies the vacuole and its membrane according to its own needs. Endodyogeny occurs at about six hours post infection, and continues until, after about 72–80 hours, a pseudocyst has developed, which contains hundreds of newly formed tachyzoites. Host cell lysis occurs, and tachyzoites are set free and infect neighbouring cells. Altogether, these events closely resemble the processes taking place during host cell entry and intracellular proliferation of *T. gondii* tachyzoites (Kasper and Mineo, 1994).

N. caninum tachyzoites also bound to both paraformaldehyde- and glutaraldehyde-fixed BAE cells, but attachment was much more efficient to living cells. This can be explained by considering the plasma membrane of living cells as a highly dynamic structure, with additional receptors probably engaged during incubation with the parasites. However, *N. caninum* also invades BAE cells fixed with 3% paraformaldehyde, provided that no glutaraldehyde was used. These findings are somewhat surprising, although *Trypanosoma cruzi*, for example, was also reported to be capable of infecting paraformaldehyde prefixed host cells *in vitro* (Schenkman *et al.*, 1991). When viewed by TEM, the mechanisms of penetration of living and prefixed BAE cells appeared to be somewhat different: penetration of prefixed BAE cells was accompanied by dramatic changes at the point of entry, with the plasmalemma being disrupted and clusters of irregular vesicles appearing around the advancing parasite. However, these observations suggested that invasion can take place independently of any energy expenditure on the part of the host cell.

Several other lines of evidence also suggested that the invasion process required only parasite energy, and was largely independent from the target cell metabolism. Treatment of tachyzoites using glycolytic and mitochondrial inhibitors such as deoxyglucose, sodium azide, oligomycin and antimycin reduced the rate of both adhesion and invasion to BAE cells, while treatments of BAE cells had no effect. Thus, metabolic energy is necessary to trigger invasion after an initial, probably low affinity, contact between parasite and host cell surface molecules. *De novo* protein synthesis on the part of the parasite did not apparently play an important role during adhesion and invasion, since cycloheximide treatment of parasites before incubation with BAE cells had no effect (Hemphill *et al.*, 1996).

Based on these criteria, the adhesion/invasion process of host cells by *N. caninum* was very similar to equivalent processes described in other Apicomplexa such as *Plasmodium* and *Toxoplasma* (Tait and Sacks, 1988). However, while for these parasites evidence has been obtained that specific target cell surface molecules serve as parasite receptors (Holder, 1994; Kasper and Mineo, 1994), possible receptors on the host cell surface which would interact with *Neospora* ligands have not been identified.

Thus, a series of experiments were performed in order to gain some basic information on the molecular nature of possible host cell surface receptors. In many cell types, cell surface adhesive molecules include glycolipids (Karlsson *et al.*, 1992), glycosaminoglycans (Lander, 1993), and glycoproteins (Oebrink, 1993). Modifications carried out on cell surface carbohydrates would therefore probably influence the adhesive properties of cell surfaces. Periodate oxidation at acidic pH had previously been used as a tool to cleave carbohydrate residues. Removal of carbohydrates from the endothelial cell surface by using a whole range of periodate concentrations at low and neutral pH significantly increased the adhesion efficiency of *N. caninum* tachyzoites. In addition, it was shown that enzymatic treatments, namely neuraminidase and hyaluronidase digestions of BAE cell surfaces, also promoted the attachment of *N. caninum* tachyzoites to BAE cells.

Similar results had previously been obtained when sialic acid residues were removed from the surface of macrophages before infection with *T. gondii* (de Carvalho *et al.*, 1993). Hyaluronidase treatment of host cells did also enhance the penetration of Hela cells by tachyzoites of *T. gondii* as previously reported by Lycke and Strannegard (1965) and Norrby (1971). It is likely that removal of sialic acid residues and acid mucopolysaccharides facilitate the interaction of surface components of parasite and host cells participating in the interaction. However, this effect is likely to be dependent on the type of host cell and parasite isolate investigated, since alterations in surface glycosylation, heparin and chondroitin sulphate did not modify invasion of CHO cells by the PTg B strain of *T. gondii* (Mack *et al.*, 1994).

The role of potential carbohydrate residues on the *N. caninum* cell surface in the adhesion to, and invasion of, BAE host cells was also assessed. Parasites were treated with either sodium periodate or tunicamycin, an inhibitor of N-glycosylation, before incubation with host cells. Neither of these treatments had any effect. However, protease treatments of tachyzoites had a negative impact on adhesion and invasion, suggesting that removal of proteins or protein fragments from the parasite surface also removed or altered the molecules responsible for mediating *N. caninum* interaction with the host cell monolayer. Thus it is likely that proteinous components present on the tachyzoite surface would act as ligands mediating host-cell recognition and invasion.

The current literature on *T. gondii* adhesion to host cells underlines the importance of four points. (i) *T. gondii* tachyzoites possess surface ligands which bind the ECM protein laminin with high affinity. Parasite-bound laminin then promotes the attachment of the tachyzoites to the host cell laminin receptor (Kasper and Mineo, 1994). (ii) *T. gondii* tachyzoite–host cell interactions could be mediated by lectin-binding proteins (De Carvalho and De Souza (1997). (iii) *T. gondii* tachyzoites possess five major surface antigens, named SAG1–5 (reviewed by Tomavo, 1996), and an additional, recently identified, surface protein named SRS1 (SAG1-related sequence 1; Hehl *et al.*, 1997). SAG1 (p30) and SAG3 (p43) have been shown to serve as attachment factors, mediating the contact between parasite and cell surface membrane during host cell adhesion and invasion (Smith, 1995). In addition, it has been proposed that SAG2 is involved in apical reorientation and detachment once the tachyzoites have established physical contact with the host cell surface membrane (Smith, 1995). (iv) Recent evidence indicates strongly that the microneme proteins MIC1 and MIC2 take part in the attachment of the parasite to the host cell surface membrane (Fourmaux *et al.*, 1996; Wan *et al.*, 1997; Carruthers and Sibley, 1997).

The effects of cytoskeletal drugs on adhesion and invasion of BAE cells by *N. caninum* tachyzoites were also assessed. Treatments of both parasites and

BAE cells with taxol and nocodazole, which are microtubule-active drugs in higher eukaryotic cells, had no effect either on adhesion or on invasion. However, pretreatment of tachyzoites with the actin inhibitor cytochalasin D were very effective in inhibiting parasite entry, indicating that actin-myosin based movements, similar to those demonstrated in *T. gondii* (Dobrowolski and Sibley, 1996) are involved during penetration of host cells by *N. caninum*.

BAE cell monolayers were also treated with cytochalasin D before the addition of *N. caninum* tachyzoites. Treatment of BAE cells with this drug prevented invasion by *N. caninum*. However, at the same time, the number of *N. caninum* tachyzoites which adhered to the host cell surface was significantly higher than in control experiments. This effect indicates that adhesion and invasion are two distinct process. Although after cytochalasin D treatment of BAE cells the parasites could still adhere to the host cells, the signal or signals which would trigger the invasive process within the parasite were not present any more; they had probably been altered or dislocated by cytochalasin D treatment. However, these experiments are not all conclusive, and they should be performed using cytochalasin-D resistant *N. caninum* tachyzoites similar to the *T. gondii* cytochalasin-D resistant mutants generated by Dobrowolski and Sibley (1996).

5.3. Identification and Characterization of Intracellular and Cell Surface-Associated *N. caninum* Tachyzoite Proteins

A considerable amount of work has been invested in identifying and characterizing *N. caninum* tachyzoite proteins, which take part in the complex interactions between parasite and host. In addition, associated comparative studies, especially in relation to *T. gondii*, were performed in order to find out whether antigenic differences between these parasites would be useful in developing a diagnostic assay for *N. caninum*. Several approaches were used, such as raising monoclonal antibodies, generating polyclonal antisera directed against the whole parasite, immunoscreening of cDNA expression libraries with sera from infected cattle, or subcellular fractionation of parasites and preparation of affinity-purified, monospecific antibodies directed against particular parasite antigens.

5.3.1. *The Use of Monoclonal Antibodies (mAbs)*

Several mAbs directed against *N. caninum* tachyzoites have been raised so far. One of these is a murine antibody which was originally reported by Cole et al. (1993) to be useful for the diagnosis of *N. caninum* infections by immunohistochemistry. Subsequently, immunoblotting demonstrated that this monoclonal antibody was directed against eight major and several minor *N. caninum* antigens with molecular weights ranging between 31 and 97.4 kDa (Cole et al., 1994). In addition, reactivity with a *T. gondii* tachyzoite antigen with a relative molecular weight of 107 kDa was also observed. Immunogold labelling showed that the antibody recognized an epitope associated with micronemes, dense granules, basal portions of the rhoptries, and the intravacuolar tubular network within the parasitophorous vacuole. In *T. gondii* tachyzoites, micronemes and basal portions of the rhoptries were also stained (Cole et al., 1994).

Baszler et al. (1996) generated an mAb which recognized a carbohydrate epitope of a 65 kDa surface antigen in *N. caninum* tachyzoites. This antibody is used in a competitive ELISA assay for the serological diagnosis of neosporosis.

A chicken mAb raised against *Eimeria acervulina* sporozoites had previously been shown to recognize the conoid of these parasites and to inhibit sporozoite invasion of lymphocytes *in vitro*. This antibody also stained the conoid from six other species of *Eimeria* and the apical complexes of both *T. gondii* and *N. caninum* tachyzoites, indicating that the mAb identified a conserved epitope on the conoid which is important in host cell invasion by apicomplexan parasites (Sasai et al., 1998).

Seven distinct surface antigens of *N. caninum* tachyzoites of molecular weights of between 17 and 56 kDa were identified using surface labelling of parasites (Schares and Conraths, 1996). All but one (a 56 kDa antigen) could be immunoprecipitated using sera from *N. caninum* infected animals. Surface protease digestion by chymotrypsin and trypsin showed that the 42 and 43 kDa antigens were sensitive to these proteolytic enzymes, while surface antigens of 29, 32 and 35 kDa exhibited only limited sensitivity. Most of the epitopes of the surface antigens showed conformational dependency, and the 32 kD antigen appeared to be a glycoprotein. Subsequently, mAbs directed against *N. caninum* tachyzoites were raised in order to analyse proteins with respect to their function in host cell invasion (Schares et al., 1997). These antibodies were characterized using immunoblotting, immunoprecipitation, indirect immunofluorescence and immunogold electronmicroscopy. Several of these antibodies were directed against three of the above-mentioned surface antigens of *N. caninum* with molecular weights (as determined by SDS-PAGE under non-reducing conditions) of 17, 42 and 43 kDa. Another antibody directed against a dense granule antigen of 35 kDa was also generated (Schares et al., 1997).

Another set of mAbs directed against *N. caninum* iscom antigen (see Section 4.2.1) were characterized by Björkman and Hemphill (1998). *Neospora* iscoms have earlier been used as antigen in ELISA systems for demonstration of antibodies to *N. caninum* in sera from dogs and cattle (Björkman *et al.*, 1994a; Björkman *et al.* 1997). The method used for construction of the iscoms was designed to select for membrane antigens. However, it was not known if the incorporated antigens were of extra- or intracellular origin. Six monoclonal antibodies raised to *N. caninum* iscoms were used in Western blots in order to test their reactivity with non-reduced and reduced *N. caninum* extracts, as well as with *N. caninum* iscoms. These antibodies recognized antigens with apparent molecular weights of 18 kDa, 30 and 32 kDa, 41 kDa and 65 kDa. *N. caninum* iscom proteins of similar molecular weights were also recognized when sera from experimentally infected rabbits and naturally infected dogs and cattle were used as probes in Western blots (Björkman *et al.*, 1994a, 1997). Antibodies directed against the 30/32 kDa doublet, the 18 kDa antigen and the 41 kDa band bound to the tachyzoite surface, suggesting that the corresponding reactive epitopes were accessible from the outside. As methanol-permeabilized tachyzoites were labelled by immunofluorescence, additional intracellular staining could be demonstrated, revealing a punctuated pattern with spots of more intense labelling distributed all over the cytoplasm of *N. caninum* tachyzoites. This type of staining is indicative for dense granule organelles, since micronemes and rhoptries would be found mainly at the anterior end of the tachyzoites (Dubey and Lindsay, 1996). Indeed, immunogold on-section labelling of *N. caninum* tachyzoites embedded in LR-White resin confirmed the presence of the 30/32 kDa doublet both on the surface and within the dense granules of the parasite, suggesting that the respective protein or proteins could play an important role during the initial interaction with the host cell surface membrane. The other antibodies could not be used for immunogold electronmicroscopy, probably because corresponding epitopes were denatured or masked during processing. Western-blotting experiments showed that the epitope recognized on the 18 kDa antigen was at least partially composed of carbohydrate residues, as indicated by the lack of staining when the blots had been treated with sodium periodate. The mAbs reacting with the 65 kDa protein failed to detect any epitopes in Western blots with non-reduced antigen of both crude *N. caninum* extracts and iscoms. The 65 kDa protein was apparently not exposed on the tachyzoite cell surface, but could be detected at the apical pole by immunofluorescence after permeabilization of tachyzoites. The mAbs directed against *N. caninum* iscoms did not react with *T. gondii* tachyzoites, neither by immunoblotting nor by immunofluorescence. Thus further investigations are needed to determine the usefulness of these mAbs as tools for the immunohistochemical detection of *N. caninum* in paraffin-embedded tissues (N. Fuchs, A. Hemphill and C. Björkman, unpublished).

5.3.2. Approaches Involving Polyclonal Antisera

Barta and Dubey (1992) characterized an anti-*N. caninum* hyperimmune rabbit serum by Western blot analysis and immunoelectron microscopy. They identified approximately 20 immunodominant antigens, with molecular weights ranging between 16 and 80 kDa. Parasite antigens separated by non-reducing SDS-PAGE were recognized more intensely than antigens separated under reducing conditions. These antigens were localized predominantly within the dense granules, in the micronemes, at the posterior end of the rhoptries and on the parasitophorous vacuole membrane. No labelling of the *N. caninum* tachyzoite surface could be detected (Barta and Dubey, 1992).

Bjerkås *et al.* (1994) showed that immune sera from a wide range of animal species exhibited a similar recognition pattern when visualized by immunoblotting, with five major (17, 27, 29, 30, and 46 kDa) and several minor *N. caninum* antigenic bands. As determined by immunoelectron-microscopy using monospecific antibodies, the 17 kDa antigen was found to be localized within the body part of the rhoptries, while the 29 and 30 kDa antigens were distributed over the parasitophorous vacuole network, the dense granules and on the parasitophorous vacuole membrane. Again, no tachyzoite surface staining was found. In addition, it was demonstrated that a polyclonal rabbit hyperimmune serum directed against *N. caninum* tachyzoites did not exhibit labelling of external membranes. In contrast, a rabbit anti-*T. gondii* hyperimmune serum exhibited a marked surface staining of *T. gondii* tachyzoites (Bjerkås *et al.*, 1994).

Another approach, leading to the identification of two dense granule proteins in *N. caninum* tachyzoites, involved immunoscreening of an *N. caninum* tachyzoite cDNA library with sera from *N. caninum*-infected cows. Subcloning of the two cDNAs and expression of respective recombinant proteins in *E. coli* led to the identification of these antigens as useful candidates to detect anti-*N. caninum* antibodies by ELISA (Lally *et al.*, 1996a). More recently, these cDNA clones were further analysed. One of them encodes a dense granule protein of 33 kDa molecular weight; thus the protein was named NCDG1 (Lally *et al.*, 1997). The second cDNA clone also coded for a dense granule protein, although of approximately 36 kDa molecular weight. It is named NCDG2 (Lidell *et al.*, 1998). Sequence analysis of a full-length cDNA clone encoding NCDG1 revealed that it contained three hydrophobic regions, namely a putative signal sequence at the N-terminus with a putative cleavage site, and two additional ones. The third hydrophobic region represented a putative transmembrane region (Lally *et al.*, 1997). Taken together, the predicted amino acid sequence of NCDG1 shared structural similarity with other dense granule proteins from *T. gondii* (Cesbron-Delauw, 1994). NCDG2 appears to be closely related to the *T. gondii* dense granule protein GRA6, with 47% nucleotide sequence identity (Liddell *et al.*, 1998).

Another approach for identifying molecules of *N. caninum* tachyzoites involved in the physical interaction with host cells was applied in our laboratory. As extracellular parasites interact with the host cell surface through membrane components, we biochemically fractionated purified tachyzoites employing the non-ionic detergent Triton-X-114. This separation step resulted in a fraction containing predominantly membrane proteins. Analysis by SDS-PAGE and subsequent immunoblotting using a polyclonal rabbit anti-*N. caninum* antiserum demonstrated that, within this Triton-X-114 fraction, a reproducible banding pattern could be observed, with major reactive bands of approximately 43, 36 and 33 kDa molecular weight. The polyclonal antiserum was affinity-purified on these three bands, and the three resulting affinity-purified antibodies were subsequently shown to be uniquely directed against their corresponding antigens. According to their molecular weights as determined by SDS-PAGE under reducing conditions, the three proteins were named Nc-p43, Nc-p36 and Nc-p33. The affinity-purified antibodies were used to further characterize these three proteins by imunofluorescence, immunogold electronmicroscopy, immunoscreening of a *N. caninum* cDNA library and analysis of the corresponding gene segments.

Immunoblotting and immunofluorescence staining of *N. caninum* tachyzoites demonstrated that Nc-p43 and Nc-p36 were major tachyzoite surface proteins, with Nc-p36 being a glycoprotein (Hemphill and Gottstein, 1996; Hemphill *et al.*, 1997a). Furthermore, we showed that Nc-p43 was expressed in both the tachyzoite and the bradyzoite stage of *N. caninum*, while Nc-p36 was exclusively expressed in tachyzoites (Fuchs *et al.*, 1998; Sonda *et al.*, 1998). Thus, specific antibodies directed against Nc-p36 could represent valuable tools for investigations on the stage conversion during different phases of infections with *N. caninum*, and for the differential and specific immunohistochemical detection of the tachyzoite stage of *N. caninum* in infected tissue samples. Both proteins were antigenically distinct from the surface proteins present on *T. gondii* tachyzoites and bradyzoites (Fuchs *et al.*, 1998). It is not known whether some of the surface proteins described by Schares and Conraths (1996) are identical with Nc-p43 and Nc-p36, since comparative studies have not yet been performed. However, comparative immunoblotting using polyclonal anti-Nc-p43 and anti-Nc-p36 antibodies and the mAbs against the 30/32 kD doublet and the 41 kDa antigens identified by Björkman and Hemphill (1998) suggested that Nc-p43 and Nc-p36 are most likely recognized by these two mAbs respectively (Björkman and Hemphill, unpublished observations).

Initial immunogold electronmicroscopy employing the original affinity-purified antibodies indicated that Nc-p43 was located on the tachyzoite surface, the dense granules and within the rhoptries of the parasite (Hemphill, 1996), while Nc-p36 was found to be associated with both the tachyzoite surface and the dense granules (Hemphill *et al.*, 1997a). However, as these localization studies were subsequently performed using antibodies affinity-purified on recombinant peptide fragments of Nc-p43 (recNc-p43) and Nc-p36 (recNc-p36), it could be demonstrated that Nc-p43 was actually associated with the cell surface and with the parasite dense granules (Hemphill *et al.*, 1997b), and Nc-p36 was found exclusively on the surface of *N. caninum* tachyzoites (Sonda *et al.*, 1998).

Both Nc-p43 and Nc-p36 are implicated in parasite adhesion to the host cell surface membrane. It has been shown that affinity-purified anti-Nc-p43 antibodies inhibit host cell invasion *in vitro*, indicating that Nc-p43 is one of the ligands which mediate the physical contact between parasite and host cell surface (Hemphill, 1996). Evidence for a participation of Nc-p36 in the invasion process is based on immunogold electronmicroscopy of invading tachyzoites (Hemphill *et al.*, 1997a) and the striking amino acid sequence similarity (76.3% similarity with 51.3% identities) between Nc-p36 and p30 (SAG1), the major *T. gondii* tachyzoite surface protein which has previously been shown to serve as an attachment factor (reviewed by Tomavo, 1996). Nc-p36 and p30 (SAG1) are closely related with respect to localization, stage-specific expression and their amino acid sequence, but they exhibit antigenic differences (Sonda *et al.*, 1998).

The amino acid sequence corresponding to the cDNA clone coding for Nc-p43 was also determined. In relation to p30 (SAG1), this sequence exhibited an overall similarity of 43% with 30% identities, and with respect to SAG3 (Cesbron-Delauw *et al.*, 1994), a 43 kDa surface protein in *T. gondii* tachyzoites which is also involved in host cell attachment (Tomavo, 1996), the overall similarity was 44%, but identities were found with only 23% of all amino acids. However, some amino acid motifs were well conserved in these three proteins. Particularly interesting was the almost identical positioning of cysteine residues within the aligned sequences (Hemphill *et al.*, 1997b; Sonda *et al.*, 1998). Cysteins are involved in the formation of secondary structures of a polypeptide because of the formation of disulphide bridges. Thus, although the surface proteins SAG1, SAG3, Nc-p43 and Nc-p36 exhibited distinct antigenic and biochemical differences, their localization and their similarities in a putative secondary structure could reflect their functional relationship.

In contrast to Nc-p43 and Nc-p36, the third protein was identified as a dense granule-associated protein of 33 kDa molecular weight, and was named Nc-p33 (Hemphill *et al.*, 1998). Isoelectric focusing and subsequent immunoblotting demonstrated that the affinity-purified antibodies recognized only a single, most likely not post-translationally modified, gene product. Nc-p33 was found in two isolates of *N. caninum* (NC-1 and

Liverpool), but could not be detected in *T. gondii* tachyzoites. Immunogold EM revealed that Nc-p33 constituted a dense granule-associated protein, and Western blotting demonstrated that Nc-p33 was most likely identical to the recently described antigen NCDG1 (Lally *et al.*, 1997). Shortly after invasion, this dense granule protein was targeted to the parasitophorous vacuole membrane, and, at later times after infection, was also found on the parasitophorous vacuolar network. This suggested that Nc-p33 could play a functional role in the modification of the parasitophorous vacuole and its membrane.

5.4. Immunology of *N. caninum* Infections

Although the immunological host response to *T. gondii* infections has been extensively studied (reviewed by Darcy and Santoro, 1994; Sher *et al.*, 1995; Candolfi *et al.*, 1996; Gazzinelli *et al.*, 1996; Hunter *et al.*, 1996; Nagasawa *et al.*, 1996; Alexander *et al.*, 1997; Innes, 1997), the immune response to *Neospora* has not been equally well elucidated and characterized. There is still a considerable lack of knowledge about the role of the humoral and the cellular immune response and on the role of immune cytokines during infection with *N. caninum*. In contrast to *T. gondii*, *N. caninum* has not been found in natural infections, either in mice or rats, hosts which would represent the most suitable models for immunological laboratory investigations, and the establishment of murine models for *Neospora* infections and disease has been hampered by some difficulties (Dubey and Lindsay, 1993). However, in some experiments *N. caninum* infection in mice has been associated with acute primary pneumonia, myositis, encephalitis, gangioradiculoneuritis and pancreatitis (Lindsay and Dubey, 1990), with some mouse strains appearing more susceptible to infection of the CNS than others (Lindsay *et al.*, l995a; Long *et al.*, 1998). A higher number of *N. caninum* tachyzoites is necessary for infection when compared to certain virulent strains of *T. gondii* such as RH, which are highly lethal at very small inoculum doses (Howe and Sibley, 1995).

Khan *et al.* (1997) have investigated the cellular immune response to *N. caninum* infections in inbred A/J mice. These mice exhibited no clinical signs of neosporosis and no significant histological evidence upon infection. Splenocytes obtained from infected mice proliferated *in vitro* in response to both *N. caninum* and *T. gondii* soluble antigens, suggesting the presence of cross-reactive immune determinants. However Lindsay *et al.* (1990) had shown that a previous infection of mice with *N. caninum* does not protect against challenge by *T. gondii* RH strain tachyzoites, demonstrating that these two species are distinct biologic entities and not immunologically closely related isolates. At day seven after infection of mice with *N. caninum* tachyzoites, a transient (two to three days) lymphocyte hyporesponsivness was observed (Kahn et al., 1997). A similar, although more prolonged, phenomenon had previously been demonstrated during acute *T. gondii* infection in mice with both virulent and avirulent strains (Howe and Sibley, 1995). In *N. caninum* infections, this immunosuppressive effect was restored at day ten, and this remained until day 14 p.i., when the study finished. The hyporesponsivness to parasite antigen and mitogen was principally due to the induction of nitric oxide, which has been shown to be an important regulatory molecule in infections by *Plasmodium* (Rockett et al., 1994) and *T. gondii* (Gazzinelli et al., 1994; Kahn et al., 1995). Treatment of spleen cells with nitric oxide synthetase inhibitor did partially restore this effect (Kahn et al., 1997).

In acute murine toxoplasmosis, the cytokine IL-10 has been found to be responsible for the downregulation of lymphocyte proliferation responses (Kahn et al., 1995). In *N. caninum* infections, this is apparently not the case (Kahn et al., 1997). The principal mechanism for murine protection against *N. caninum* is likely to involve IL-12 and Interferon (IFN) gamma. These two cytokines have also been found to be essential for protection against a number of infectious agents, including *T. gondii* (Hunter et al., 1995). Previous *in vitro* observations had already suggested that IFN gamma is critically involved in the growth of *N. caninum* tachyzoites: treatment of ovine fibroblasts with ovine recombinant IFN gamma for 24 hours before infection with *N. caninum* significantly inhibited intracellular multiplication of the parasite (Innes et al., 1995). In conclusion, the mechanisms of host protection against *N. caninum* appear to be similar in many respects to the immune response elicited by *T. gondii*.

Although the study described above focused on the first 14 days of infection, a more recently published paper (Long et al., 1998) describes a comparison of intracerebral parasite load, lesion development and systemic cytokines in different mouse strains (BALB/c, C57BL/6, B.10.D2) infected with *N. caninum* at six weeks post infection. BALB/c and C57BL/6 mice were highly susceptible (as defined by central nervous system lesions and parasite load) to the development of *N. caninum*-induced encephalitis, whereas B10.2 mice were found to be resistant. Resistance in the latter was associated with high IFN gamma : IL-4 ratio from antigen-stimulated splenocytes.

Recently, Kasper and Kahn (1998) demonstrated that vaccination of mice with intact *N. caninum* tachyzoites protects these mice against a lethal challenge from *T. gondii*, and that this protection is mediated by antigen-cross-reactive $CD8^+$ T cells obtained from spleens of *N. caninum*-vaccinated mice. These observations differ significantly from those in earlier studies undertaken by Lindsay and co-workers (1990). However, these studies employed the very virulent *T. gondii* RH strain, and it appears that protection is highly dependent on the strain used for the challenge. This was confirmed in a study recently undertaken by Lindsay et al. (1998), where the effect of vaccination of mice with N. caninum was investigated with respect to oral challenge with T. gondii oocysts. These experiments indicated that infection with N. caninum provides some protection against fatal infection with T. gondii oocysts of a moderately pathogenic strain, but not tachyzoites of a highly pathogenic strain. The protection achieved by N. caninum vaccination is much less than that provided by previous exposure to T. gondii (Lindsay et al., 1998).

Recently, Éperon et al. (1998) established a mouse model for N. caninum infections comprising B-cell-deficient antibody knock-out mice (C57BL/6 mice with a transgenic mutation in the transmembrane exon of the IgM μ chain gene; μMT mice) and the corresponding parental strain C57BL/6. Parental and μMT mice were infected with N. caninum tachyzoites, or were immunized with dead parasites, and mice were sacrificed at various times after infection. Brains, hearts, lungs, livers, spleens and kidneys were investigated using histology, immunohistochemistry and PCR. PCR detection (Müller et al., 1996) revealed that more organs were also more heavily infected in μMT mice compared to wild-type mice, and immunohistochemistry of brain tissues of μMT mice using the polyclonal anti-N. caninum antiserum demonstrated a high infection density with multifocal necrotic lesions, which were absent in the brains of wild-type mice. In both wild-type and μMT mice infected with living N. caninum tachyzoites, spleen cells stimulated in vitro with N. caninum antigen produced high levels of IFN-gamma and IL-10 in comparison with immunized mice. The μMT spleen cells produced less IL-10 than wild-type splenocytes; the phenomenon became more evident with increasing time of infection. Consequently, the susceptibility of μMT mice may be partially related to decreased production of IL-10 in spleen cells (Éperon et al., 1998).

Other animal model systems to study the immune response to N. caninum infections are currently used. Experimental infection of sheep has shown that these are highly susceptible to N. caninum infection (McAllister et al., 1996b; Buxton et al., 1997a). Buxton et al. (1997b) reported on experimental infections of sheep, showing that N. caninum inoculated during pregnancy killed all foetuses, while inoculation before pregnancy did not cause any mortality, but did provide some protection against subsequent challenge with N. caninum during pregnancy. The live T. gondii vaccine Toxovax (Intervet, BV) did not protect ewes against challenge with N. caninum during pregnancy, in spite of the preliminary observations that the two parasites exhibit cross-reactivity in T-cell responses (Buxton et al., 1997b).

Cellular and humoral immune responses in cattle are being investigated (Williams et al., 1997; Innes et al., 1997). Following experimental infection of heifers, antigen-specific proliferation of lymphocytes was recorded from day eight to day 77 p.i. High levels of IFN gamma were measured. IgM responses occurred at day 12 p.i., and IgGl as well as IgG2 responses were detected at day 29 p.i., peaking at day 35, with the IgG2 response higher and more sustained. Since bovine B cells treated with IFN gamma *in vitro* produce high levels of IgG2 (Estes *et al.*, 1994), these results may indicate that a Thl response was induced upon experimental infection with *N.*

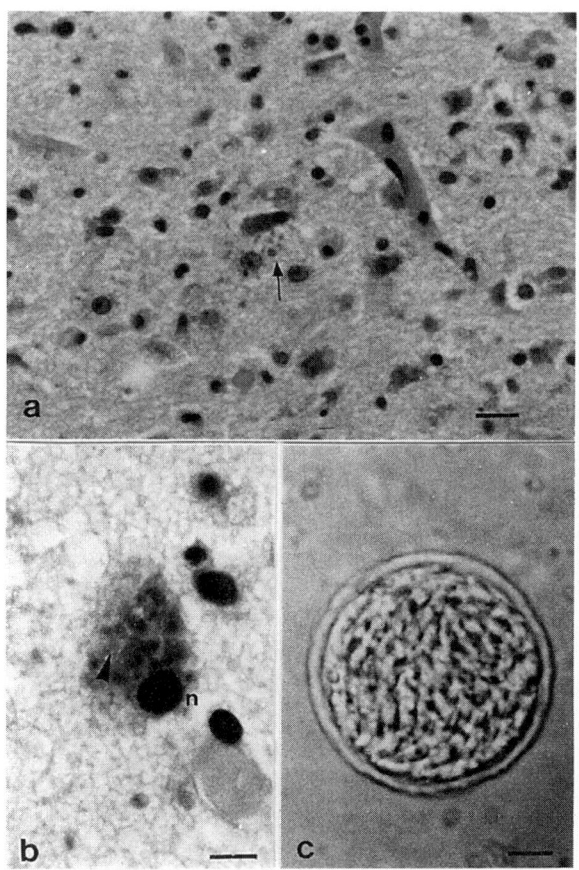

Figure 2 Neospora caninum tachyzoites and bradyzoites. a. Histological section (hemotoxylin eosin-stain) of brain tissue of a μMT mouse experimentally infected with the NC-1 isolate of *N. caninum*. The arrow points towards a group of tachyzoites. Scale bar = 25 μm. b. Higher magnification view of a pseudocyst containing numerous *N. caninum* tachyzoites (arrowhead). Hemotoxylin eosin stain. n = host cell nucleus. Scale bar = 7.5 μm. c. *Neospora* NC-Liverpool-isolate tissue cyst containing bradyzoites. Nomarski illumination. Note the thick cyst wall. Scale bar = 17.5 μm. a, b courtesy of Dr Simone Éperon; c courtesy of Dr Milton McAllister.

caninum (Williams *et al.*, 1997b). Innes *et al.* (1997) also detected cell proliferation responses to a crude *N. caninum* lysate from days six to eight p.i. Phenotypic analysis of these cells activated *in vitro* showed that the majority were CD^{4+} T-cells. Supernatants from these activated cells contained IFN gamma, and these T-cell supernatants were able to inhibit multiplication of *N. caninum* tachyzoites.

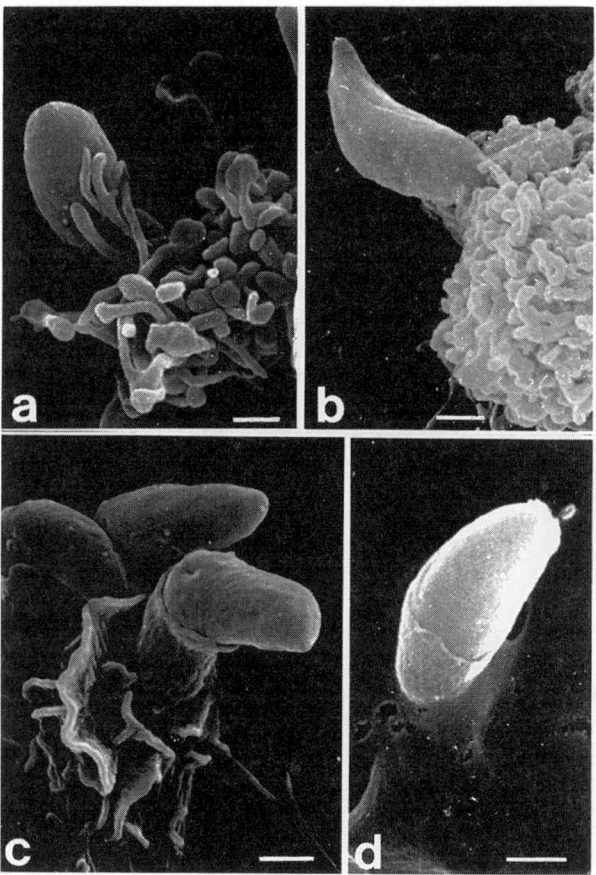

Figure 3 Adhesion and host cell entry by *Neospora caninum*. a, b. Tachyzoites establishing the initial contact between parasite and host cell surface membrane. *N. caninum* tachyzoites enter cells either passively, e.g. by being phagocytosed by a bovine macrophage (c), or by active invasion of non-phagocytic cells as shown in d. Scale bars = 0.5 μm.

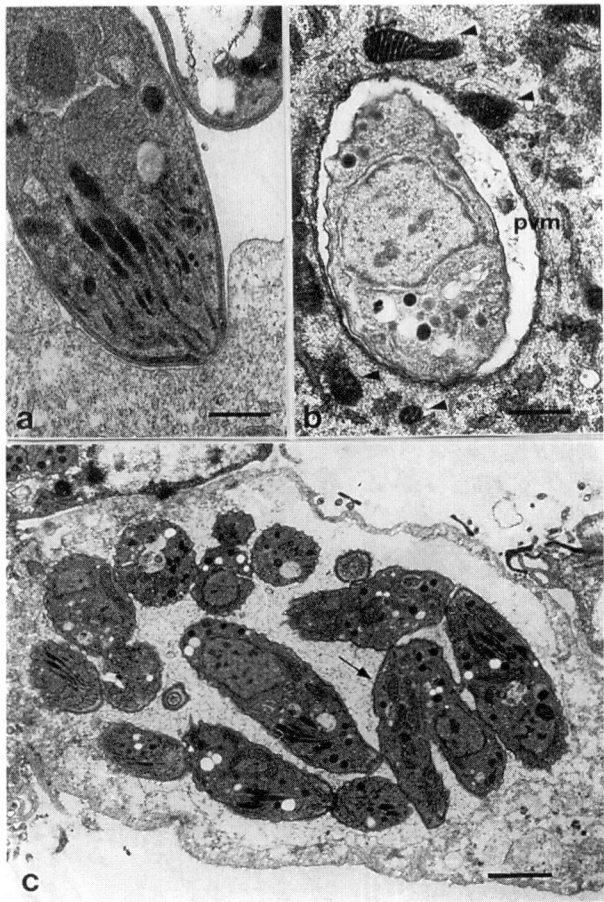

Figure 4 TEM of host cell invasion and intracellular proliferation of *Neospora caninum* tachyzoites within bovine aorta endothelial cells. a. Parasites enter their host cells apical end first. Scale bar = 0.3 μm. b. Tachyzoite within a parasitophorous vacuole at 30 minutes post invasion. Extensive secretion of parasite products into the lumen of the vacuole takes place. The parasitophorous vacuole membrane (pvm) is clearly visible. Note the host cell mitochondria which accumulate around the parasitophorous vacuole (arrowheads). Scale bar = 0.3 μm. c. Pseudocyst containing numerous tachyzoites at 48 hours post-invasion. Parasites divide by endodyogeny (arrow). The lumen of the vacuole is filled with a membranous network (arrowhead). Scale bar = 1 μm.

6. CONCLUDING REMARKS

Since the discovery of *N. caninum* ten years ago, this parasite has attracted considerable attention, not only as an important causative agent of abortion in cattle and neuromuscular disease in dogs but also as a complementary model system to *T. gondii* for investigating the basic biology of intracellular parasitism. Major efforts have been invested in the development of sensitive and specific molecular tools for the detection of *N. caninum* within tissues and body fluids. These newly developed diagnostic tools will contribute largely towards assessing the relevance of neosporosis as a potential risk factor for animal and also human health. They also play a major role during molecular-epidemiological investigations. Most likely molecular approaches (such as transfection of parasites using techniques previously developed for *T. gondii*) will be increasingly used to elucidate molecular pathogenetic events in the course of neosporosis, but also to prepare potential new immunotherapeutic tools for future vaccination against infection or disease mediated by *N. caninum*.

ACKNOWLEDGEMENTS

Many thanks are addressed to Norbert Müller, Nicole Fuchs, Sabrina Sonda (Institute of Parasitology, University of Bern) and Camilla Björkman (Swedish University of Agricultural Sciences, Uppsala) for many suggestions and critical reading of the manuscript, and Banû Yürüker for moral support. Simone Éperon and Milton McAllister are gratefully acknowledged for providing the micrographs shown in Figure 2. Moreover, it is a great pleasure to thank Bruno Gottstein (Institute of Parasitology, University of Bern) for his constant encouragement and enthusiastic support. Many thanks also to those colleagues who have generously provided data on still unpublished observations. This work is financed by the Swiss National Science Foundation (project grant No. 31.46846.96).

REFERENCES

Achbarou, A., Mercereau-Puijalon, O., Autheman, J. M., Fortier, B., Camus, D. and Dubremetz, J.F. (1991a). Characterization of microneme proteins of *Toxoplasma gondii*. *Molecular and Biochemical Parasitology* **47**, 223–233.
Achbarou, A., Mercereau-Puijalon O., Sadak, A. and Dubremetz, J.F. (1991b).

Differential targeting of dense granule proteins in the parasitophorous vacuole of *Toxoplasma gondii*. *Parasitology* **103**, 321–329.

Adams, J.H., Hudson, D.E., Torii, M., Ward, G.E., Wellems, T.E., Aikawa, M. and Miller, L.H. (1990). The Duffy receptor family of *Plasmodium knowlesi* is located within the micronemes of invasive merozoites. *Cell* **63**, 141–153.

Aikawa, M., Torii, M., Sjoelander, A., Berzins, K., Perlman, P. and Miller, L.H. (1990). Pf155/RESA antigen is localized in dense granules of *Plasmodium falciparum* merozoites. *Experimental Parasitology* **71**, 326–329.

Alexander, J., Scharton-Kersten, T.M., Yap, G., Roberts, C.W., Liew, F.Y. and Sher, A. (1997). Mechanisms of innate resistance to *Toxoplasma gondii*. *Philosophical Transaction of the Royal Society of London* **352**, 1355–1359.

Anderson, M.L., Blanchard, P.C., Barr, B.C., Dubey, J.P., Hoffman, R.L. and Conrad, P.A. (1991). *Neospora*-like protozoan infection as a major cause of abortion in California dairy cattle. *Journal of the American Veterinary Medical Association* **198**, 241–244.

Anderson, M.L., Palmer, C.W., Thurmond, M.C., Picanso, J.P., Blanchard, P.C., Breitmeyer, R.E., Layton, A.W., McAllister, M., Dafl, B., Kinde, H., Read, D.H., Dubey, J.P., Conrad, P.A. and Barr, B.C. (1995). Evaluation of abortions in cattle attributable to neosporosis in selected dairy herds in California. *Journal of the American Veterinary Medical Association* **207**, 1206–1210.

Augustine, P.C. (1989). The *Eimeria*: cellular invasion and host-cell parasite interactions. In: *Coccidia and intestinal coccidiomorphs*. Collogues de INTRA 49. (P. Yvore, ed.) pp. 205–215. Tours, France.

Baker, D.G., Morishita, T.Y., Brooks, D.L., Shen, S.K., Lindsay, D.S. and Dubey, J.P. (1995). Experimental oral inoculations in birds to evaluate potential definitive host of *Neospora caninum*. *Journal of Parasitology* **81**, 783–785.

Bannister, L. H. and Dluzewski A.R. (1990). The ultrastructure of red cell invasion in malaria infections: a review. *Blood Cell* **16**, 257–92.

Barber, J.S. and Trees, A.J. (1996). Clinical aspects of twenty-seven cases of neosporosis in dogs. *Veterinary Record* **139**, 439–443.

Barber, J., Trees, A.J., Owen, M. and Tennant, B. (1993). Isolation of *Neospora caninum* from a British dog. *Veterinary Record* **133**, 531–532.

Barber, J.S., Holmdahl, O.J.M., Owen, M.R., Guy, F., Uggla, A. and Trees, A.J. (1995). Characterization of the first European isolate of *Neospora caninum* (Dubey, Carpenter, Speer, Topper and Uggla). *Parasitology* **111**, 563–568.

Barr, B.C., Anderson, M.L., Dubey, J.P. and Conrad, P.A. (1991). *Neospora*-like protozoal infection associated with bovine abortions. *Veterinary Pathology* **28**, 110–116.

Barr, B.C., Conrad, P.A., Breitmeyer, R., Sverlow, K., Anderson, M.L., Reynolds, J., Chauvet, A.E., Dubey, J.P. and Ardans, A.A. (1993). Congenital *Neospora* infection in calves born from cows that had previously aborted *Neospora*-infected fetuses: Four cases (1990–1992). *Journal of the American Veterinary Medical Association* **202**, 113–117.

Barr, B.C., Conrad, P.A., Sverlow, K.W., Tarantal, A.F. and Hendrickx, A.G. (1994a). Experimental fetal and transplacental *Neospora* infection in the non-human primate. *Laboratory Investigations* **71**, 236–242.

Barr, B.C., Rowe, J.D., Sverlow, K.W., BonDurant, R.H., Ardans, A.A., Oliver, M.N. and Conrad, P.A. (1994b). Experimental reproduction of bovine fetal *Neospora* infection and death with a bovine *Neospora* isolate. *Journal of Veterinary Diagnosis and Investigations* **6**, 207–215.

Barr, B.C., Anderson, M.L., Sverlow, K.W. and Conrad, P.A. (1995). Diagnosis of bovine fetal *Neospora* infection with an indirect fluorescent antibody test. *Veterinary Record* **137**, 611–613.

Barta, J.R. and Dubey, J.P. (1992). Characterization of anti-*Neospora caninum* hyperimmune rabbit serum by Western blot analysis and immunoelectron microscopy. *Parasitology Research* **78**, 689–694.

Baszler, T.V., Knowles, D.P., Dubey, J.P., Gay, J.M., Mathison, B.A. and McElwain, T.F. (1996). Serological diagnosis of bovine neosporosis by *Neospora caninum* monoclonal antibody-based competitive inhibition ELISA. *Journal of Clinical Microbiology* **34**, 1423–1428.

Bauer, C., Dubremetz, J.F. and Enzeroth, R. (1995). Characterization of surface antigens of *Eimeria nieschulzi* (Sporozoa, Eimeriidae) merozoites. *Parasitology Research* **81**, 230–234.

Beckers, C.J., Wakefield, T. and Joiner, K.A. (1997). The expression of *Toxoplasma* proteins in *Neospora caninum* and the identification of a gene encoding a novel rhoptry protein. *Molecular and Biochemical Parasitology* **89**, 209–223.

Bjerkås, I. and Dubey, J.P. (1991). Evidence that *Neospora caninum* is identical to the *Toxoplasma*-like parasite of Norwegian dogs. *Acta Veterinaria Scandinavica* **32**, 407–410.

Bjerkås, I., Mohn, S.F. and Presthus, J. (1984). Unidentified cyst-forming sporozoon causing encephalomyelitis and myositis in dogs. *Zeitschrift für Parasitenkunde* **70**, 271–274.

Bjerkås, I., Jenkins, M.C. and Dubey, J.P. (1994). Identification and characterization of *Neospora caninum* tachyzoite antigens useful for diagnosis of neosporosis. *Clinical Diagnostical Laboratory Immunology* **1**, 214–221.

Björkman, C. and Hemphill, A. (1998). Characterization of *Neospora caninum* iscom antigens using monoclonal antibodies. *Parasite Immunology* **20**, 73–80.

Björkman, C. and Lunden, A. (1998). Application of iscom antigen preparations in ELISA for diagnosis of *Neospora* and *Toxoplasma* infections. *International Journal of Parasitology* **28**, 187–193.

Björkman, C., Lunden, A., Holmdahl, O.J.M., Barber, J., Trees, A.J. and Uggla, A. (1994a). *Neospora caninum* in dogs: detection of antibodies by ELISA using an iscom antigen. *Parasite Immunology* **16**, 643–648.

Björkman, C., Lunden, A. and Uggla, A. (1994b). Prevalence of antibodies to *Neospora caninum* and *Toxoplasma gondii* in Swedish dogs. *Acta Veterinaria Scandinavica* **35**, 445–447.

Björkman, C., Johansson, O., Stenlund, S., Holmdahl, J. and Uggla, A. (1996). *Neospora* species infection in a herd of dairy cattle. *Journal of the American Veterinary Medical Association* **188**, 1441–1444.

Björkman, C., Holmdahl, J. and Uggla, A. (1997). An indirect enzyme linked immunoassay (ELISA) for demonstration of antibodies to *Neospora caninum* in serum and milk of cattle. *Veterinary Parasitology* **68** 251–256.

Bonhomme, A., Pingret, L. and Pinon, J.M. (1992). Review: *Toxoplasma gondii* cellular invasion. *Parasitologia* **34**, 31–43.

Boothroyd, J.C., Black, M., Bonnefoy, S., Hehl, A., Knoll, L.J., OrtegaBarria, E. and Tomavo, S. (1997). Genetic and biochemical analysis of development in *Toxoplasma gondii*. *Philosophical Transactions of the Royal Society of London Series B—Biological Sciences* **352**, 1347–1354.

Braun-Breton, C. and Perreira Da Silva, L.H. (1993). Malaria proteases and red blood cell invasion. *Parasitology Today* **9**, 92–96.

Brezin, A.P., Egwuago, C.E., Silveira, C., Thulliez, P., Martins, M.C., Mahdi, R.M.,

Belfort, R. and Nussenblatt, R.B. (1991). Analysis of aqueous humor in ocular toxoplasmosis. *New England Journal of Medicine* **324**, 699.

Brindley, P.J., Gazzinelli, R.T., Denkers, E.Y., Davis, S.W., Dubey, J.P., Belfon, R., Marüns, M.-C., Silveiran C., Jamra, L., Waters, A.P. and Sher, A. (1993). Differentiation of *Toxoplasma gondii* from closely related coccidia by riboprint analysis and a surface antigen gene polymerase chain reaction. *American Journal of Tropical Medicine and Hygiene* **48**, 447–456.

Bryan, L.A., Cajedhar, A.A., Dubey, J.P. and Haines, D.M. (1994). Bovine neonatal encephalomyelitis associated with a *Neospora sp.* protozoan. *Canadian Veterinary Journal* **35**, 111–113.

Burg, J.L., Grover, C.M., Pouletty, P. and Boothroyd, J.C. (1989). Direct and sensitive detection of a pathogenic protozoan, *Toxoplasma gondii*, by polymerase chain reaction. *Journal of Clinical Microbiology* **27**, 1787–1792.

Buxton, D., Maley, S.W., Thomson, K.M., Trees, A.J. and Innes, E.A. (1997a). Experimental infection of non-pregnant and pregnant sheep with *Neospora caninum*. *Journal of Comparative Pathology* **117**, 1–16.

Buxton, D., Wright, S., Maley, S.W., Lunden, A., Vermeulen, A.N. and Innes, E.A. (1997b). Protective immunity to *Neospora caninum*. *VIIth International Coccidiosis Conference and European Union COST820 Workshop*. September 1–5. Keble College, Oxford, England.

Candolfi, E., Villard, O., Thouvenin, M. and Kien, T.T. (1996). Role of nitric oxide-induced immune suppression in toxoplasmosis during pregnancy and in infection by a virulent strain of *Toxoplasma gondii*. *Current Topics in Microbiology and Immunology* **219**, 141–154.

Carruthers, V.B. and Sibley, L.D. (1997). Sequential secretion from three distinct organelles of *Toxoplasma gondii* accompanies invasion of human fibroblasts. *European Journal of Cell Biology* **73**, 114–123.

Cesbron-Delauw, M.F. (1994). Dense granule organelles of *Toxoplasma gondii*: the role in the host-parasite relationship. *Parasitology Today* **10**, 239–246.

Cesbron-Delauw, M.F., Tomavo, S., Beauchamps, P., Fourmaux, M.P., Camus, D., Capron, A. and Dubremetz, J.F. (1994). Similarities between the primary structures of two distinct major surface proteins of *Toxoplasma gondii*. *Journal of Biological Chemistry* **269**, 16217–16222.

Cesbron-Delauw, M.F., Lecordier, L. and Mercier C. (1996). Role of secretory dense granule organelles in the pathogenesis of toxoplasmosis. *Current Topics in Microbiology and Immunology* **219**, 59–65.

Charif, H., Darcy, F., Torpier, G., Cesbron-Delauw, M.F. and Capron, A. (1990). *Toxoplasma gondii:* characterization and localization of antigens secreted from tachyzoites. *Experimental Parasitology* **71**, 114–124.

Cheng, Q. and Saul, A. (1994). Sequence analysis of the apical membrane antigen I (AMA-1) of *Plasmodium vivax*. *Molecular and Biochemical Parasitology* **65**, 183–187.

Chobotar, B., Danforth, H.D. and Entzeroth, R. (1993). Ultrastructural observations of host cell invasion by sporozoites of *Eimeria papillata in vivo*. *Parasitology Research* **79**, 15–23.

Cole, R.A., Lindsay, D.S., Dubey, J.P. and Blagburn, B.L. (1993). Detection of *Neospora caninum* in tissue sections using a murine monoclonal antibody. *Journal of Veterinary Diagnosis and Investigations* **5**, 579–584.

Cole, R.A., Lindsay, D.S., Dubey, J.P., Toivio-Kinnucan, M.A. and Blagburn, B.L. (1994). Characterization of a murine monoclonal antibody generated against *Neospora caninum* by western blot analysis and immunoelectron microscopy. *American Journal of Veterinary Research* **55**, 1717–1722.

Cole, R.A., Lindsay, D.S., Blagburn, B.L. and Dubey, J.P. (1995a). Vertical transmission of *Neospora caninum* in mice. *Journal of Parasitology* **81**, 730–732.

Cole, R.A., Lindsay, D.S., Blagburn, B.L., Sorjonen, D.C. and Dubey, J.P. (1995b). Vertical transmission of *Neospora caninum* in dogs. *Journal of Parasitology* **81**, 208–211.

Conrad, P.A., Barr, B.C., Sverlow, K.W., Anderson, M., Daft, B., Kinde, H., Dubey, J.P., Munson, L. and Ardans, A. (1993a). In vitro isolation and characterization of a *Neospora* sp. from aborted bovine fetuses. *Parasitology* **106**, 239–249.

Conrad, P.A., Sverlow, K., Anderson, M., Rowe, J., BonDurant, R., Tuter, G., Breitmeyer, R., Palmer, C., Thummond, M., Ardans, A., Dubey, J.P., Duhamel, G. and Barr, B. (1993b). Detection of serum antibody responses in cattle with natural or experimental *Neospora* infections. *Journal of Veterinary Diagnosis and Investigations* **5**, 572–578.

Cuddon, P., Lin, D.S., Bowman, D.D., Lindsay, D.S., Miller, T.K., Duncan, J.D., De Lahunta, A., Cummings, J., Suter, M., Cooper, B., King, J.M. and Dubey, J.P. (1992). *Neospora caninum* infection in English Springer Spaniel littermates: diagnostic evaluation and organism isolation. *Journal of Veterinary Internal Medicine* **6**, 325–332.

Dannatt, L., Guy, F., and Trees, A.J. (1995). Abortion due to *Neospora* species in a dairy herd. *Veterinary Record* **137**, 566–567.

Darcy, F. and Santoro F. (1994). Toxoplasmosis. In *Parasitic Infections and the Immune System* (F. Kierszenbaum, ed.). pp. 163–190. New York: Academic Press.

De Braganca, K., Peschka, B., Peters, B. and Seitz, H.M. (1996). How long does the obligate intracellular parasite *Toxoplasma gondii* survive in cell free medium? *Proceedings of the 10th Japanese–German Cooperative Symposium on Protozoan Diseases.* September 11–15. Hannover, Germany.

De Carvalho, L. and De Souza, W. (1997). Carbohydrate-containing molecules found in rhoptries are released during active penetration of tachyzoites of *Toxoplasma gondii* into host cells. *Biology of the Cell* **21**, 129–135.

De Carvalho, L., Yan, C.Y.I. and De Souza, W. (1993). Effect of various digestive enzymes on the interaction of *Toxoplasma gondii* with macrophages. *Parasitology Research* **79**, 114–118.

Dobrowolski, J. and Sibley, L.D. (1996). *Toxoplasma* invasion of mammalian cells is powered by the actin skeleton of the parasite. *Cell* **84**, 933–939.

Dobrowolski, J. and Sibley, L.D. (1997). The role of the cytoskeleton in host cell invasion by *Toxoplasma gondii*. *Behring Institut Mitteilungen* **99**, 90–96.

Dobrowolski, J.M., Niesman, I.R. and Sibley, L.D. (1997). Actin in the parasite *Toxoplasma gondii* is encoded by a single copy gene, ACT1 and exists primarily in a globular form. *Cell Motility and the Cytoskeleton* **37**, 253–262.

Dubey, J.P. (1993). *Toxoplasma, Neospora, Sarcocystis,* and other tissue cyst-forming coccidia of humans and animals. In: *Parasitic Protozoa* (J.P. Kreier, ed.) Vol. 6, pp. 1–158. New York: Academic Press.

Dubey, J.P. and Beattie, C.P. (1988). *Toxoplasmosis of Animals and Man.* Boca Raton, FL: CRC Press.

Dubey, J.P. and De Lahunta, A. (1993). Neosporosis associated congenital limb deformities in a calf. *Applied Parasitology* **34**, 229–233.

Dubey, J.P. and Lindsay, D.S. (1989a). Transplacental *Neospora caninum* infection in cats. *Journal of Parasitology* **75**, 765–771.

Dubey, J.P. and Lindsay, D.S. (1989b). Transplacental *Neospora caninum* infection in dogs. *American Journal of Veterinary Research* **50**, 1578–1579.

Dubey, J.P. and Lindsay, D.S. (1990a). Neosporosis in dogs. *Veterinary Parasitology* **36**, 147–151.
Dubey, J.P. and Lindsay, D.S. (1990b). *Neospora caninum* induced abortion in sheep. *Journal of Veterinary Diagnosis and Investigations* **2**, 230–233.
Dubey, J.P. and Lindsay, D.S. (1993). Neosporosis. *Parasitology Today* **9**, 452–458.
Dubey, J.P. and Lindsay, D.S. (1996). A review of *Neospora caninum* and neosporosis. *Veterinary Parasitology* **67**, 1–59.
Dubey, J.P. and Porterfield, M.L. (1990). *Neospora caninum* (Apicomplexa) in an aborted equine foetus. *Journal of Parasitology* **76**, 732–734.
Dubey, J.P., Carpenter, J.L., Speer, C.A., Topper, M.J. and Uggla, A. (1988a). Newly recognized fatal protozoan disease of dogs. *Journal of the American Veterinary Medical Association* **192**, 1269–1285.
Dubey, J.P., Hanel, A.L., Lindsay, D.S. and Topper, M.J. (1988b). Neonatal *Neospora caninum* infection in dogs: isolation of the causative agent and experimental transmission. *Journal of the American Veterinary Medical Association* **193**, 1259–1263.
Dubey, J.P., Hartley, W.J., Lindsay, D.S. and Topper, M.J. (1990a). Fatal congenital *Neospora caninum* infection in a lamb. *Journal of Parasitology* **76**, 127–130.
Dubey, J.P., Lindsay, D.S. and Lipscomb, T.P. (1990b). Neosporosis in cats. *Veterinary Pathology* **27**, 335–339.
Dubey, J.P., Acland, H.M. and Hamir, A.N. (1992). *Neospora caninum* (Apicomplexa) in a stillborn goat. *Journal of Parasitology* **78**, 532–534.
Dubey, J.P., Hamir, A.N., Shen, S.K., Thulliez, P. and Rupprecht, C.E. (1993). Experimental *Toxoplasma gondii* infection in raccoons (Procyon lotor). *Journal of Parasitology* **79**, 548–552.
Dubey, J.P., Metzger, F.L.J., Hattel, A.L., Lindsay, D.S. and Fritz, D.L. (1995). Canine cutaneous neosporosis: clinical improvement with clindamycin. *Veterinary Dermatology* **6**, 37–43.
Dubey, J.P., Lindsay, D.S., Adams, D.S., Gay, I.M., Baszler, T.V., Blagburn, B.L. and Thulliez, P. (1996a). Serologic responses of cattle and other animals infected with *Neospora caninum*. *American Journal of Veterinary Research* **57**, 320–336.
Dubey, J.P., Morales, J.A., Villalobos, P., Lindsay, D.S., Blagburn, B.L. and Topper, M.J. (1996b). Neosporosis-associated abortion in a dairy goat. *Journal of the American Veterinary Medical Association* **208**, 263–265.
Dubey, J.P., Rigoulet, J., Lagourette, P., George, C., Longeart, L. and Le Net, J.L. (1996c). Fatal transplacental neosporosis in a deer (*Ceraus eldi siumensis*) from a zoo. *Journal of Parasitology* **82**, 338–339.
Dubey, J.P., Jenkins, M.C., Adams, D.S., McAllister, M.M., Anderson-Sprecher, R., Baszler, T.V., Kwok, O.C., Lally, N.C., Björkman, C., and Uggla, A. (1997). Antibody responses of cows during an outbreak of neosporosis evaluated by indirect fluorescence antibody test and different enzyme-linked immunosorbent assay. *Journal of Parasitology* **83**, 1063–1069.
Dubremetz, J.F. (1998). Host cell invasion by *Toxoplasma gondii*. *Trends in Microbiology* **6**, 27–30.
Dubremetz, J.F. and McKerrow, J.H. (1995). Invasion mechanisms. In: *Biochemistry and Molecular Biology of Parasites*. (J.J. Marr and M. Mueller, eds). pp. 307–322. New York: Academic Press.
Dubremetz, J.F. and Schwartzman, J.D. (1993). Subcellular organelles of *Toxoplasma gondii* and host cell invasion. *Research in Immunology* **144**, 31–33.
Ellis, J., Luton, K., Baverstock, P.R., Brindley, P.J., Nimmo, K.A. and Johnson,

A.M. (1994). The phylogeny of *Neospora caninum*. *Molecular and Biochemical Parasitology* **64**, 303–311.

Ellis, J., McMillan, D., Croan, D., Reddacliff, L. and Harper, P.A.W. (1997). Detection of *Neospora caninum* and *Toxoplasma gondii* DNA by single tube nested PCR. *Journal of Microbiological Methods* **30**, 237.

Éperon, S., Hemphill, A. and Gottstein, B. (1998). *Neospora caninum:* susceptibility versus resistance parameters in B-cell deficient mice and corresponding wild-type mice. *30th Annual Meeting of the USGEB*. March 5–6. Lausanne, Switzerland.

Estes, D.M., Closser, N.M. and Allen, G.K. (1994). IFN-gamma stimulates IgG2 production from bovine B cells costimulated with anti-mu and mitogen. *Cellular Immunology* **154**, 287–295.

Fischer, H.G., Stachelhaus, S., Sahm, M., Meyer, H.E. and Reichmann, G. (1998). GRA7, and excretory 29 kDa *Toxoplasma gondii* dense granule antigen released by infected host cells. *Molecular and Biochemical Parasitology* **91**, 251–262.

Fourmaux, M.N., Achbarou, A., Mercereau-Puijalon, O., Biderre, C., Briche, I., Loyens, A., Odberg-Ferragut, C., Camus, D. and Dubremetz, J.F. (1996). The MIC1 microneme protein of *Toxoplasma gondii* contains a duplicated receptor-like domain and binds to host cell surface. *Molecular and Biochemical Parasitology* **83**, 201–210.

Foussard, F., Leriche, M.A. and Dubremetz, J.F. (1991). Characterization of the lipid content of *Toxoplasma gondii* rhoptries. *Parasitology* **102**, 367–370.

Fuchs, N., Sonda, S., Gottstein, B. and Hemphill, A. (1998). Differential expression of cell surface- and dense granule-associated *Neospora caninum* proteins in tachyzoites and bradyzoites. *Journal of Parasitology* **84**, 753–758.

Galinski, M.R. and Barnwell, J.W. (1996). *Plasmodium vivax:* merozoites, invasion of reticulocytes and considerations for malaria vaccine development. *Parasitology Today* **12**, 20–28.

Gazzinelli, R.T., Hayashi, S., Wysocka, M., Carrera, L., Kuhn, R., Müller, W., Roberge, F., Trinchieri, G. and Sher, A. (1994). Role of IL-12 in the initiation of cell-mediated immunity by *Toxoplasma gondii* and its regulation by IL-10 and nitric oxide. *Journal of Eukaryotic Microbiology* **41**, 92.

Gazzinelli, R.T., Amichay, D., Sharton-Kersten, T., Grunwald, E., Farber, J.M. and Sher, A. (1996). Role of macrophage-derived cytokines in the induction and regulation of cell-mediated immunity to *Toxoplasma gondii*. *Current Topics in Microbiology and Immunology* **219**, 127–139.

Gomez Marin, J.E., Bonhomme, A., Guenounou, M. and Pinon, J.M. (1996). Role of interferon-gamma against invasion by *Toxoplasma gondii* in a human monocytic cell line (THP1): involvement of the parasite's secretory phospholipase A2. *Cellular Immunology* **169**, 218–225.

Gottstein, B., Hentrich, B., Wyss, R., Thür, B., Busato, A., Stärk, K.D.C. and Müller, N. (1998). Molecular and immunodiagnostic investigations on bovine neosporosis in Switzerland. *International Journal of Parasitology* **28**, 679–691.

Gray, M.L., Harmon, B.G., Sales, L. and Dubey, J.P. (1996). Visceral neosporosis in a 10-year-old horse. *Journal of Veterinary Diagnosis and Investigations* **8**, 130–133.

Grimwood, J. and Smith, J.E. (1995). *Toxoplasma gondii:* redistribution of tachyzoite surface protein during host cell invasion and intracellular development. *Parasitology Research* **81**, 657–661.

Grimwood, J., Mineo, J.R. and Kasper, L.H. (1996). Attachment of *Toxoplasma gondii* to host cells is host cell cycle dependent. *Infection and Immunity* **64**, 4099–4104.

Guo, Z.-G. and Johnson, A.M. (1995). Genetic comparison of *Neospora caninum* with *Toxoplasma* and *Sarcocystis* by random amplified polymorphic DNA-polymerase chain reaction. *Parasitology Research* **81**, 365–370.

Hay, W.H., Shell, L.G., Lindsay, D.S. and Dubey, J.P. (1990). Diagnosis and treatment of *Neospora caninum* infections in a dog. *Journal of the American Veterinary Medical Association* **197**, 87–89.

Hehl, A., Krieger, T. and Boothroyd, J.C. (1997). Identification and characterization of SRS1, a *Toxoplasma gondii* surface antigen upstream of and related to SAG1. *Molecular and Biochemical Parasitology* **89**, 271–282.

Hemphill, A. (1996). Subcellular localization and functional characterization of Nc-p43, a major *Neospora caninum* tachyzoite surface protein. *Infection and Immunity* **64**, 4279–4287.

Hemphill, A. and Gottstein, B. (1996). Identification of a major surface protein on *Neospora caninum* tachyzoites. *Parasitology Research* **82**, 497–504.

Hemphill, A., Gottstein, B. and Kaufmann, H. (1996). Adhesion and invasion of bovine endothelial cells by *Neospora caninum*. *Parasitology* **112**, 183–197.

Hemphill, A., Fuchs, N., Sonda, S., Gottstein, B. and Hentrich, B. (1997a). Identification and partial characterization of a 36 kD surface protein on *Neospora caninum* tachyzoites. *Parasitology* **115**, 371–380.

Hemphill, A., Felleisen, R., Connolly, B., Gottstein, B., Hentrich, B. and N. Müller, N. (1997b). Characterization of a cDNA clone encoding Nc-p43, a major *Neospora caninum* surface protein. *Parasitology* **115**, 581–590.

Hemphill, A., Gajendran, N., Sonda, S., Fuchs, N., Gottstein, B., Hentrich, B. and Jenkins, M. (1998). Identification and characterization of a dense granule-associated protein in *Neospora caninum* tachyzoites. *International Journal of Parasitology* **28**, 429–438.

Ho, M.S., Barr, B.C., Marsh, A.E., Anderson, M.L., Rowe, I.D., Tarantal, A.F., Hendrickx, A.G., Sverlow, K., Dubey, J.P. and Conrad, P.A. (1996). Identification of bovine *Neospora* parasites by PCR amplification and specific small subunit rRNA sequence probe hybridization. *Journal of Clinical Microbiology* **34**, 1203–1208.

Ho, M.S., Barr, B.C., Tarantal, A.F., Lai, L.T., Hendrickx, A.G., Marsh, A.E., Sverlow, K.W., Packham, A.E. and Conrad, P.A. (1997a). Detection of *Neospora* from tissues of experimentally infected rhesus macaques by PCR and specific DNA probe hybridization. *Journal of Clinical Microbiology* **35**, 1740–1745.

Ho, M.S., Barr, B.C., Rowe, J.D., Anderson, M.L., Sverlow, K.W., Packham, A., Marsh, A.E. and Conrad, P.A. (1997b). Detection of *Neospora* sp. from infected bovine tissues by PCR and probe hybridization. *Journal of Parasitology* **83**, 508–514.

Holder, A.A. (1994). Proteins on the surface of the malaria parasite and cell invasion. *Parasitology* **108**, S5–S18.

Holmdahl, O.J.M. and Mattsson, J.G. (1996). Rapid and sensitive identification of *Neospora caninum* by *in vitro* amplification of the internal transcribed spacer 1. *Parasitology* **112**, 177–182.

Holmdahl, O.J.M., Mattsson, J.G., Uggla, A. and Johansson, K.E. (1994). The phylogeny of *Neospora caninum* and *Toxoplasma gondii* based on ribosomal RNA sequences. *Federation of the European Microbiology Societies Microbiology Letters* **119**, 187–192.

Holmdahl, O.J.M., Björkman, C. and Uggla, A. (1995). A case of *Neospora* associated bovine abortion in Sweden. *Acta Veterinaria Scandinavica* **36**, 279–281.

Homan, W.L., Limper, L., Verlaan, M., Borst, A., Vercammen, M. and van Knapen, F. (1997). Comparison of the internal transcribed spacer, ITS 1, from *Toxoplasma gondii* isolates and *Neospora caninum*. *Parasitology Research* **83**, 285–289.

Howe, D.K. and Sibley, L.D. (1995). *Toxoplasma gondii* comprises three clonal lineages: correlation of parasite genotype with human disease. *Journal of Infectious Diseases* **172**, 1561–1566.

Howe, D.K. and Sibley, L.D. (1997). Development of molecular genetics for *Neospora caninum:* a complementary system to *Toxoplasma gondii. Methods* **13**, 123–133.

Howe, D.K., Mercier, C., Messina, M. and Sibley, L.D. (1997). Expression of *Toxoplasma gondii* genes in the closely related apicomplexan parasite *Neospora caninum*. *Molecular and Biochemical Parasitology* **86**, 29–36.

Ho-Yen, D.O., Joss, A.W.L., Balfour, A.H., Smyth, E.T.M, Baird, D. and Chatterton, J.M.W. (1992). Use of polymerase chain reaction to detect *Toxoplasma gondii* in human blood samples. *Journal of Clinical Pathology* **45**, 910–913.

Huldt, G. (1981). Workshop no. 3. Serodiagnosis of parasitic infections. *Parasitology* **82**, 49–55.

Hunter, C.A., Candolfi, E., Subauste, C., VanCleave, V. and Remington, J.S. (1995). Studies on the role of interleukin.12 in acute murine toxoplasmosis. *Immunology* **84**, 16–20.

Hunter, C.A., Suzuki, Y., Subauste, C.S. and Reminton, J.S. (1996). Cells and cytokines in resistance to *Toxoplasma gondii. Current Topics in Microbiology and Immunology* **219**, 113–125.

Innes, E.A. (1997). Toxoplasmosis: comparative species susceptibility and host immune response. *Comparative Immunology, Microbiology and Infectious Diseases* **20**, 131–138.

Innes, E.A., Panton, W.R.M., Marks, J., Trees, A.J., Holmdahl, J. and Buxton, D. (1995). Interferon gamma inhibits the intracellular multiplication of *Neospora caninum*, as shown by incorporation of 3H uracil. *Journal of Comparative Pathology* **113**, 95–100.

Innes, E.A., Marks, J., Lunden, A., McLean-Tooke, A., Maley, S.W., Wright, S. and Buxton, D. (1997). Cell-mediated immune responses to *Neospora caninum*. *VIIth International Coccidiosis Conference and European Union COST820 Workshop*. September 1–5. Keble College, Oxford, England.

Jacobs, D., Dubremetz, J.F., Loyens, A., Bosman, F. and Saman, E. (1998). Identification and heterologous expression of a new dense granule protein (GRA7) from *Toxoplasma gondii*. *Molecular and Biochemical Parasitology* **91**, 237–249.

Jardine, J.E. (1996). The ultrastructure of bradyzoites and tissue cysts of *Neospora caninum* in dogs: absence of distinguishing morphological features between parasites of canine and bovine origin. *Veterinary Parasitology* **62**, 231–240.

Jenkins, M.C., Wouda, W. and Dubey, J.P. (1997). Serological response over time to recombinant *Neospora caninum* antigens in cattle after a neosporosis induced abortion. *Clinical and Diagnostic Laboratory Immunology* **4**, 270–274.

Joiner, K.A., Fuhrman, S.A., Miettinen, H.M., Kasper, L.H. and Mellman, I. (1990). *Toxoplasma gondii:* Fusion competence of parasitophorous vacuoles in Fc-receptor-transfected fibroblasts. *Science* **249**, 641–646.

Joiner, K.A. and Dubremetz, J.F. (1993). *Toxoplasma gondii:* a protozoan for the nineties. *Infection and Immunity* **61**, 1169–1172.

Kahn, I.A., Matsuura, T. and Kasper, L.H. (1995). IL-10 mediates immunesuppression following primary infection with *Toxoplasma gondii* in mice. *Parasite Immunology* **17**, 185–195.

Kahn, I.A., Schwartzman, J.D., Fonseka, S. and Kasper, L.H. (1997). *Neospora caninum*: role for immune cytokines in host immunity. *Experimental Parasitology* **85**, 24–34.

Karlsson, K.A., Milh, M.A., Anggstroem, J., Bergstroem, J., Dezfoolian, H., Lanne, B., Leonardson, I. and Teneberg, S. (1992). Membrane proximity and internal binding in the microbial recognition of host cell glycolipids: a conceptual discussion. In: *Molecular recognition in host–parasite interactions. FEMS Symposium No. 61.* (T.K. Korhonen, H. Tapani and P.H. Maekelae eds). pp. 115–132. New York: Plenum Press.

Kasper, L.H., and Mineo, J.R. (1994). Attachment and invasion of host cells by *Toxoplasma gondii*. *Parasitology Today* **10**, 184–188.

Kasper, L.H and Kahn, I.A. (1998). Antigen-specific CD8$^+$ T cells protect against lethal toxoplasmosis in mice infected with *Neospora caninum*. *Infection and Immunity* **66**, 1554–1560.

Kaufmann, H., Yamage, M., Roditi, I., Dobbelaere, D., Dubey, J.P., Holmdahl, O.J.M., Trees, A. and Gottstein, B. (1996). Discrimination of *Neospora caninum* from *Toxoplasma gondii* and other apicomplexan parasites by hybridization and PCR. *Molecular and Cellular Probes* **10**, 289–297.

Koudela, B., Svoboda, M., Björkman, C. and Uggla, A. (1998). Neosporosis in dogs — the first case report in the Czech republic. *Veterinary Medicine-Czech* **43**, 33–36.

Lally, N.C., Jenkins, M.C. and Dubey, J.P. (1996a). Evaluation of two *Neospora caninum* recombinant antigens for use in an ELISA for the diagnosis of bovine neosporosis. *Clinical and Diagnostic Laboratory Immunology* **3**, 275–279.

Lally, N.C., Jenkins, M.C. and Dubey, J.P. (1996b). Development of polymerase chain reaction assay for the diagnosis of neosporosis using the *Neospora caninum* 14-3-3 gene. *Molecular and Biochemical Parasitology* **75**, 169–178.

Lally, N.C., Jenkins, M., Lidell, S. and Dubey, J.P. (1997). A dense granule protein (NCDG1) gene from *Neospora caninum*. *Molecular and Biochemical Parasitology* **87**, 239–243.

Lander, A.D. (1993). Proteoglycans. In: *Guidebook to Extracellular Matrix and Adhesion Proteins.* (T. Kreis and R. Vale eds). pp. 12–16. Oxford: University Press.

Lecordier, L., Moleon-Borodowsky, I., Dubremetz, J.F., Tourvieille, B., Mercier, C., Deslée, D., Capron, A. and Cesbron-Delauw, M.F. (1995). Characterization of a dense granule antigen of *Toxoplasma gondii* (GRA6) associated to the network of the parasitophorous vacuole. *Molecular and Biochemical Parasitology* **70**, 85–94.

Leriche, M.A. and Dubremetz, J.F. (1991). Characterization of the protein contents of rhoptries and dense granules of *Toxoplasma gondii* tachyzoites by subcellular fractionation and monoclonal antibodies. *Molecular and Biochemical Parasitology* **45**, 249–260.

Liddell, S., Jenkins, M.C. and Dubey, J.P. (1997). Development of a mouse model for neosporosis: PCR detection and quantification of *Neospora caninum*. *VIIth International Coccidiosis Conference and European Union COST820 Workshop.* September 1–5. Keble College, Oxford, UK.

Liddell, S., Lally, N.C., Jenkins, M.C. and Dubey, J.P. (1998). Isolation of the cDNA encoding a dense granule associated antigen (NCDG2) of *Neospora caninum*. *Molecular and Biochemical Parasitology* **93**, 153–158.

Lindsay, D.S. and Dubey, J.P. (1989a). Evaluation of anti-coccidial drugs' inhibition of *Neospora caninum* development in cell cultures. *Journal of Parasitology* **75**, 990–992.

Lindsay, D.S. and Dubey, J.P. (1989b). Immunohistochemical diagnosis of *Neospora caninum* in tissue sections. *American Journal of Veterinary Research* **50**, 1981–1983.
Lindsay, D.S. and Dubey, J.P. (1989c). *Neospora caninum* (Protozoa: Apicomplexa) infections in mice. *Journal of Parasitology* **75**, 772–779.
Lindsay, D.S. and Dubey, J.P. (1990). Effects of sulfadiazine and amprolium on *Neospora caninum* (Protozoa: Apicomplexa) infections in mice. *Journal of Parasitology* **76**, 177–179.
Lindsay, D.S., Blagburn, B.L. and Dubey, J.P. (1990). Infection of mice with *Neospora caninum* (Protozoa: Apicomplexa) does not protect against challenge with *Toxoplasma gondii*. *Infection and Immunity* **58**, 2699–2700.
Lindsay, D.S., Speer, C.A., Toivio-Kinaucan, M.A., Dubey, J.P. and Blagburn, B.L. (1993). Use of infected cultured cells to compare ultrastructural features of *Neospora caninum* from dogs and *Toxoplasma gondii*. *American Journal of Veterinary Research* **54**, 103–106.
Lindsay, D.S., Rippey, N.S., Cole, R.A., Parsons, L.C., Dubey, J.P., Tidwell, R.R. and Blagburn, B.L. (1994). Examination of the activities of 43 chemotherapeutic agents against *Neospora caninum* tachyzoites in cultured cells. *American Journal of Veterinary Research* **55**, 976–981.
Lindsay, D.S., Lenz, S.D., Cole, R.A., Dubey, J.P. and Blagburn, B.L. (1995a). Mouse model for central nervous system *Neospora caninum* infections. *Journal of Parasitology* **81**, 313–315.
Lindsay, D.S., Rippey, N.S., Powe, T.A., Sartin, E.A., Dubey, J.P. and Blagburn, B.L. (1995b). Abortion, fetal death, and stillbirth in pregnant pygmy goats inoculated with tachyzoites of *Neospora caninum*. *American Journal of Veterinary Research* **56**, 1176–1180.
Lindsay, D.S., Butler, J.M., Rippey, N.S. and Blagburn, B.L. (1996a). Demonstration of synergistic effects of sulfonamides and dihydrofolate reductase/thymidylate synthase inhibitors against *Neospora caninum* tachyzoites in cultured cells, and characterization of mutants resistant to pyrimethamine. *American Journal of Veterinary Research* **57**, 68–72.
Lindsay, D.S., Kelly, E.J., McKown, R., Stein, F.J., Plozer, J., Herman, J., Blagburn, B.L. and Dubey, J.P. (1996b). Prevalence of *Neospora caninum* and *Toxoplasma gondii* antibodies in coyotes (*Canis latrans*) and experimental infections of coyotes with *Neospora caninum*. *Journal of Parasitology* **82**, 657–659.
Lindsay, D.S., Steinberg, H., Dubielzig, R.R., Semrad, S.D., Konkle, D.M., Miller, P.E. and Blagburn, B.L. (1996c). Central nervous system neosporosis in a foal. *Journal of Veterinary Diagnosis and Investigations* **8**, 507–510.
Lindsay, D.S., Butler, J.M. and Blagburn, B.L. (1997). Efficacy of decoquinate against *Neospora caninum* tachyzoites in cell culture. *Veterinary Parasitology* **68**, 35–40.
Lindsay, D.S., Lenz, S.D., Dykstra, C.C., Blagburn, B.L. and Dubey, J.P. (1998). Vaccination of mice with *Neospora caninum*: response to oral challenge with *Toxoplasma gondii* oocysts. *Journal of Parasitology* **84**, 311–315.
Loevgren, K., Uggla, A. and Morein, B. (1987). A new approach to the preparation of a *Toxoplasma gondii* membrane antigen for use in ELISA. *Journal of Veterinary Medicine* **34**, 274–282.
Long, M.T. and Baszler, T.V. (1996). Fetal loss in Balb/c mice infected with *Neospora caninum*. *Journal of Parasitology* **82**, 608–611.
Long, M.T., Baszler, T.V. and Mathison, B.A. (1998). Comparison of intracerebral

parasite load, lesion development, and systemic cytokines in mouse strains infected with *Neospora caninum*. *Journal of Parasitology* **84**, 316–324

Louie, K., Sverlow, K.W., Barr, B.C., Anderson, M.L. and Conrad, P.A. (1997). Cloning and characterization of two recombinant *Neospora* protein fragments and their use in serodiagnosis of bovine neosporosis. *Clinical Diagnostic Laboratory Immunology* **4**, 692–699.

Lycke, E. and Strannegard, O. (1965). Enhancement by lysozyme and hyaluronidase of the penetration by *Toxoplasma gondii* into cultured host cells. *British Journal of Experimental Pathology* **46**, 189–199.

Lycke, E. and Norrby, R. (1966). Demonstration of a factor of *Toxoplasma gondii* enhancing the penetration of *Toxoplasma* parasites into cultured host cells. *British Journal of Experimental Pathology* **47**, 248–256.

Mack, D., Kasper, L. and McLeod, R. (1994). Alterations in cell surface glycosylation, heparin and chondroitin sulfate do not modify invasion of CHO cells by the PTg B strain of *Toxoplasma gondii*. *Journal of Eukaryotic Microbiology* **41**, 14S.

Marsh, A.E., Bar, B.C., Sverlow, K., Ho, M., Dubey, J.P. and Conrad, P.A. (1995). Sequence analysis and comparison of ribosomal DNA from bovine *Neospora* to similar coccidial parasites. *Journal of Parasitology* **81**, 530–535.

Marsh, A.E., Barr, B.C., Madigan, J.E. and Conrad, P.A. (1996). *In vitro* cultivation and characterization of a *Neospora* isolate obtained from a horse with protozoal myeloencephalitis. *Procedures of the American Society of Parasitology and The Society of Protozoologists*. June 11–15. Tucson, Arizona.

Mauël, J. (1996). Intracellular survival of protozoan parasites with special reference to *Leishmania* spp., *Toxoplasma gondii* and *Trypanosoma cruzi*. *Advances in Parasitology* **38**, 1–53.

Mayhew, I.G., Smith, K.C., Dubey, J.P., Gatward, L.K. and McGlennon, N.J. (1991). Treatment of encephalomyelitis due to *Neospora caninum* in a litter of puppies. *Journal of Small Animal Practice* **32**, 609–612.

McAllister, M.M., Hoffmann, E.M., Hietala, S.K., Conrad, P.A., Anderson, M.L. and Salman, M.O. (1996a). Evidence suggesting a point source exposure in an outbreak of bovine abortion due to neosporosis. *Journal of Veterinary Diagnosis and Investigations* **8**, 355–357.

McAllister, M.M., McGuire, A.M., Jolley, W.R., Lindsay, D.S., Trees, A.J. and Stobart. R.H. (1996b). Experimental neosporosis in pregnant ewes and their offspring. *Veterinary Pathology* **33**, 647–655.

McAllister, M.M., Parmley, S.F., Weiss, L.M., Welch, V.J. and McGuire, A.M. (1996c). An immunohistochemical method for detecting bradyzoite antigen (BAG5) in *Toxoplasma gondii*-infected tissues cross-reacts with a *Neospora caninum* bradyzoite antigen. *Journal of Parasitology* **82**, 354–355.

McGuire, A.M., McAllister, M.M., Jolley, W.R. and Anderson-Sprecher, R.C. (1997a). A protocol for the production of *Neospora caninum* tissue cysts in mice. *Journal of Parasitology* **83**, 647–651.

McGuire, A.M., McAllister, M.M. and Jolley, W.R. (1997b). Separation and cryopreservation of *Neospora caninum* tissue cysts from murine brains. *Journal of Parasitology* **83**, 319–321.

Moen, A.R., Wouda, W. and van Werven, T. (1995). Clinical and sero-epidemiological follow-up study in four dairy herds with an outbreak of *Neospora* abortion. *Procedures of the Dutch Society for Veterinary Epidemiology and Economics*. December 13. Lelystad.

Mondragon, R. and Frixione, E. (1996). Ca(2+) dependence of conoid extrusion

in *Toxoplasma gondii* tachyzoites. *Journal of Eukaryotic Microbiology* **43**, 120–127.
Mordue, D.G. and Sibley, L.D. (1997). Intracellular fate of vacuoles containing *Toxoplasma gondii* is determined at the time of formation and depends on the mechanism of entry. *Journal of Immunology* **159**, 4452–4459.
Morisaki, J.H., Heuser, J.E. and Sibley, D.L. (1995). Invasion of *Toxoplasma gondii* occurs by active penetration of the host cell. *Journal of Cell Science* **108**, 2457–2464.
Müller, N., Zimmermann, V., Hentrich, B. and Gosttstein, B. (1996). Diagnosis of *Neospora caninum* and *Toxoplasma gondii* infection by PCR and DNA hybridization immunoassay. *Journal of Clinical Microbiology* **34**, 2850–2852.
Müller, N., McAllister, M., Fuchs, N., Hentrich, B., Gottstein, B. and Hemphill, A. (1997). Differential immunohistochemical- and PCR-based detection of *Neospora caninum* and *Toxoplasma gondii* bradyzoites in paraffin-embedded brain tissue from infected mice. *VIIth International Coccidiosis Conference and European Union COST820 Workshop*. September 1–5. Keble College, Oxford, England.
Nagasawa, H., Himeno, K. and Suzuki, N. (1996). Mini review: protective immunity in toxoplasmosis. *Applied Parasitology* **37**, 284–292.
Norrby, R. (1971). Immunological study on the host cell penetration factor of *Toxoplasma gondii*. *Infection and Immunity* **3**, 278–286.
Obendorf, D.L., Murray, N., Veldhuis, G., Munday, B.L. and Dubey, J.P. (1995). Abortion caused by neosporosis in cattle. *Austrian Veterinary Journal* **72**, 117–118.
Oebrink, B. (1993). Cell adhesion and cell–cell contact proteins. In: *Guidebook to the Extracellular Matrix and Adhesion Proteins*. (T. Kreis and R. Vale eds). pp. 109–114. Oxford: University Press.
Odin, M. and Dubey, J.P. (1993). Sudden death associated with *Neospora caninum*-myocarditis in a dog. *Journal of the American Veterinary Medical Association* **203**, 831–833.
Ossorio, P.N., Schwartzman, J.D. and Boothroyd, J.C. (1992). A *Toxoplasma gondii* rhoptry protein associated with host cell penetration has unusual charge asymmetry. *Molecular and Biochemical Parasitology* **50**, 1–16.
Paré, J., Hietala, S.K. and Thurmond, M.C. (1995). An enzyme-linked immunosorbent assay (ELISA) for serological diagnosis of *Neospora* sp. infection in cattle. *Journal of Veterinary Diagnosis and Investigations* **7**, 352–359.
Parish, S.M., Maag-Miller, L., Besser, T.E., Weidner, J.P., McElwain, T., Knowles, D.P. and LeaUhers, C.W. (1987). Myelitis associated with protozoal infection in newborn calves. *Journal of the American Veterinary Medical Association* **191**, 1599–1600.
Parmley, S.F., Goebel, F.D. and Remington, J.S. (1992). Detection of *Toxoplasma gondii* in cerebrospinal fluid from AIDS patients by polymerase chain reaction. *Journal of Clinical Microbiology* **30**, 1127–1133.
Pasvol, G., Carlsson, J. and Clough, B. (1992). Recognition of molecules on red cells for malarial parasites. In: *Molecular Recognition in Host–Parasite Interactions. FEMS Symposium No. 61*. (T.K. Korhonen, T. Hovi and P.H. Maekelae eds). pp. 173–200. New York: Plenum Press.
Payne, S. and Ellis, J. (1996). Detection of *Neospora caninum* DNA by the polymerase chain reaction. *International Journal of Parasitology* **26**, 347–351.
Perkins, M.E. (1992). Rhoptry organelles of apicomplexan parasites. *Parasitology Today* **8**, 28–32.
Rockett, K.A., Auburn, M.M., Rockett, A.J., Cowden, W.B. and Clark, I.A. (1994).

Possible role of nitric oxide in malarial immunosuppression. *Parasite Immunology* **16**, 243–249.
Romand, S., Thulliez, P., and Dubey, J.P. (1998). Direct agglutination test for serologic diagnosis of *Neospora caninum* infection. *Parasitology Research* **60**, 50–53.
Saffer, L. and Schwartzman, J.D. (1991). A soluble phospholipase of *Toxoplasma gondii* associated with host cell penetration. *Journal of Protozoology* **38**, 454–460.
Saffer, L., Mercereau-Puijalon, O., Dubremetz, J.F. and Schwartzman, J.D. (1992). Localization of a *Toxoplasma gondii* rhoptry protein by immunoelectron microscopy during and after host cell penetration. *Journal of Protozoology* **39**, 526–530.
Sam-Yellowe, T.Y. (1996). Rhoptry organelles of the apicomplexa: their role in host cell invasion and intracellular survival. *Parasitology Today* **12**, 308–316.
Sasai, K., Lillehoj, H.S., Hemphill, A., Matsuda, H., Hanioka, V., Fukata, T., Baba, E. and Arakawa, A. (1998). A chicken monoclonal antibody identifies a common epitope which is present on the apicomplexan coccidian parasites. *Journal of Parasitology* **84**, 654–656.
Savva, D., Morris, J.C., Johnson, J.D. and Holliman, R.E. (1990). Polymerase chain reaction for detection of *Toxoplasma gondii*. *Journal of Medical Microbiology* **32**, 25–31.
Sawada, M., Park, CH., Morita, T., Shimada, A., Umemura, T. and Haritani, M. (1997). Pathological findings of nude mice inoculated with bovine *Neospora*. *Journal of Veterinary Medical Science* **59**, 947–948.
Schares, G. and Conraths, F.J. (1996). Characterization of surface antigens of *Neospora caninum* tachyzoites. *Proceedings COST820. Vaccines against Animal Coccidiosis*. October 10–12, Copenhagen, Denmark.
Schares, G., Dubremetz, J.F., Loyens, A., Bärwald, A. and Conraths, F.J. (1997). Characterization of *Neospora caninum* antigens by monoclonal antibodies. *VIIth International Coccidiosis Conference and European Union COST820 Workshop*. September 1–5. Keble College, Oxford, England.
Schenkman, S., Robbins, E.S. and Nussenzweig, V. (1991). Attachment of *Trypanosoma cruzi* requires parasite energy and invasion can be independent of the target cell cytoskeleton. *Infection and Immunity* **59**, 645–654.
Shaw, M.K. (1997). The same but different: the biology of *Theileria* sporozoite entry into bovine cells. *International Journal of Parasitology* **27**, 457–474.
Sher, A., Denkers, E.Y., and Gazzinelli, R.T. (1995). Induction and regulation of host cell-mediated immunity by *Toxoplasma gondii*. *Ciba Foundation Symposia* **195**, 95–104.
Sibley, L.D. (1995). Invasion of vertebrate cells by *Toxoplasma gondii*. *Trends in Cell Biology* **5**, 129–133.
Sibley, L.D., Adams, L.B. and Krahenbuhl, J. (1993). Macrophage interactions in toxoplasmosis. *Research in Immunology* **144**, 38–40.
Sibley, L.D., Niesman, I.R., Asai, T. and Takeuchi, T. (1994). *Toxoplasma gondii*: secretion of a potent nucleoside triphosphate hydrolase into the parasitophorous vacuole. *Experimental Parasitology* **79**, 301–311.
Sibley, L.D., Niesman, I.R., Parmley, S.F. and Cesbron-Delauw, M.F. (1995). Regulated secretion of multilamellar vesicles leads to formation of a tubuvesicular network in host-cell vacuoles occupied by *Toxoplasma gondii*. *Journal of Cell Science* **108**, 1669–1677.
Sim, B.K.L., Orlandi, P.A., Haynes, J.D., Klotz, F.W., Carter, J.M., Camus, D., Zegans, M.E. and Chulay, J.D. (1990). Primary structure of the 175K *Plasmodium*

falciparum erythrocyte binding antigen and identification of a peptide which elicits antibodies that inhibit malaria merozoite invasion. *Journal of Cell Biology* **111**, 1877–1884.

Sim, B.K.L., Toyoshima, T., Haynes, J.D. and Aikawa, M. (1992). Localization of the 175-kilodalton erythrocyte binding antigen in micronemes of *Plasmodium falciparum* merozoites. *Molecular and Biochemical Parasitology* **51**, 157–160.

Smith, J.E. (1995). A ubiquitous intracellular parasite: the cellular biology of *Toxoplasma gondii*. *International Journal of Parasitology* **25**, 1301–1309.

Soldati, D., Kim, K., Kampmeier, J., Dubremetz, J.F. and Boothroyd, J.C. (1995). Complementation of a *Toxoplasma gondii* ROP1 knock-out mutant using phleomycin selection. *Molecular and Biochemical Parasitology* **74**, 87–97.

Sonda, S., Fuchs, N., Conolly, B., Fernandez, P., Gottstein, B. and Hemphill, A. (1998). The major 36 kDa *Neospore caninum* tachyzoite surface protein is closely related to the major *Toxoplasma gondii* surface antigen 1. *Molecular and Biochemical Parasitology*.

Speer, C.A. and Dubey, J.P. (1989). Ultrastructure of tachyzoites, bradyzoites and tissue cysts of *Neospora caninum*. *Journal of Protozoology* **36**, 458–463.

Stenlund, S., Björkman, C., Holmdahl, O.J.M., Kindahl, H. and Uggla, A. (1997a). Characterization of a Swedish bovine isolate of *Neospora caninum*. *Parasitology Research* **83**, 214–219.

Stenlund, S., Kindahl, H., Uggla, A. and Björkman, C. (1997b). *Neospora caninum* antibody levels in pregnant cattle. *VIIth International Coccidiosis Conference and European Union COST820 Workshop*. September 1–5. Keble College, Oxford, England.

Sundermann, C.A., Estridge, B.H., Branton, M.S., Bridgman, C.R. and Lindsay, D.S. (1977). Immunohistochemical diagnosis of *Toxoplasma gondii*: potential for crossreactivity with *Neospora caninum*. *Journal of Parasitology* **83**, 440–443.

Suss-Toby, E., Zimmerberg, J. and Ward, G.E. (1996). *Toxoplasma* invasion: the parasitophorous vacuole is formed from host cell plasma membrane and pinches off via a fission pore. *Procedures of the National Academy of Science USA* **93**, 8413–8418.

Tait, A. and Sacks, D.L. (1988). The cell biology of parasite invasion and survival. *Parasitology Today* **8**, 228–234.

Tenter, A.M. and Johnson, A.M. (1997). Phylogeny of the tissue-cyst forming coccidia. *Advances in Parasitology* **39**, 70–141.

Thilsted, J.P. and Dubey, J.P. (1989). Neosporosis-like abortions in a herd of dairy cattle. *Journal of Veterinary Diagnosis and Investigations* **1**, 205–209.

Thornton, R.N., Gajadhar, A. and Evans, J. (1994). *Neospora* abortion epidemic in a dairy herd. *New Zealand Veterinary Journal* **42**, 190–191.

Thurmond, M.C. and Hietala, S.K. (1997). Effect of *Neospora caninum* infection on milk production in first-lactation dairy cows. *Journal of the American Veterinary Medical Association* **210**, 672–674.

Tomavo, S. (1996). The major surface proteins of *Toxoplasma gondii*: structures and functions. *Current Topics in Microbiology and Immunology* **219**, 45–54.

Wan, K.L., Carruthers, V.B., Sibley, L.D. and Ajioka, J.W. (1997). Molecular characterization of an expressed sequence tag locus of *Toxoplasma gondii* encodes the micronemal protein MIC2. *Molecular and Biochemical Parasitology* **84**, 203–214.

Wastling, J.M., Nicoll, S. and Buxton, D. (1993). Comparison of two gene amplification methods for the detection of *Toxoplasma gondii* in experimentally infected sheep. *Journal of Clinical Microbiology* **38**, 360–365.

Weiss, L.M. and Ma, Y. (1997). *In vitro* development of bradyzoites and cysts of *Neospora caninum*. *Molecular Parasitology Meeting VIII*. September 24–28. Marine Biological Laboratory, Woods Hole, MA, USA.

Werk, R. (1985). How does *Toxoplasma gondii* enter host cells? *Reviews in Infectious Diseases* **7**, 449–457.

Werner-Meier, R. and Entzeroth, R. (1997). Diffusion of microinjected markers across the parasitophorous vacuole membrane in cells infected with *Eimeria nieschulzi* (Coccidia, Apicomplexa). *Parasitology Research* **83**, 611–613.

Williams, D.J.L., McGarry, J., Guy, F., Barber, J.S. and Trees, A.J. (1996). A novel ELISA for detection of *Neospora*-specific antibodies in cattle. *Veterinary Record* **140**, 328–331.

Williams, D.J.L., Guy, C., Taylor, K., Davison, H.C., McGarry, J., Guy, F. and Trees, A.J. (1997). Cellular and humoral immune responses in cattle infected with *Neospora caninum*. *VIIth International Coccidiosis Conference and European Union COST820 Workshop*. September 1–5. Keble College, Oxford, England.

Wouda, W., Moen, A.R., Visser, I.J.R. and VanKnappen, F. (1997). Bovine fetal neosporosis: a comparison of epizootic and sporadic abortion cases and different age classes with regards to lesion severity and immunohistochemical identification in brain, heart and liver. *Journal of Veterinary Diagnosis and Investigations* **9**, 180–185.

Wouda, W., Dubey, J.P. and Jenkins, M.C. (1997). Serological diagnosis of bovine fetal neosporosis. *Journal of Parasitology* **83**, 545–547.

Yaeger, M.J., Shawd-Wessels, S. and Leslie-Steen, P. (1994). *Neospora* abortion storm in a midwestern dairy. *Journal of Veterinary Diagnosis and Investigations* **6**, 506–508.

Yamage, M., Flechtner, O. and Gottstein, B. (1996). *Neospora caninum*: specific oligonucleotide primers for the detection of brain 'cyst' DNA of experimentally infected nude mice by the polymerase chain reaction (PCR). *Journal of Parasitology* **82**, 272–279.

Yamane, I., Kokuho, T., Shimura, K., Eto, M., Shibahara, T., Haritani, M., Ouchi, Y., Sverlow, K. and Conrad, P.A. (1997). *In vitro* isolation and characterization of a bovine *Neospora* species in Japan. *Research in Veterinary Science* **63**, 77–80.

Youn, J.H., Nam, H.W., Kim, D.J., Park, Y.M., Kim, W.K., Kim, W.S. and Choi, W.Y. (1991). Cell cycle-dependent entry of *Toxoplasma gondii* into synchronized HL-60 cells. *Kisaengch'ung Hak Chapchi* **29**, 121–128.

NOTE ADDED IN PROOF

McAllister *et al.* (1998) have recently reinvestigated whether dogs could be a final host of *N. caninum*. They fed dogs with tissue cysts in infected mouse tissue, and subsequently examined dog faeces for 30 days for the presence of oocysts. Oocysts could be found in the faeces of three out of four infected dogs. These oocysts were subspherical, measuring 10–11 microns in diameter, were shed unsporulated, and sporulation occurred within 3 days. Sporulated oocysts contained two sporocysts with four sporozoites each. Oocyst-containing canine faecal extracts were proven to be orally infected for outbred, inbred, and gamma interferon knockout mice using a variety of diagnostic tests. Based on this study, dogs are a definitive host of *N. caninum*.

McAllister, M.M., Dubey, J.P., Lindsay, D.S., Jolley, W.R., Wills, R.A. and McGuire, A.M. (1998). Dogs are definitive hosts of *Neospora caninum*. *International Journal for Parasitology* **28**, 1473–1478.

PROTEASES OF PROTOZOAN PARASITES

Philip J. Rosenthal

Department of Medicine, Box 0811, San Francisco General Hospital, University of California, San Francisco CA 94143–0811, USA

Abstract	106
1. Introduction	106
2. Classification of Proteases	106
3. Survey of Identified Protozoan Proteases	108
4. Protease Inhibitors as Antiprotozoan Drugs	110
5. *Leishmania*	110
5.1. *Leishmania* metalloprotease (leishmanolysin)	111
5.2. *Leishmania* cysteine protease	114
5.3. Other proteases of *Leishmania*	117
6. African Trypanosomes	117
6.1. *T. brucei* cysteine protease (trypanopain)	117
6.2. Other proteases of African trypanosomes	118
7. *Trypanosoma cruzi*	119
7.1. Cysteine protease of *T. cruzi* (cruzain)	120
7.2. Other proteases of *T. cruzi*	122
8. Malaria Parasites	122
8.1. Plasmodial proteases mediating erythrocyte invasion and rupture	123
8.2. Plasmodial proteases mediating haemoglobin degradation	126
8.3. Plasmodial proteases as potential chemotherapeutic targets	130
9. *Entamoeba histolytica*	131
9.1. Cysteine protease of *E. histolytica*	132
9.2. Other proteases of *E. histolytica*	135
10. *Giardia lamblia*	135
11. *Cryptosporidium parvum*	136
12. *Trichomonas vaginalis*	137
13. *Toxoplasma gondii*	138
14. Summary	139
Acknowledgements	139
References	139

ABSTRACT

Proteolytic enzymes seem to play important roles in the life cycles of all medically important protozoan parasites, including the organisms that cause malaria, trypanosomiasis, leishmaniasis, amebiasis, toxoplasmosis, giardiasis, cryptosporidiosis and trichomoniasis. Proteases from all four major proteolytic classes are utilized by protozoans for diverse functions, including the invasion of host cells and tissues, the degradation of mediators of the immune response and the hydrolysis of host proteins for nutritional purposes. The biochemical and molecular characterization of protozoan proteases is providing tools to improve our understanding of the functions of these enzymes. In addition, studies in multiple systems suggest that inhibitors of protozoan proteases have potent antiparasitic effects. This review will discuss recent advances in the identification and characterization of protozoan proteases, in the determination of the function of these enzymes, and in the evaluation of protease inhibitors as potential antiprotozoan drugs.

1. INTRODUCTION

Protozoan parasites cause many of the most important infectious diseases in the world. Malaria, leishmaniasis, trypanosomiasis and amebiasis are associated with profound morbidity and extensive mortality, mostly in less developed countries. Toxoplasmosis, giardiasis, cryptosporidiosis and trichomoniasis cause significant disease in both developed and developing countries. Certain pathogens, including *Toxoplasma*, *Leishmania* and *Cryptosporidium*, are particular problems in immunocompromised individuals.

It has become increasingly clear that proteolytic enzymes play key roles in the life cycles of protozoan parasites. Many protozoan proteases have been identified and characterized. In a number of cases, important insights into the biological roles of these enzymes have been made. This review will survey available information on proteases of medically important protozoan parasites. Emphasis will be given to studies that have shed light on the functions of these enzymes, and to studies that have evaluated the possibility of inhibiting proteases in the treatment and prevention of protozoan infections.

2. CLASSIFICATION OF PROTEASES

Proteases catalyse the hydrolysis of peptide bonds, and thereby break down protein substrates. Endopeptidases (or proteinases) cleave internal peptide bonds. Exopeptidases cleave peptide bonds at or near the amino or carboxy terminus of a protein. Most proteases will hydrolyse only certain peptide bonds. In some cases highly specific enzymes hydrolyse only a particular bond within a specific protein. In other cases a protease will cleave many peptide bonds that fit certain sequence or conformational requirements. Evaluation of the substrate specificity of a protease can be very helpful in determining the biological role of the enzyme. The hydrolysis of the true substrate of a protease or that of synthetic peptide substrates can be assessed. However, interpretations based on such studies must be made with caution, as *in vitro* assessments of substrate cleavage may not predict the specificity of the protease in its true biological context.

Most proteases fall into one of four large classes (Neurath, 1989; Barrett, 1994). Three of these classes, the serine, cysteine and aspartic proteases, are named for a key active site amino acid, and metalloproteases are named for the requirement of a metal ion for catalysis. Proteases are most readily categorized into one of these four classes by inhibitor studies. Specific inhibitors of each of the four protease classes are available (Salvesen and Nagase, 1989). The use of inhibitors to help determine the biological functions of proteases will be extensively discussed in this review. Proteases are generally synthesized as proforms, which are later processed to mature, enzymatically active forms. Proform processing may occur due to autohydrolysis or the action of other proteases.

Serine proteases (Rawlings and Barrett, 1994b) utilize a catalytic triad of serine, aspartate and histidine, with catalysis proceeding after the formation of a tetrahedral transition state intermediate. Eukaryotic serine proteases include digestive enzymes such as trypsin, chymotrypsin and elastase, each of which has rather broad substrate specificity, and a large number of regulatory proteases with much more limited specificity. The limited specificity of regulatory proteases is mediated both by their inherent catalytic abilities and by the accessibility and conformation of natural substrates.

Cysteine proteases (Rawlings and Barrett, 1994a) utilize a catalytic cysteine and a thiol ester intermediate. The papain family of cysteine proteases is represented in a great many eukaryotic organisms, including protozoans. Most of these proteases act intracellularly, often at acid pH, with rather broad substrate specificity (Bond and Butler, 1987). Other papain-family proteases act extracellularly. The plant cysteine protease papain and the mammalian lysosomal cysteine proteases cathepsin B and cathepsin L are well studied enzymes that serve as models for the many protozoan proteases with similar biochemical features. Many papain family proteases can be classified as cathepsin B-like, with preference for cleavage of peptide substrates with arginine at the P2 position, or cathepsin L-like, with preference for hydrophobic amino acids at P2 (Rawlings and Barrett,

1994a). Cathepsin B- and cathepsin L-like enzymes can also be distinguished by the fact that cathepsin B exhibits carboxypeptidase activity in addition to endopeptidase activity, and by the presence of subfamily-specific sequence motifs (Berti and Storer, 1995; Turk *et al.*, 1997).

Aspartic proteases (Rawlings and Barrett, 1995b) utilize two aspartates in catalysis. Eukaryotic aspartic proteases include intracellular acid proteases, digestive enzymes and extracellular regulatory proteases with tight substrate specificity. Metalloproteases (Rawlings and Barrett, 1995a) are a diverse group of enzymes that utilize a metal ion, usually zinc, in catalysis. Many of the metalloproteases share a conserved HEXXH metal binding site. Biologically important eukaryotic metalloproteases include enzymes that degrade extracellular matrix components (e.g. elastases and collagenases) and regulatory proteases.

The proteasome is a newly identified structure that includes proteases with unique mechanisms (Rivett, 1993). This structure is a cytoplasmic complex of multiple subunits that performs three independent proteolytic functions. The three activities catalyse peptide bond cleavage on the carboxyl side of basic, hydrophobic and acidic amino acids, and are described as trypsin-like, chymotrypsin-like and peptidylglutamyl-peptide hydrolase activities respectively (Rivett, 1993). Some proteasome subunits utilize a catalytic threonine for peptide bond cleavage (Seemüller *et al.*, 1995), a previously unrecognized mechanism of proteolysis.

3. SURVEY OF IDENTIFIED PROTOZOAN PROTEASES

Many proteases of protozoan parasites have been identified and characterized (Table 1). These include members of all four major protease classes. In some cases the genes encoding important proteases have been isolated and sequenced, and molecular studies are helping to characterize protease functions. In other cases only biochemical data are available so far. Even when detailed molecular and biochemical data are available, it has often been difficult to determine if results from different laboratories concern the same or different enzymes. This situation is complicated by the fact that, for a number of protozoans, numerous enzymes with similar biochemical features are expressed. The reason for the redundancy in protease expression is generally not clear. This review will attempt to simplify the extensive literature on protozoan proteases and provide hypotheses regarding protease functions. However, as our understanding of all protozoan proteases is incomplete, statements about protease functions will probably include some oversimplifications and incorrect hypotheses.

Table 1 Major identified proteases of protozoan parasites

Parasite	Class (name)	Gene(s) cloned?	Putative functions[1]
Leishmania	Metallo (leishmanolysin, gp63)	Yes	Cleavage of complement; intracellular proteolysis
	Cysteine	Yes	Intracellular proteolysis
Trypanosoma brucei	Cysteine (trypanopain)	Yes	Parasite differentiation
	Serine oligopeptidase	No	Extracellular proteolysis
	Metallo	Yes	Cleavage of complement
	Proteasome	No	Cytosolic proteolysis
Trypanosoma cruzi	Cysteine (cruzain or cruzipain)	Yes	Parasite differentiation; invasion of host cells
	Serine oligopeptidase	Yes	Mediator of calcium signalling
	Metallo	No	Parasite differentiation
	Proteasome	No	Parasite development and differentiation
	Collagenase	No	Host cell infection
Plasmodium	Cysteine (falcipain)	Yes	Haemoglobin degradation
	Aspartic (plasmepsins I and II)	Yes	Haemoglobin degradation
	Aminopeptidase	No	Globin peptide hydrolysis
	Serine	No	Erythrocyte rupture and/or invasion
	Cysteine	No	Erythrocyte rupture and/or invasion
Entamoeba histolytica	Cysteine (amoebapain, histolysin)	Yes	Tissue invasion; complement activation
	Collagenase	No	Tissue invasion
	Proteasome	Yes	Cytosolic proteolysis
Giardia	Cysteine	Yes	Excystation
Trichomonas	Cysteine	Yes	Attachment to epithelial cells
Toxoplasma	Cysteine	Yes[2]	Intracellular proteolysis
	Serine	No	?
Cryptosporidium	Cysteine	Yes[2]	Invasion of host cells
	Aminopeptidase	No	Excystation

[1] In many cases, data supporting the putative function listed are limited.
[2] The genes have been cloned, but sequences have not yet been published or entered into databases.

Some of the protozoan proteases discussed in this review may have important biological properties independent of their enzymatic activities. For example, some of the proteases appear to be important antigenic targets of antiprotozoan immune responses, and some may mediate cellular attachment and other functions. These nonenzymatic properties of the proteases may be critically important in parasite biology, but they are beyond the scope of this review.

4. PROTEASE INHIBITORS AS ANTIPROTOZOAN DRUGS

As proteases play key roles in the life cycles of protozoan parasites, their characterization should add important insights to our understanding of the means by which these organisms cause disease. These insights may improve our ability to develop new methods for the prevention, treatment, and control of these diseases. In particular, many protozoan proteases may be appropriate targets for chemotherapy. Protease inhibitors are now used to treat a number of diseases. Inhibitors of angiotensin-converting enzyme are among the most widely used cardiovascular drugs (Jackson and Garrison, 1996). Inhibitors of the HIV protease have recently had an enormous impact on the treatment of HIV infection and AIDS (McDonald and Kuritzkes, 1997). Many of the protozoan proteases that will be discussed below are worthy of consideration as potential targets for antiprotozoal chemotherapy. This is a very important aspect of research on protozoan proteases, as for all of the protozoan diseases discussed in this review new drugs are needed. In particular, because of the problems of drug resistance, toxicity and limited efficacy, new drugs are sorely needed for the treatment of malaria, leishmaniasis, trypanosomiasis and cryptosporidiosis. Available evidence regarding the utility of protease inhibitors in the treatment of protozoan infections will be reviewed for each of the protozoans discussed below.

5. *LEISHMANIA*

Leishmaniasis is one of the most important protozoan infections of humans (Berman, 1997). Species of the genus *Leishmania* cause visceral, cutaneous and mucosal disease in many parts of the developing world. Visceral leishmaniasis (kala-azar), which causes hepatosplenomegaly and bone marrow dysfunction, is an important cause of morbidity and mortality in India, Africa and parts of South America. Cutaneous leishmaniasis causes skin ulcers which heal slowly over months or years, often with scarring. Mucosal leishmaniasis can cause severe destructive lesions of mucosal membranes of the nose and mouth. At least 100 000 cases of visceral leishmaniasis and 300 000 cases of cutaneous leishmaniasis are estimated to occur annually (Ashford et al., 1992), leading to about 1000 deaths per year (Walsh, 1989). Available treatments for leishmaniasis, including pentavalent antimonials, pentamidine and amphotericin, are limited by drug resistance, significant toxicities and high cost (Berman, 1997). Two important classes of leishmanial proteases have been identified and extensively characterized, a promastigote surface protease named leishmanolysin and a set of intracellular cysteine proteases.

5.1. *Leishmania* Metalloprotease (Leishmanolysin)

Leishmania parasites contain a major surface glycoprotein with protease activity that has been referred to as gp63, promastigote surface protease and, most recently, leishmanolysin (Bouvier *et al.*, 1985; Chang *et al.*, 1990; Olafson *et al.*, 1990; Bouvier *et al.*, 1995). Leishmanolysin is present on the parasite surface at high density (Bouvier *et al.*, 1995) and is bound to the parasite plasma membrane by a glycosyl-phosphatidylinositol anchor (Bordier *et al.*, 1986; Etges *et al.*, 1986; Schneider *et al.*, 1990). Leishmanolysin molecules from the different species of *Leishmania* are very similar (Bouvier *et al.*, 1987), and they will be discussed interchangeably below.

Leishmanolysin was first identified in promastigotes, the parasite stage that develops in the sandfly vector and then circulates in humans after inoculation by the insect (Bouvier *et al.*, 1985; Davies *et al.*, 1990). It is also present in the amastigote forms that reside within human macrophages, as RNA encoding leishmanolysin is transcribed in both promastigotes and amastigotes (Button *et al.*, 1989), and as antibodies to leishmanolysin recognized the protein in amastigotes in immunofluorescence, immunoblotting and immunoprecipitation studies (Medina-Acosta *et al.*, 1989; Frommel *et al.*, 1990). However, leishmanolysin production is markedly down-regulated at the RNA and protein levels in amastigotes, and the leishmanolysin proteolytic activity of amastigotes is less than 1% that of promastigotes (Schneider *et al.*, 1992). Leishmanolysin is apparently processed and localized differently in the two life-cycle stages. In promastigotes leishmanolysin is surface-localized, while in amastigotes most of the protein appears to be localized to the flagellar pocket and is not bound to a phosphatidylinositol membrane anchor (Medina-Acosta *et al.*, 1989). Another form of leishmanolysin has also been identified in amastigote lysosomes (Ilg *et al.*, 1993). A small portion of amastigote leishmanolysin is on the parasite surface. Though a small portion of total leishmanolysin, this protein nonetheless is the most abundant amastigote surface protein (Medina-Acosta et al., 1989).

Leishmanolysin was initially studied as a surface protein and antigen of infecting parasites; it was subsequently shown to have protease activity (Etges et al., 1986; Chaudhuri and Chang, 1988). The protease has the biochemical properties of a zinc metalloprotease, including incorporation of radiolabelled zinc, inhibition by zinc chelators and reversal of this inhibition with added zinc (Chaudhuri and Chang, 1988; Bouvier et al., 1989; Chaudhuri et al., 1989). The protease cleaved albumin, haemoglobin, C3, IgG and purified rat liver lysosomal proteins (Chaudhuri and Chang, 1988) and preferentially cleaved synthetic peptide substrates with tyrosine at the P1 site, hydrophobic residues at the P1' site and basic amino acid residues at the P2' and P3' sites (Bouvier et al., 1990). Proteolytic activity was maximal at slightly basic pH for the promastigote surface form of leishmanolysin, and at acid pH for intracellular amastigote forms (Ilg et al., 1993; Bouvier et al., 1995), although the pH optimum of the enzyme was also dependent on the substrate studied (Tzinia and Soteriadou, 1991).

Cloning of the genes encoding leishmanolysin from multiple leishmanial species predicted a preproenzyme that was processed to a mature form with at least 70% amino acid identity between the species (Button and McMaster, 1988; Miller et al., 1990; Webb et al., 1991; Steinkraus and Langer, 1992; Gonzalez-Aseguinolaza et al., 1997). The sequences predicted zinc metalloproteases with conservation of zinc binding sites and other sequences typical for this class of protease (Bouvier et al., 1989; Chaudhuri et al., 1989). In most species studied, leishmanolysin is encoded by a set of tandemly linked genes located on a single chromosome (Button et al., 1989). In L. mexicana (Medina-Acosta et al., 1993) and L. chagasi (Ramamoorthy et al., 1992), three linked families of tandemly repeated genes were identified. In L. mexicana transcripts encoding all three families were present in promastigotes, but only one class, which lacked a glycosylphosphatidylinositol anchor addition site, was represented in amastigotes (Medina-Acosta et al., 1993). In L. chagasi a total of 18 genes, which could be distinguished by unique 3' untranslated regions, were identified, including constitutively expressed genes and those expressed only in highly infectious stationary or less infectious logarithmic-phase cultures (Ramamoorthy et al., 1992; Roberts et al., 1993).

Leishmanolysin activity appears to be controlled at a number of levels. Both the regulated expression of different leishmanolysin genes (Ramamoorthy et al., 1992) and post-transcriptional control of protein expression (Wilson et al., 1993; Ramamoorthy et al., 1995) mediate differential production of different leishmanolysins during the developmental cycle of the parasite. Leishmanolysin is processed from a proform to its active mature form, possibly by an autocatalytic process similar to that in other metalloproteases (Bouvier et al., 1990; Button et al., 1993; Macdonald et al., 1995). Recent studies utilizing site-directed mutagenesis of the leishmanolysin gene demonstrated that conserved residues at the zinc binding site are required for activity and stability of the protease, that asparagineglycosylation at three sites contributes to stability, and that Asn-577 acts as the glycosyl phosphatidylinositol addition site and is required for the membrane anchoring of the protease (McGwire and Chang, 1996). However, deglycosylation of leishmanolysin did not block appropriate folding, transport, and activity of the protease (Funk et al., 1994).

The function of leishmanolysin is unknown, but it is believed to act as a *Leishmania* virulence factor (Chang et al., 1990). Leishmanolysin is an abundant surface protein of promastigotes, making up about 1% of total cellular protein (Chang et al., 1990), and its abundance has been correlated with *Leishmania* infectivity in most (Kweider et al., 1987; Kink and Chang, 1988; Wilson et al., 1989; Seay et al., 1996), but not all (Murray et al., 1990) comparisons of cultured parasite strains. Leishmanolysin appears to participate in the binding between *Leishmania* promastigotes and macrophages (Russell and Wilhelm, 1986; Chang et al., 1990; Liu and Chang, 1992), though this property may be independent of the proteolytic activity of the molecule. Leishmanolysin may also act to cleave C3 into C3b and other products, thus supporting the complement receptor-mediated endocytosis of promastigotes by macrophages (Chaudhuri and Chang, 1988). *Leishmania* expressing wild-type leishmanolysin more efficiently cleaved C3, activated complement and bound to complement receptors than did mutants lacking functional leishmanolysin (Brittingham et al., 1995). In addition, leishmanolysin may protect circulating promastigotes by cleaving membranolytic complement components. *Leishmania* expressing wild-type leishmanolysin on their surface fixed only small amounts of the terminal complement components, and were much more resistant to lysis by complement than were mutants lacking functional leishmanolysin (Brittingham et al., 1995). In more recent studies, mutants lacking six of seven genes encoding the *L. major* leishmanolysins were resistant to complement-mediated lysis (Joshi et al., 1998). Taken together, these results suggest that leishmanolysin may allow *Leishmania* to utilize the opsonic properties of complement to mediate entry into macrophages while avoiding the lytic effects of activation of the complete complement pathway.

Intracellular amastigote leishmanolysin may support organisms within macrophage lysosomes by cleaving microbicidal factors (Chang et al., 1990), by degrading proteins to provide nutrients, or by preventing intracellular degradation of the parasites (Chaudhuri et al., 1989; Seay et al., 1996). Leishmanolysin-deficient *Leishmania* mutants have markedly decreased survival in macrophages. Transfection of wild-type leishmanolysin into these mutants significantly improved survival (McGwire and Chang, 1994).

The presence of leishmanolysin at the surface of the related, but non-parasitic, organisms *Crithidia* and *Herpetomonas* (Etges, 1992; Inverso *et al.*, 1993; Schneider and Glaser, 1993) suggests that the enzyme has roles in addition to its support of the intracellular amastigote and interaction with circulating immune mediators. For example, it may contribute to the survival of promastigotes in the sandfly midgut (Etges, 1992).

Leishmanolysin has recently been crystallized (Schlagenhauf *et al.*, 1995), although structure co-ordinates have not yet been released. The determination of the structure of this protease should aid in the evaluation of its biological role and in the development of specific inhibitors. Current research directed toward the inhibition of other metalloproteases (Brown and Giavazzi, 1995) may lead to compounds worthy of study as drugs to treat leishmaniasis.

5.2. *Leishmania* Cysteine Protease

A number of distinct cysteine protease activities have been identified in *Leishmania* (Coombs, 1982; Pupkis and Coombs, 1984; Robertson and Coombs, 1990). These activities mostly reside within lysosome-like organelles of amastigotes known as megasomes (Pupkis *et al.*, 1986). Recent molecular studies have helped sort out the activities and roles of the multiple leishmanial cysteine proteases.

Cysteine protease genes of *L. mexicana* can be grouped into a single-copy *lmcpa* gene (Mottram *et al.*, 1992) multiple *lmcpb* genes, which are found in an array of 19 nonidentical 2.8 kilobase repeat units (Souza *et al.*, 1992) and a single copy *lmcpc* gene (Robertson and Coombs, 1994). The *lmcpa* and *lmcpb* genes encode cathepsin L-like proteases, while the *lmcpc* gene is predicted, based on sequence comparisons, to encode a cathepsin B-like enzyme. The proteases are alternatively categorized biochemically as gelatinase activities A, B, C and E, which are encoded by *lmcpb* genes, and separated based on differences in size, glycosylation, charge and substrate preferences (Robertson and Coombs, 1994), and activity D, which is encoded by the *lmcpc* gene (Robertson and Coombs, 1993). The product of the *lmcpa* gene does not appear to have gelatinase activity, although it does bind active site-directed inhibitors (Mottram *et al.*, 1992). Evaluations of both mRNA (Souza *et al.*, 1992) and protease activity (Pral *et al.*, 1993; Robertson and Coombs, 1994) have shown that most of these proteases are expressed principally in amastigote forms, but lower levels of expression are detected in promastigotes, particularly metacyclic promastigotes, the life-cycle stage that is infective to humans and that was previously shown to express unique cysteine proteases (Bates *et al.*, 1994). Recent studies have shown that the first two copies of the *lmcpb* tandem array (*lmcpb1* and *lmcpb2*) are expressed primarily in metacyclic promastigotes, that the next 16 genes in the array are expressed in amastigotes, and that the last gene in the array is not transcribed and is probably a pseudogene (Mottram *et al.*, 1997).

The products of the *lmcpb* genes and homologous proteases from *L. major* (Sakanari *et al.*, 1997) and *L. pifanoi* (Traub-Cseko *et al.*, 1993) share typical papain family proforms and cathepsin L-like mature domains, but they also contain unusual carboxy-terminal extensions. Such extensions have been seen only in the trypanasomatid protozoans (including *T. cruzi* and *T. brucei*; see other sections of this review) and in some plant cysteine proteases. However, the carboxy-terminal extensions of trypanosomatid and plant cysteine proteases do not have marked similarities in sequence, other than sharing a richness in cysteines that may be important in determining secondary structure (Mottram *et al.*, 1997). The function of the carboxy terminal extensions of leishmanial cysteine proteases is unknown. The extensions may determine life-cycle stage-dependent expression of proteases, as *lmcpb1* and *lmcpb2*, the two genes that are uniquely expressed primarily in metacyclic promastigotes, have truncated extensions (Mottram *et al.*, 1997). The carboxy-terminal extensions do not appear to be necessary for localization to the megasome, as the products of *lmcpb1*, *lmcpb2* (Mottram *et al.*, 1997), and a *L. pifanoi lmcpa* homologue (Duboise *et al.*, 1994), all of which lack full carboxy-terminal extensions, localize to this organelle. In studies of the *L. pifanoi lmcpa* homologue, a number of processing steps were identified. Processing was inhibited by cysteine protease inhibitors, suggesting that autohydrolysis was the mode of processing to active enzyme. Most of the enzyme localized to megasomes, although it was also present in the flagellar pocket and plasma membrane of promastigotes (Duboise *et al.*, 1994).

L. mexicana (Robertson and Coombs, 1993; Bart *et al.*, 1995) and *L. major* (Sakanari *et al.*, 1997) amastigotes also express cathepsin B-like proteases (activity D), which differ from the *lmcpb* gene-encoded proteases in substrate specificity and protein sequence. The gene encoding this protease is single copy in *L. mexicana* (Bart *et al.*, 1997).

The specific roles of each of the leishmanial cysteine proteases and the reason that so many protease genes are expressed are unexplained. Studies of the targeted disruption of a number of the protease genes in *L. mexicana* are now being carried out, however, and so an improved understanding of the functions of these enzymes is developing. Disruption of the *lmcpa* gene did not noticeably alter parasite viability, growth rate or pathogenicity in a murine model (Souza *et al.*, 1994). Disruption of the entire 19-gene *lmcpb* tandem array did not eliminate the viability of cultured parasites, proving that the *lmcpb* gene products are not essential for parasite viability (Mottram *et al.*, 1996). However, mutant parasites were markedly less virulent than wild-type parasites, and had an 80% reduction in infectivity to macrophages. This decreased infectivity was not due to decreased macrophage invasion, but rather to decreased intracellular survival. When evaluated in an animal model, the mutant parasites produced subcutaneous lesions in BALB/C mice, though at a decreased rate compared to wild-type parasites. The mutant parasites did not produce lesions in less susceptible mouse strains (C57BL/6, SvEv 129; Mottram, J.C., personal communication). The product of a single *lmcpb* gene re-expressed in the mutant was enzymatically active and restored infectivity to macrophages to wild-type levels. Mutants in which both the *lmcpb* array and the single *lmcpa* gene were disrupted had a macrophage infectivity phenotype similar to that of the *lmcpb* mutant, although they were unable to form lesions in mice (Mottram, J.C., personal communication). Lastly, mutants with a disrupted *lmcpc* gene, which encodes the cathepsin B-like leishmanial protease, have recently been produced (Bart *et al.*, 1997). The mutant parasites grew normally in promastigote and amastigote cultures, but infectivity to macrophages was reduced by 50%, and skin lesions in mice were somewhat smaller after infection with mutant compared to wild-type parasites. The ultimate cysteine protease gene disruption, in which all of the known (*lmcpa*, *lmcpb*, and *lmcpc*) leishmanial cysteine protease genes have been disrupted, has not yet been reported. In summary, available results suggest that the cysteine proteases are important, but not essential, for the infectivity of *L. mexicana* to macrophages, and that among the proteases the products of the *lmcpb* gene array are probably the most important in mediating parasite virulence.

One putative function for leishmanial cysteine proteases is the degradation of immune molecules. Amastigotes of *L. amazonensis* residing within activated macrophages internalized major histocompatibility complex class II molecules (De Souza Leao *et al.*, 1995). The degradation of the class II molecules was inhibited by cysteine protease inhibitors, leading to the accumulation of the proteins in leishmanial megasomes (De Souza Leao *et al.*, 1995). Thus, amastigote cysteine proteases appear to degrade internalized class II molecules, possibly as a means of evading the host immune response.

Two inhibitors of cysteine proteases, antipain and leupeptin, were potent inhibitors of the growth of *L. mexicana* amastigotes in mouse peritoneal macrophages (Coombs and Baxter, 1984). Recently, both nonpeptide hydrazides identified via a molecular modelling approach (Selzer *et al.*, 1997) and pseudopeptide substrate analogues were shown to inhibit *L. major* cysteine proteases, to block parasite growth at concentrations that had no toxic effects on cultured mammalian cells and to ameliorate the pathology associated with experimental infection in mice (Selzer *et al.*, 1997; Selzer, P. and McKerrow, J.H., personal communication). The inhibitors appeared to act by interfering with parasite lysosomal function, as inhibitor-treated parasites accumulated undigested debris in lysosomes and in the flagellar pocket. The results suggest that cysteine protease inhibitors have promise as antileishmanial drugs.

5.3. Other Proteases of *Leishmania*

One study identified a cytosolic calcium-dependent protease activity of *L. donovani* promastigotes, termed caldonopain (Bhattacharya *et al.*, 1993). The detailed biochemical features and function of this protease are unknown.

6. AFRICAN TRYPANOSOMES

Organisms of the *Trypanosoma brucei* complex cause African trypanosomiasis, a severe febrile illness that is usually fatal (Kirchhoff, 1995). Recent control efforts against African trypanosomiasis have been largely unsuccessful (Kuzoe, 1993). About 20 000 cases occur each year, and about 50 million Africans are at risk of infection (Kirchhoff, 1995). Available therapies for African trypanosomiasis are inadequate. Suramin, pentamidine, arsenicals and eflornithine all show less than optimal efficacy, and all cause significant toxicity (Kirchhoff, 1995).

6.1. *T. brucei* Cysteine Protease (Trypanopain)

A major cysteine protease, now named trypanopain, has been identified in *T. brucei* (North *et al.*, 1983; Lonsdale-Eccles and Mpimbaza, 1986; Lonsdale-Eccles and Grab, 1987; Robertson *et al.*, 1990; Troeberg *et al.*, 1996). Much of the study of this enzyme has been in *T. brucei brucei*, the subspecies of the parasite that infects cattle, but based on biochemical and molecular studies a similar enzyme also appears to be present in the human parasites *T. brucei gambiense* (Lonsdale-Eccles and Mpimbaza, 1986) and *T. brucei rhodesiense* (Pamer *et al.*, 1990). As is the case with *Leishmania*, more than one cysteine protease appears to be present in *T. brucei*; four similar but independent activities were identified in *T. brucei* lysates using substrate gel electrophoresis techniques (Robertson *et al.*, 1990). However, as information on individual enzymes is limited, trypanopain activity will be discussed as that of a single protease. Trypanopain is a papain-family cysteine protease that is located in the lysosome of trypanosomes (Lonsdale-Eccles and Grab, 1987). It has an acid pH optimum and the substrate preference of cathepsin L, and it is inhibited by standard cysteine protease inhibitors (Lonsdale-Eccles and Mpimbaza, 1986; Lonsdale-Eccles and Grab, 1987; Troeberg et al., 1996).

Genes encoding trypanopain have been identified and characterized. The genes predict fairly typical papain family proteases, except for a 108-amino-acid carboxy-terminal extension that is rich in prolines (Mottram et al., 1989; Pamer et al., 1990). When the protease was heterologously expressed in E. coli, the pro sequence, but not the carboxy-terminal extension, was required for activity (Pamer et al., 1991). The T. brucei genome contains more than 20 copies of the trypanopain gene arranged in a long tandem array (Mottram et al., 1989).

The function of trypanopain is not known. Unlike the related trypanosomatids Leishmania and T. cruzi, T. brucei replicates extracellularly and does not invade host cells. Trypanopain may play a role in the development of bloodstream trypanosomes, as trypanopain activity increased as parasites differentiated from long slender to short stumpy forms (Pamer et al., 1989). The protease may inhibit parasite opsonization by degrading antibody-bound trypanosome antigens, as protease inhibitors decreased the processing of internalized antibody complexes by T. brucei (Russo et al., 1993). However, trypanopain is probably not active in the bloodstream, as it is effectively inhibited by the circulating cysteine protease inhibitor cystatin (Troeberg et al., 1996). Immune responses to trypanopain may also be important, as humoral responses to an analogous protease of the veterinary parasite T. congolense correlated with natural protection against bovine trypanosomiasis (Authie et al., 1993).

Cysteine protease inhibitors may be appropriate new drugs for treating African trypanosomiasis. In in vitro studies, peptide protease inhibitors lysed infectious bloodstream forms but not noninfectious forms of T. brucei (Ashall et al., 1990). In recent studies, a number of peptidyl chloromethyl ketone, diazomethyl ketone and fluoromethyl ketone trypanopain inhibitors killed cultured bloodstream forms of T. brucei at micromolar concentrations (Troeberg, L., personal communication). Studies using biotinylated derivatives of two of these inhibitors verified that trypanopain was their likely target. In addition, a series of chalcone and related nonpeptide reversible inhibitors of trypanopain killed cultured bloodstream forms of T. brucei at micromolar concentrations, and were active against the parasite in a murine model (Troeberg, L., personal communication).

6.2. Other Proteases of African Trypanosomes

A second identified T. brucei protease is a serine protease, referred to as the serine oligopeptidase (Huet et al., 1992; Kornblatt et al., 1992; Troeberg *et al.*, 1996). This enzyme appears to be an atypical trypsin-like serine protease, as it cleaves typical trypsin substrates, but is not inhibited by some trypsin inhibitors (Troeberg *et al.*, 1996). The enzyme is not inhibited by the cysteine protease inhibitor E-64, but is inhibited by other cysteine protease inhibitors that act by modifying cysteine residues, and it is stimulated by reducing agents (Troeberg *et al.*, 1996). These data suggest that, although it is a serine protease, the serine oligopeptidase contains a cysteine residue that must be reduced for optimal activity. Purified serine oligopeptidase cleaved a number of small peptide substrates, but not larger molecules (Troeberg *et al.*, 1996). The function of the protease is unknown. It may act extracellularly to cleave circulating peptide hormones and perhaps mediate some of the clinical features of African trypanosomiasis.

T. brucei has recently been shown to contain genes encoding homologues of leishmanolysin, the *Leishmania* surface metalloprotease (El-Sayed and Donelson, 1997). The metalloprotease active site sequence HEXXH, 19 cysteines and 10 prolines are conserved between the leishmanial and trypanosomal genes. Transcription of the *T. brucei* genes is equal between insect and mammalian parasite forms, but mRNAs accumulate to much higher levels in the bloodstream forms. The function of the newly identified trypanosomal leishmanolysin homologues is unknown, but their expression in obligate extracellular parasites suggests a different role from that proposed for leishmanolysin in leishmanial amastigotes. One possible role, similar to that discussed for leishmanial promastigotes, is protection against complement-mediated cell lysis.

A cytosolic multicatalytic protease complex or proteasome was recently purified from bloodstream and insect stages of *Trypanosoma brucei* (Hua *et al.*, 1996; Lomo *et al.*, 1997; To and Wang, 1997). As is typical for mammalian proteasomes (Rivett, 1993), the complex demonstrated chymotrypsin-like, trypsin-like and peptidylglutamylpeptide-hydrolysing activities against synthetic substrates (Hua *et al.*, 1996; Lomo *et al.*, 1997). The proteasome probably plays an important role in cytosolic protein turnover in trypanosomes.

7. *TRYPANOSOMA CRUZI*

T. cruzi infection causes American trypanosomiasis (Chagas disease) in much of Central and South America (Kirchhoff, 1993a,b). Chagas disease causes an initial, usually mild, febrile illness that is followed years later by a serious heart and gastrointestinal tract disease that can manifest as cardiomyopathy, megaesophagus and megacolon. It has recently been estimated that more than 16 million people are infected with *T. cruzi* in

South America, and that up to 50 000 people die each year of Chagas disease (Kirchhoff, 1993a). Therapy for this disease is unsatisfactory. Nifurtimox and benznidazole have limited efficacy for the treatment of acute Chagas disease. Only about 50% of patients treated for acute disease are cured, and serious toxicity is common (Kirchhoff, 1993a). Antiparasitic therapy for chronic Chagas disease is not available.

7.1. Cysteine Protease of *T. cruzi* (Cruzain)

The cysteine protease of *T. cruzi*, named cruzipain (Cazzulo *et al.*, 1989, 1990) or cruzain (Eakin *et al.*, 1992) has been well characterized. The enzyme is a papain-family cysteine protease with an acidic pH optimum, a cathepsin L-like substrate specificity and inhibition by typical cysteine protease inhibitors (Cazzulo *et al.*, 1989, 1990). Cruzain is inhibited by a number of the circulating cysteine protease inhibitors known as cystatins (Stoka *et al.*, 1995), and its substrate specificity was more rigorously defined based on its hydrolysis of substrates containing conserved cystatin sequences (Lalmanach *et al.*, 1996; Serveau *et al.*, 1996) and other typical cysteine protease cleavage sites (Del Nery *et al.*, 1997b). The specificities of cruzain and mammalian cysteine proteases for peptide substrates differed significantly, suggesting that specific inhibitors of cruzain can be developed (Serveau *et al.*, 1996).

Cruzain was localized to the lysosomes of *T. cruzi* epimastigotes (the insect stage of the parasite) by subcellular fractionation (Bontempi *et al.*, 1989), and immunocytochemistry studies identified cruzain in the lysosomes of epimastogotes, amastigotes (the intracellular human stage) and trypomastigotes (the circulating human stage), on the surfaces of epimastigotes and amastigotes and in the flagellar pocket of trypomastigotes (Souto-Padron *et al.*, 1990). The enzyme also appears to be secreted by trypomastigotes (Yokoyama-Yasunaka *et al.*, 1994). Biochemical studies showed epimastigotes contained much greater protease activity than amastigotes or trypomastigotes (Campetella *et al.*, 1990; Bonaldo *et al.*, 1991). As cruzain mRNAs are present at approximately the same steady-state levels in each of the parasite life cycle stages, the developmental regulation of cruzain activity appears to occur at the translational or post-translational level (Tomas and Kelly, 1996).

The gene encoding cruzain was identified independently by three laboratories (Cazzulo *et al.*, 1989; Eakin *et al.*, 1990; Murta *et al.*, 1990). Multiple copies of the gene are present in tandem arrays on more than one chromosome (Campetella *et al.*, 1992; Eakin *et al.*, 1992; Tomas and Kelly, 1996). The gene is expressed in all developmental forms of *T. cruzi* (Eakin *et al.*, 1992). The cruzain gene encodes a papain-family cysteine protease with a typical pro-sequence and an unusual carboxy-terminal extension similar to that of *T. brucei* and *Leishmania* cysteine proteases (Eakin *et al.*, 1992). The carboxy-terminal extension differs from that of *T. brucei* in that a polyproline stretch is replaced in *T. cruzi* by polythreonine (Eakin *et al.*, 1992). Both the prosequence and the carboxy-terminal extension of cruzain are cleaved by autohydrolysis (Aslund *et al.*, 1991; Hellman *et al.*, 1991; Eakin *et al.*, 1992). The function of the carboxy-terminal extension is unknown; it was not required for correct protein folding or enzymatic activity (Eakin *et al.*, 1992, 1993). Potential functions include targeting to intracellular sites or anchoring to the cell membrane.

The function of cruzain is unknown. A number of protease inhibitors blocked adherence to and invasion of fibroblasts (Piras *et al.*, 1985) and invasion of Vero cells (Franke de Cazzulo *et al.*, 1994), and anti-cruzain antibody blocked invasion of macrophages (Souto-Padron *et al.*, 1990) by trypomastigotes, suggesting that cruzain and other proteases may mediate these processes. Peptide cysteine protease inhibitors lysed parasites (Ashall *et al.*, 1990a; Franke de Cazzulo *et al.*, 1994) and, at lower concentrations, blocked amastigote development (Meirelles *et al.*, 1992; Franke de Cazzulo *et al.*, 1994) and transformation from epimastagotes to trypomastigotes (metacyclogenesis) (Bonaldo *et al.*, 1991; Franke de Cazzulo *et al.*, 1994), trypomastigotes to amastigotes (Harth *et al.*, 1993), and amastigotes to trypomastigotes (Meirelles *et al.*, 1992; Harth *et al.*, 1993). *T. cruzi* engineered to overexpress cruzain had an enhanced ability to undergo metacyclogenesis and relative resistance to the deleterious effects of a diazomethyl ketone cysteine protease inhibitor (Tomas *et al.*, 1997). These results suggest that cruzain plays a role both in intracellular protein degradation and also in parasite remodelling during transformation between parasite stages. Cruzain may also mediate inflammation coincident with *T. cruzi* infection. The protease has been shown to generate the proinflammatory peptide bradykinin by cleavage of kininogen (Del Nery *et al.*, 1997a).

Although the specific function of cruzain remains uncertain, the demonstration of antiparasitic effects of multiple inhibitors of the enzyme suggests that it may be an ideal target for new drugs to treat Chagas disease. Cruzain is the only protozoan cysteine protease for which a structure has been determined (McGrath *et al.*, 1995). This structure should be valuable in the prediction of inhibitors of the enzyme (Gillmor *et al.*, 1997). Extensive efforts to develop peptide-based (Harth *et al.*, 1993) and nonpeptide (Li *et al.*, 1996) inhibitors of cruzain are currently ongoing (McKerrow *et al.*, 1995). In studies with fluoromethyl ketone cysteine protease inhibitors, concentrations of inhibitors that interrupted the *T. cruzi* life cycle had no observable toxicity to macrophages, fibroblasts or epithelial cells in culture (Harth *et al.*, 1993). Fluorescence microscopy showed that a biotinylated form of a fluoromethyl ketone inhibitor was selectively taken up by intracellular parasites and did not effectively reach host lysosomes (McGrath et al., 1995), perhaps, in part, explaining the selective effectiveness of the inhibitors against cruzain and not host cysteine proteases. In recent studies, fluoromethyl ketone and vinyl sulfone inhibitors of cruzain markedly altered the morphology of the *T. cruzi* Golgi apparatus, inhibited the transport of cruzain to lysosomes and arrested parasite growth (Engel et al., 1998). These results offer additional support to the consideration of cruzain as a promising target for new therapies for Chagas disease.

7.2. Other Proteases of *T. cruzi*

Metalloprotease activity was identified in cultured parasites undergoing metacyclogenesis, and this process was blocked by metalloprotease inhibitors (Bonaldo et al., 1991). However, metalloprotease expression was highly variable among different *T. cruzi* isolates (Lowndes et al., 1996).

A *T. cruzi* alkaline peptidase activity has also been identified (Ashall, 1990). This activity was present in all life-cycle stages and, based on its biochemical properties, including a lack of inhibition by E-64, it appears to be analogous to the serine oligopeptidase activity of *T. brucei* (Ashall, 1990; Ashall et al., 1990b; Santana et al., 1992; Troeberg et al., 1996). The enzyme was purified and shown to mediate the calcium signalling in host cells that is required for cell entry by trypomastigotes (Burleigh and Andrews, 1995). The protease did not directly mediate calcium fluxes, but presumably acted via the hydrolysis of another molecule. The gene encoding the serine oligopeptidase was characterized and shown to be homologous to serine prolyl oligopeptidases from other organisms (Burleigh et al., 1997).

An alkaline *T. cruzi* collagenase activity was identified in extracts of amastigotes, trypomastigotes and epimastigotes (Santana et al., 1997). The purified enzyme hydrolysed human type I and type IV collagen (Santana et al., 1997). The *T. cruzi* collagenase may play a role in optimizing the efficiency of host cell infection by trypomastigotes.

A proteasome of *T. cruzi* has recently been identified (Gonzalez et al., 1996). The proteasome inhibitor lactacystin inhibited the transformation of trypomastigotes to amastigotes and intracellular amastigote development, suggesting that proteasome activity is required for these processes (Gonzalez et al., 1996).

8. MALARIA PARASITES

Malaria is the most deadly protozoan infection of humans. Hundreds of millions of cases of malaria occur annually, and infections with *Plasmodium falciparum*, the most virulent human malaria parasite, lead to over one million deaths each year (Walsh, 1989). Despite extensive control efforts, the incidence of malaria is not decreasing in most endemic areas of the world, and in some areas it is clearly increasing (Oaks *et al.*, 1991). A major reason for the persistence of the severe malaria problem is the increasing resistance of parasites to available chemotherapeutic agents. Resistance to chloroquine, the most widely used antimalarial in the last 50 years, is now very common, and other available antimalarials are limited by resistance, high cost and toxicity (Olliaro *et al.*, 1996).

Most studies of malaria parasites utilize erythrocytic stage parasites, which are the stages responsible for human disease and also the easiest to study in culture and in animal models. Proteases of erythrocytic stage parasites are probably required for a number of functions. Serine and/or cysteine proteases appear to mediate the invasion of erythrocytes by free merozoites, which initiates the erythrocytic cycle and the rupture of erythrocytes by mature schizonts that completes this cycle. Cysteine and aspartic proteases mediate the crucial process of haemoglobin degradation by intracellular trophozoites. Nonerythrocytic stages of malaria parasites are difficult to study, and no specific information regarding proteases of these stages is yet available.

Numerous older reports described neutral and acidic proteases of human (*P. falciparum*), monkey (*P. knowlesi*), rodent (*P. berghei*, *P. chabaudi*, *P. yoelii*, *P. vinckei*) and avian (*P. lophurae*) malaria parasites. Older studies are often difficult to interpret, however, because of inconsistencies with protease assays and the difficulty of distinguishing host and parasite enzyme activities (see McKerrow *et al.*, 1993, for a review of the older literature). More recent studies have made efforts to control for the activities of host enzymes, but it remains difficult to relate biochemical data from different laboratories with specific gene products. This discussion will concentrate on newer studies, with emphasis on the proposed functions of the malarial proteases that have been identified and on the evaluation of inhibitors of malarial proteases as potential antimalarial drugs.

8.1. Plasmodial Proteases Mediating Erythrocyte Invasion and Rupture

During erythrocyte invasion, merozoites secrete the contents of the rhoptry and microneme organelles as they enter the erythrocyte within a parasitophorous vacuole (Perkins, 1989). The mechanism of rupture of the erythrocyte by mature schizonts is poorly understood, but the secretion of a portion of rhoptry and microneme contents has also been observed at the time of rupture in *P. knowlesi* (Bannister and Mitchell, 1989). During both erythrocyte invasion and rupture, an important barrier to malaria parasites, in addition to the erythrocyte plasma membrane, is the erythrocyte cytoskeleton. Most likely, the rhoptry and microneme contents that are released during erythrocyte invasion and rupture include proteases that degrade cytoskeletal proteins.

Studies of the effects of protease inhibitors on the invasion and rupture of erythrocytes by malaria parasites suggest roles for both serine and cysteine proteases in these processes. In *P. knowlesi*, chymostatin (an inhibitor of serine proteases), but not leupeptin (an inhibitor of cysteine proteases and trypsin-like serine proteases) (Hadley *et al.*, 1983), and in *P. chabaudi* a number of serine protease inhibitors (Braun-Breton *et al.*, 1992), specifically blocked the invasion of erythrocytes by isolated merozoites. Pretreatment of human erythrocytes with the serine protease chymotrypsin (Dluzewski *et al.*, 1986) or pretreatment of murine erythrocytes with an Mr 68 000 *P. chabaudi* serine protease (Braun-Breton *et al.*, 1992) reversed the inhibitory effects on erythrocyte invasion of serine protease inhibitors, further suggesting that serine protease activity is required for invasion. Cysteine protease activity may also be required, as peptide inhibitors of a *P. falciparum* Mr 68 000 cysteine protease (Mayer *et al.*, 1991) and specific inhibitors of calpain (Olaya and Wasserman, 1991) appeared to inhibit invasion in *P. falciparum*. Other investigators also showed that in *P. knowlesi* (Banyal *et al.*, 1981) and *P. falciparum* (Dejkriengkraikhul and Wilairat, 1983; Dluzewski *et al.*, 1986; Lyon and Haynes, 1986) chymostatin and leupeptin blocked erythrocyte rupture and/or invasion. Chymostatin may act by inhibiting the degradation of cytoskeletal proteins by a chymotrypsin-like protease. The incubation of erythrocytes with chymotrypsin led to cleavage of the erythrocyte cytoskeletal protein band 3 and localized disruption of the erythrocyte membrane, possibly mediating the formation of the parasitophorous vacuole during parasite invasion (McPherson *et al.*, 1993). Considering the process of erythrocyte rupture, in both *P. knowlesi* (Hadley *et al.*, 1983) and *P. falciparum* (Lyon and Haynes, 1986), the incubation of parasites with either chymostatin or leupeptin was followed by the accumulation of erythrocytes containing unreleased merozoites. In these erythrocytes schizonts apparently completed development into merozoites, but the rupture of the erythrocyte cytoskeleton or other structures required to free the merozoites was blocked by the serine and cysteine protease inhibitors.

Multiple proteins of mature schizonts and merozoites are proteolytically processed immediately before or during erythrocyte rupture and invasion, suggesting that proteolytic fragments have roles in these processes. The best characterized processing is that of the major surface protein of merozoites (MSP-1), which is processed to form a surface multi-component complex (Blackman *et al.*, 1990; Cooper and Bujard, 1992). A final processing event yields a small carboxy-terminal portion of MSP-1, which is retained on the merozoite, and appears to be necessary for successful erythrocyte invasion (Blackman et al., 1991). This final processing event appears to be due to the action of a membrane-bound calcium-dependent serine protease (Blackman and Holder, 1992; Blackman et al., 1993). This protease has not yet been identified. The processing of P126, a parasitophorous vacuole protein also known as SERA (see below), was altered by the cysteine protease inhibitor leupeptin (Debrabant and Delplace, 1989). A parasite phosphoprotein is hydrolysed in schizonts such that parasite protein kinase activity is lost; this cleavage was inhibited by cysteine protease inhibitors (Wiser, 1995). The *P. falciparum* acidic basic repeat antigen (ABRA) undergoes proteolysis that is inhibited by chymostatin (Weber et al., 1988); this processing may be due to autohydrolysis, suggesting that the antigen is a serine protease (Nwagwu et al., 1992).

A number of proteases have been identified as potentially involved in erythrocyte invasion and rupture by malaria parasites.

(a) An M_r 68 000 cysteine protease was identified in schizonts and merozoites of *P. berghei*, *P. chabaudi* and *P. falciparum* (Bernard and Schrevel, 1987; Schrevel et al., 1988; Grellier et al., 1989; Mayer et al., 1991). Antisera directed against the Mr 68 000 protease localized primarily to the merozoite apex (Bernard and Schrevel, 1987), suggesting that the protease might be released from the rhoptry organelle during invasion.

(b) An M_r 35–40 000 cysteine protease of mature schizonts and an M_r 75 000 serine protease of merozoites were identified in highly synchronized *P. falciparum* parasites (Rosenthal et al., 1987). The stage-specific activities of these proteases suggest that they may also be involved in erythrocyte invasion or rupture.

(c) An M_r 76 000 serine protease of *P. falciparum* schizonts and merozoites was shown to be bound in an inactive form to the schizont/merozoite membrane by a glycosyl-phosphatidylinositol anchor, and to be activated by phosphatidylinositol-specific phospholipase C during the merozoite stage (Braun-Breton et al., 1988), suggesting a role in erythrocyte invasion. This protease and a *P. chabaudi* analogue cleaved the erythrocyte cytoskeletal protein band 3 (Braun-Breton et al., 1992; Roggwiller et al., 1996).

(d) An M_r 37 000 protease of *P. falciparum* and *P. berghei* that was inhibited by both chymostatin and leupeptin, but was not easily categorized as to catalytic class, also hydrolysed erythrocyte cytoskeletal proteins (spectrin and band 4.1), suggesting potential roles in both erythrocyte invasion and rupture (Deguercy et al., 1990). At this time, the genes encoding the schizont and merozoite protease activities described above have not been identified. Thus, it is not possible to determine if similar activities identified by different laboratories represent the same enzyme. However, recent progress in a number of laboratories and the ongoing *P. falciparum* genome project suggest that the sequences of key schizont and merozoite proteases will soon be available.

A well characterized *P. falciparum* antigen may represent a schizont protease. The serine-rich antigen (referred to independently as P126 (Delplace *et al.*, 1987), SERA (Bzik *et al.*, 1988), and SERP (Knapp *et al.*, 1989)) is located in the parasitophorous vacuole of mature schizonts (Delplace *et al.*, 1987; Knapp *et al.*, 1989) and on the surface of free merozoites (Perkins and Ziefer, 1994). Portions of SERA and the related homologue SERPH (Knapp *et al.*, 1991) have limited sequence homology to cysteine proteases, particularly near highly conserved active site residues. However, in SERA the predicted active site cysteine is replaced by a serine residue, while in SERPH the catalytic cysteine is conserved. Whether SERA or SERPH have protease activity is unknown, but the identification of these proteins in the parasitophorous vacuole of schizonts suggests potential roles for these perhaps multifunctional proteins in the proteolysis of erythrocyte cytoskeletal or parasite proteins. SERA is itself processed in a manner that is in part inhibited by leupeptin (Debrabant and Delplace, 1989), suggesting that hydrolysis (perhaps autohydrolysis) may be required for protease activity.

Host proteases may also play a role in erythrocyte rupture by *P. falciparum*. The circulating serine protease urokinase was recently shown to bind to the surface of *P. falciparum*-infected erythrocytes, and the depletion of urokinase from parasite culture medium inhibited erythrocyte rupture by mature schizonts (Roggwiller *et al.*, 1997). This inhibition was reversed by the addition of exogenous urokinase.

8.2. Plasmodial Proteases Mediating Haemoglobin Degradation

Extensive evidence suggests that the degradation of haemoglobin is necessary for the growth of intraerythrocytic malaria parasites (Scheibel and Sherman, 1988; McKerrow *et al.*, 1993; Rosenthal and Meshnick, 1996; Francis *et al.*, 1997b). These parasites apparently require haemoglobin as a source of free amino acids, as the parasites have a limited capacity for synthesizing amino acids, and the quantity of free amino acids within erythrocytes is not sufficient for parasite needs. Evidence that malaria parasites degrade haemoglobin into free amino acids includes the following.

(a) The haemoglobin content of infected erythrocytes decreases 25–75% during the life cycle of erythrocytic parasites (Ball *et al.*, 1948; Groman, 1951; Roth *et al.*, 1986).

(b) The concentration of free amino acids is greater in infected than in uninfected erythrocytes (Sherman and Mudd, 1966).
(c) The composition of the amino acid pool of infected erythrocytes is similar to the amino acid composition of haemoglobin (Cenedella et al., 1968; Barry, 1982; Zarchin et al., 1986).
(d) The infection of erythrocytes containing radiolabelled haemoglobin is followed by the appearance of labelled amino acids in parasite proteins (Fulton and Grant, 1956; Sherman and Tanigoshi, 1970; Theakston et al., 1970).
(e) Only five amino acids, most of which are in limited supply in human haemoglobin, are required to support normal growth of cultured parasites (Divo et al., 1985; Francis et al., 1994).
(f) Cysteine protease inhibitors that block haemoglobin degradation block the development of cultured parasites (Rosenthal et al., 1988).

In *P. falciparum*, haemoglobin degradation occurs predominantly in trophozoites and early schizonts. Trophozoites ingest erythrocyte cytoplasm via a specialized organelle known as the cytostome, and transport the cytoplasm within vesicles to a large central food vacuole (Aikawa, 1988). In the food vacuole haemoglobin is broken down into heme, which is a major component of malarial pigment (Slater, 1992), and globin, which is hydrolysed to its constituent free amino acids. The food vacuole is an acidic organelle that appears to be analogous to lysosomes (Krogstad and Schlesinger, 1987). The *P. falciparum* food vacuole contains at least three proteases, the cysteine protease falcipain (Rosenthal et al., 1988; Rosenthal and Nelson, 1992; Gluzman et al., 1994) and the aspartic proteases plasmepsin I and plasmepsin II (Goldberg et al., 1991; Vander Jagt et al., 1992; Dame et al., 1994; Gluzman et al., 1994; Hill et al., 1994), and each of these enzymes probably participates in haemoglobin degradation.

The aspartic proteases plasmepsin I and plasmepsin II have recently been well characterized (Goldberg et al., 1991; Vander Jagt et al., 1992; Dame et al., 1994; Gluzman et al., 1994; Hill et al., 1994). Earlier reports of plasmodial aspartic protease activity probably all referred to these enzymes (Aissi et al., 1983; Sherman and Tanigoshi, 1983; Sato et al., 1987; Bailly et al., 1991). Both proteases are located in the food vacuole, have acid pH optima, and share significant sequence homology with other aspartic proteases. The enzymes are biochemically similar, but not identical; in evaluations of globin cleavage, substrate preferences of the two enzymes were distinct (Gluzman et al., 1994; Luker et al., 1996). The plasmepsins are synthesized as proenzymes, which are integral membrane proteins that are cleaved to soluble forms under acidic conditions (Francis et al., 1997a). The synthesis and processing of both plasmepsins peaks in trophozoites, but plasmepsin I is also synthesized, processed, and presumably active in young ring-stage parasites (Francis *et al.*, 1997a). Analogues of the *P. falciparum* plasmepsins have recently been identified in the human malaria parasites *P. vivax* and *P. malariae* (Westling *et al.*, 1997). Only one aspartic protease has been identified to date in each of these species; the biochemical properties of the proteases were quite similar to those of the *P. falciparum* plasmepsins.

Falcipain is a *P. falciparum* papain-family cysteine protease with biochemical properties and sequence similar to that of cathepsin L and many of the other protozoan cysteine proteases discussed in this review (Rosenthal *et al.*, 1988; Rosenthal *et al.*, 1989; Rosenthal and Nelson, 1992; Salas *et al.*, 1995; Francis *et al.*, 1996). Falcipain has an acidic pH optimum and a cathepsin L-like substrate specificity, and its activity is enhanced by reducing agents (Rosenthal *et al.*, 1989). The gene encoding falcipain is single copy (Rosenthal and Nelson, 1992). The falcipain message is most abundant in rings (Rosenthal and Nelson, 1992; Francis *et al.*, 1996), but the activity of this protease is maximal in trophozoites (Rosenthal *et al.*, 1987). Falcipain has an unusually long pro-sequence for a papain-family enzyme (Rosenthal and Nelson, 1992); the function of the amino-terminal portion of the pro-sequence, which has no homology with other papain-family proteases, is unknown. Partial or complete sequences of falcipain analogues from nine other plasmodial species, including all species that infect humans, are now available (Rosenthal, 1993, 1996; Rosenthal *et al.*, 1994). The predicted mature forms of the proteases of the different species are highly conserved (Rosenthal, 1996).

Each of the three plasmodial food vacuole proteases that have been identified probably participates in globin hydrolysis. Each enzyme degraded haemoglobin *in vitro* (Rosenthal *et al.*, 1988; Gluzman *et al.*, 1994; Salas *et al.*, 1995), and the hydrolysis of haemoglobin by parasite and food vacuole extracts was inhibited in an additive manner by cysteine and aspartic protease inhibitors (Rosenthal *et al.*, 1988; Goldberg *et al.*, 1990). However, the precise roles of the three food vacuole proteases in haemoglobin degradation are unclear, in part because of uncertainties regarding the redox state of the food vacuole. In one set of studies, performed under nonreducing conditions, plasmepsin I was found to readily cleave native haemoglobin at a hinge region of the molecule, plasmepsin II to prefer denatured haemoglobin as a substrate, and falcipain to cleave only denatured globin, suggesting that haemoglobin degradation requires an ordered pathway of enzymatic events initiated by plasmepsin I (Goldberg *et al.*, 1990, 1991; Gluzman *et al.*, 1994). It was further shown, in a series of *in vitro* experiments, that falcipain degraded haemoglobin only after treatment of the substrate with reducing agents, acid-acetone or an aspartic protease (Francis *et al.*, 1996). However, direct studies of the chemical composition and reducing state of the food vacuole have not been done. Since erythrocytes contain millimolar concentrations of reduced glutathione

(Mills and Lang, 1996), which is presumably transported into the food vacuole with haemoglobin, the food vacuole may well be a reducing environment, providing optimal conditions for proteolysis by falcipain (Rosenthal et al., 1988). In studies utilizing reducing conditions, falcipain readily degraded native haemoglobin (Salas et al., 1995; Francis et al., 1996).

Studies with cultured parasites have suggested that falcipain is required for the degradation of native haemoglobin. Treatment of cultured parasites with a number of inhibitors of falcipain caused a block in haemoglobin degradation such that parasites developed food vacuoles filled with undegraded globin (Dluzewski et al., 1986; Rosenthal et al., 1988; Bailly et al., 1992; Rosenthal, 1995). Falcipain also appears to play a role in initial steps of haemoglobin degradation, as the cysteine protease inhibitor E-64, but not the aspartic protease inhibitor pepstatin, inhibited the dissociation of the haemoglobin tetramer (Gamboa de Domínguez and Rosenthal, 1996), the release of haeme from globin (Gamboa de Domínguez and Rosenthal, 1996), the initial processing of α- and β-globin (Kamchonwongpaisan et al., 1997), and the formation of the haemoglobin breakdown product haemozoin (Asawamahasakda et al., 1994). The simplest explanation for the inhibition of haemoglobin degradation by cysteine protease inhibitors is that falcipain directly processes native haemoglobin. Another possibility is that falcipain activity is required for the processing of another enzyme (e.g. plasmepsin I) that initiates the hydrolysis of haemoglobin. However, although the aspartic protease inhibitor pepstatin caused gross morphological abnormalities (Bailly et al., 1992; Rosenthal, 1995), aspartic protease inhibitors have not been shown to block haemoglobin degradation or haemozoin formation by cultured parasites (Bailly et al., 1992; Asawamahasakda et al., 1994; Rosenthal, 1995; Gamboa de Domínguez and Rosenthal, 1996; Kamchonwongpaisan et al., 1997).

It is not clear whether globin is fully hydrolysed to free amino acids in the food vacuole or if peptides derived from globin are transported from the food vacuole and then additionally processed in the cytosol. Recent in vitro evidence suggests the latter possibility. Incubation of haemoglobin with P. falciparum food vacuole lysates generated multiple discrete peptide fragments ranging in size from eight to 49 amino acids (Kolakovich et al., 1997), suggesting that falcipain, plasmepsins I and II and possibly other proteases generated multiple cleavages on the globin chains. Free amino acids were not identified, however, suggesting that additional proteolytic activity was required for the complete processing of globin. Exopeptidase activity in the cytosol may generate free amino acids after globin peptides are transported from the food vacuole. Aminopeptidase activity has been identified in parasite but not food vacuole lysates (Gyang et al., 1982; Vander Jagt et al., 1984; Curley et al., 1994; Kolakovich et al., 1997). This activity has a neutral pH optimum, consistent with a cytosolic site of action. Preferred substrates had amino-terminal alanine or leucine, two amino acids that are particularly abundant in human haemoglobin (Curley et al., 1994). These results suggest that a cytosolic aminopeptidase participates in the processing of globin to amino acids. In summary, although the precise roles of the three food vacuole proteases and the cytosolic aminopeptidase remain uncertain, it appears that multiple plasmodial proteases act in a concerted manner to hydrolyse globin to free amino acids.

8.3. Plasmodial Proteases as Potential Chemotherapeutic Targets

As noted above, the antimalarial effects of peptide protease inhibitors have been studied by a number of groups. Both leupeptin and chymostatin have been shown to inhibit erythrocyte invasion and rupture by malaria parasites. Peptide ethylamide inhibitors of the *P. falciparum* Mr 68 000 schizont cysteine protease inhibited erythrocyte invasion when incubated with mature schizont-stage parasites, but high (millimolar) concentrations of the peptides were required for inhibition (Mayer et al., 1991). Peptide inhibitors of calpain, a calcium-dependent cysteine protease that has not been identified in *Plasmodium*, blocked erythrocyte invasion at nanomolar concentrations (Olaya and Wasserman, 1991).

The cysteine protease inhibitor E-64 and the aspartic protease inhibitor pepstatin have both been shown to block *P. falciparum* development (Dluzewski et al., 1986; Rosenthal et al., 1988; Vander Jagt et al., 1989; Bailly et al., 1992; Rosenthal, 1995). The two inhibitors acted synergistically *in vitro* (Bailly et al., 1992), suggesting that optimal chemotherapy might inhibit both classes of proteases. Numerous peptide fluoromethyl ketone inhibitors of falcipain inhibited *P. falciparum* growth at nanomolar concentrations (Rosenthal et al., 1989, 1991; Rockett et al., 1990). The inhibition of falcipain correlated with biological effects, suggesting that the inhibitors were acting by directly inhibiting the protease (Rosenthal et al., 1991). One fluoromethyl ketone inhibitor of falcipain blocked *P. vinckei* cysteine protease activity *in vivo* after a single subcutaneous dose, and when administered for four days cured 80% of murine malaria infections (Rosenthal et al., 1993). More recently, vinyl sulphone inhibitors of falcipain, which appear to have minimal toxicity, have also been shown to have nanomolar antimalarial effects (Rosenthal et al., 1996). Thus, despite the theoretical limitations of potentially rapid degradation *in vivo* and inhibition of host proteases, peptide-based protease inhibitors appear to be promising candidate antimalarial drugs.

In order to identify nonpeptide inhibitors, the structure of the *P. falciparum* trophozoite cysteine protease was modelled based on the predicted protease sequence and the known structures of other papain-family proteases, and potential inhibitors were identified by computer modelling techniques (Ring *et al.*, 1993), Screening of potential nonpeptide inhibitors identified a low micromolar lead compound (Ring *et al.*, 1993), and subsequent synthesis and screening identified chalcones (Li *et al.*, 1995) and phenothiazines (Domínguez *et al.*, 1997) that inhibited falcipain and blocked the development of cultured parasites at nanomolar-low micromolar concentrations.

Aspartic protease inhibitors are also under study as potential antimalarials. Three different projects have identified compounds that inhibit plasmepsin I or plasmepsin II and block parasite development. SC-50083, a peptidomimetic compound, inhibited haemoglobin hydrolysis by plasmepsin I and blocked parasite metabolism at low micromolar concentrations (Francis *et al.*, 1994). The most effective of a set of inhibitors with modified statine motifs inhibited plasmepsin II at nanomolar concentrations and showed modest effects against cultured parasites (Silva *et al.*, 1996). Two other peptidomimetic compounds were low-nanomolar inhibitors of plasmepsin I and inhibited parasite metabolism at high nanomolar concentrations (Moon *et al.*, 1997). Drug development efforts should benefit from the recent determination of the structure of plasmepsin II (Silva *et al.*, 1996).

Effective antimalarial protease inhibitors should ideally inhibit parasite proteases but not analogous host proteases. In this regard, studies of aspartic protease inhibitors demonstrated specificity between the inhibition of plasmepsins I and II (Francis *et al.*, 1994; Moon *et al.*, 1997) and between these enzymes and cathepsin D (Silva *et al.*, 1996). Cysteine protease inhibitors under study have not been rigorously evaluated for specificity, but it is encouraging that a number of amino acids predicted to surround the active site of falcipain are conserved in homologues from nine other plasmodial species, but are not conserved in host cysteine proteases (Rosenthal *et al.*, 1994; Rosenthal, 1996). Additional specificity may be provided by differences between the transport of compounds to the food vacuole and to host lysosomes, where the analogous host proteases are located. In this regard, the increased permeability of erythrocytes infected with mature parasites may act to direct delivery of protease inhibitors to the parasite targets (Elford *et al.*, 1995).

9. *ENTAMOEBA HISTOLYTICA*

E. histolytica infects the human gut, where it causes colitis and dysentery (Ravdin, 1995). The parasite can also penetrate the gut wall and migrate to other organs, particularly the liver, where it causes abscesses. Both intestinal and extraintestinal manifestations of amebiasis can cause fatal disease.

Approximately 40 million episodes of amebiasis resulting in 70 000 deaths were recently estimated to occur annually (Walsh, 1989). However, only a small fraction (probably about 10%) of those infected with parasites identified as *E. histolytica* develop clinical disease. An important recent finding has been that there are two morphologically identical species of *Entamoeba* that infect humans. *E. histolytica* is pathogenic, and commonly causes disease, while *E. dispar*, which is more prevalent, causes only an asymptomatic carrier state (Diamond and Clark, 1993). Among potential *E. histolytica* virulence factors are a number of proteases. Recent studies suggest that the expression of cysteine proteases appears to play a key role in the determination of the pathogenicity of an infecting parasite.

9.1. Cysteine Protease of *E. histolytica*

Cysteine protease activity in *E. histolytica* was reported by a number of investigators more than a decade ago (Scholze and Werries, 1984; Lushbaugh *et al.*, 1985; Keene *et al.*, 1986; Luaces and Barrett, 1988). Purified proteases were reported to have cathepsin B- or cathepsin L-like activities. Until recently it has been difficult to distinguish various cysteine protease activities biochemically, and so activities reported in older papers may represent more than one enzyme. More recently, molecular analysis has begun to clarify the activities and biological roles of specific gene products.

A number of *E. histolytica* cysteine proteases have been purified and characterized (Scholze and Tannich, 1994). From older studies, it appears that at least two similar but distinct cysteine protease activities are present in *E. histolytica*. The activities were distinguished by migration on SDS-PAGE gels, and referred to as amoebapain (Scholze and Schulte, 1988; Schulte and Scholze, 1989) and histolysin (Luaces and Barrett, 1988). Other reports probably identified these same or very similar enzymes (Lushbaugh *et al.*, 1985; Keene *et al.*, 1986). All reported activities of the *E. histolytica* cysteine proteases were similar. They are probably all of similar size (23–27 kDa) and are not glycosylated. Larger species identified in some studies probably represent multimers (Keene *et al.*, 1986; Ostoa-Saloma *et al.*, 1989). The enzymes are fairly typical papain-family cysteine proteases, although pH optima were broad, ranging from about pH5 to pH9 (Scholze and Tannich, 1994). The substrate specificity of *E. histolytica* cysteine proteases has generally been reported as cathepsin B-like, with preference for peptide substrates with arginine at the P1 and P2 positions (Lushbaugh *et al.*, 1985; Keene *et al.*, 1986; Luaces and Barrett, 1988; Scholze and Tannich, 1994), although molecular analysis of protease genes has identified only cathepsin L-like sequences. Amoebapain was localized to both the amoeba cell surface and to subcellular pinocytotic vesicles (Scholze *et al.*, 1992), and surface-associated cysteine protease activity of *E. histolytica* has been described (Avila and Calderon, 1993).

Recent molecular studies have helped to clarify our understanding of the multiple *E. histolytica* cysteine proteases. Six distinct *E. histolytica* cysteine protease genes have been identified (Tannich *et al.*, 1991, 1992; Reed *et al.*, 1993; Bruchhaus *et al.*, 1996). Each of these genes predicts a protein with homology to papain-family cysteine proteases; identity among the *E. histolytica* cysteine protease genes is 43–87% (Bruchhaus *et al.*, 1996). All of the genes predict proteins with 34–39% identity with papain, 36–46% identity with cathepsin L, and conservation of a prosequence typical of cathepsin L- but not cathepsin B-like enzymes (Bruchhaus *et al.*, 1996). It has been hypothesized that three cysteine proteases, encoded by *ehcp1*, *ehcp2* and *ehcp5*, account for nearly all *E. histolytica* cysteine protease activity, as these genes represented 90% of parasite cysteine protease transcripts, and three independent activities recently identified in parasite lysates were due to these three proteins (Bruchhaus *et al.*, 1996). It has also been proposed that *ehcp1* encodes amoebapain, *ehcp2* encodes histolysin and *ehcp5* encodes a third major cysteine protease that may play a key role in the determination of virulence (Bruchhaus *et al.*, 1996).

E. histolytica cysteine proteases appear to be virulence factors mediating tissue invasion. It has long been hypothesized that *E. histolytica* protease activity may mediate a cytolytic parasite activity that is required for virulence (Neal, 1960; McLaughlin and Faubert, 1977). Studies with cultured cells and protease inhibitors suggested that cysteine protease activity is required for a cytopathic effect (Montfort *et al.*, 1993). Purified *E. histolytica* cysteine proteases degraded collagens, fibronectin, laminin and a model of connective tissue extracellular matrix (Keene *et al.*, 1986; Schulte and Scholze, 1989), suggesting that they may foster the invasion of tissues by trophozoites. Supporting this possibility, purified cysteine proteases produced a cytopathic effect in cultured mammalian cells (Lushbaugh *et al.*, 1985; Keene *et al.*, 1986; Luaces and Barrett, 1988). The virulence of laboratory strains has also been correlated with the protease activity of *Entamoeba* isolates (Gadasi and Kobiler, 1983; Keene *et al.*, 1986; Luaces and Barrett, 1988; Keene *et al.*, 1990). In studies of clinical isolates, extracellular cysteine protease activity was significantly greater in patients with colitis or liver abscesses than in those with asymptomatic carriage of *Entamoeba* (Reed *et al.*, 1989a).

As noted above, the species *E. dispar* denotes organisms that are nonpathogenic, though capable of colonizing humans. *E. dispar* is believed to be nonpathogenic because of its limited invasiveness (Diamond and Clark, 1993). Recent work has utilized genetic analyses to relate older studies with laboratory strains of varied pathogenicity to human infection with *E. histolytica* and *E. dispar*. As genes encoding *E. histolytica* cysteine proteases were identified, it was noted that pathogenic and nonpathogenic strains expressed different proteases (Tannich et al., 1991; Reed et al., 1993). Reports disagreed, however, as to which genes differed in expression between pathogenic and nonpathogenic strains (Reed et al., 1993; Bruchhaus and Tannich, 1996; Mirelman et al., 1996). The most detailed analysis of the six known E. histolytica cysteine protease genes has shown that at least three of the E. histolytica proteases (EHCP2, EHCP3, also known as ACP1, and EHCP4) appear to have close homologues in E. dispar with > 90% sequence identity (Bruchhaus et al., 1996). Two highly expressed E. histolytica proteases, EHCP1 and EHCP5, are reportedly not expressed in E. dispar (Bruchhaus et al., 1996), although results with EHCP1, which is very similar in sequence to EHCP2, are controversial. EHCP5 is much less similar to the other described E. histolytica cysteine proteases (49–58% identity). It is intriguing to speculate that EHCP5 or other unique proteases may be specific virulence determinants that mediate the pathogenicity of E. histolytica. Recently EHCP5 was shown to be a membrane-associated cysteine protease and to produce a cytopathic effect on cultured cells (Jacobs et al., 1998). Thus EHCP5 may be a surface-expressed protease that degrades extracellular matrix proteins and/or host cellular proteins to mediate tissue invasion. The conclusion that EHCP5 is a unique virulence determinant for E. histolytica might be an oversimplification, however. As discussed above, other E. histolytica cysteine proteases have also been shown to produce a cytopathic effect, and the cysteine protease genes identified as those expressed in E. histolytica, and not E. dispar, have differed between studies. Although the identification of certain genes as specific to E. histolytica has been somewhat controversial, it seems clear that E. histolytica expresses an increased number of cysteine protease genes, produces more protease-specific mRNA, and secretes more active protease than does E. dispar.

E. histolytica cysteine protease activity can activate complement by degrading C3 (Reed et al., 1989b) and can also degrade the anaphylatoxic complement components C3a and C5a (Reed et al., 1995). Complement activation by this mechanism leads to the lysis of nonpathogenic, but not pathogenic, parasite isolates (Reed and Gigli, 1990). Thus the protease activity appears to contribute to the host immune response against nonpathogenic strains, but may inhibit an effective response, via degradation of C3a and C5a, against pathogenic strains. The cysteine protease activity also degrades IgA (Kelsall and Ravdin, 1993), an important mediator of immune responses against gut pathogens, and IgG (Tran et al., 1998), possibly explaining the lack of development of protective antibody responses against E. histolytica.

As one or more of the E. histolytica cysteine proteases may be virulence factors, the inhibition of these enzymes might be a promising modality of chemotherapy. Cysteine protease inhibitors blocked the cytopathic effect caused by parasite extracts and purified *E. histolytica* proteases (McGowan *et al.*, 1982; Lushbaugh *et al.*, 1984, 1985; Keene *et al.*, 1990). In recent studies with infected severe combined immunodeficiency mice, antibodies to EHCP1 identified the protease in extracellular amoebic abscess fluid, and treatment with the cysteine protease inhibitor E-64 markedly reduced liver abscess formation (Stanley *et al.*, 1995). In other studies, the compound allicin, which was extracted from garlic, blocked both the cysteine protease activity of cultured *E. histolytica* trophozoites and the ability of these parasites to destroy monolayers of mammalian cells (Ankri *et al.*, 1997).

9.2. Other Proteases of *E. histolytica*

E. histolytica also expresses collagenase activity that is apparently independent of the cysteine protease activities discussed above. Cultured *E. histolytica* trophozoites degraded type I and type III collagen (Munoz *et al.*, 1982; Rosales-Encina *et al.*, 1992). Incubation of trophozoites with collagen induced the release of electron-dense granules containing collagenase activity (Munoz *et al.*, 1990; Leon *et al.*, 1997). Collagenase activity correlated with the virulence of *E. histolytica* strains (Gadasi and Kessler, 1983; Magos *et al.*, 1992). These results suggest that one or more *E. histolytica* collagenases, which have not yet been extensively characterized, may act as virulence factors. They presumably act extracellularly to degrade extracellular matrix proteins, and thus facilitate parasite invasion of host tissues.

Two high molecular weight proteases of *E. histolytica* were identified recently (Scholze *et al.*, 1996). One of the activities was attributed to the *E. histolytica* proteasome based on the electrophoretic mobility of subunits and on its reactivity with an anti-proteasome antibody. The other activity was due to a complex of six subunits that had unique proteolytic activity, including the cleavage of substrates with aromatic P1 residues, and inhibition by chymostatin and a calpain inhibitor (Scholze *et al.*, 1996). An *E. histolytica* proteasome a-subunit gene has also recently been identified (Ramos *et al.*, 1997).

10. *GIARDIA LAMBLIA*

Giardia lamblia is the most common parasite identified as a cause of diarrhoea in humans (Ortega and Adam, 1997). Disease is caused by trophozoites, which emerge from infectious cysts after they pass through the stomach and into the small intestine. Giardiasis is treated with metronidazole, quinacrine and a number of other agents, but the use of these drugs is limited by toxicity and efficacies that are suboptimal (Ortega and Adam, 1997).

Cysteine protease activity of *G. lamblia* has been identified by a number of investigators (Hare *et al.*, 1989; Parenti, 1989; Werries *et al.*, 1991). Recently, three *G. lamblia* cysteine protease genes were characterized (Ward *et al.*, 1997). Two of the enzymes, termed CP2 and CP3, were more than 80% identical in sequence, while the third, CP1, was about 50% identical to the others. All three genes encode proteases that are cathepsin B-like in sequence. Indeed, *G. lamblia* is among the most primitive eukaryotes known (Sogin *et al.*, 1989), and phylogenetic analysis identified the *G. lamblia* cysteine proteases at the earliest known branch of the cathepsin B family (Ward *et al.*, 1997).

Giardial protease activity was localized by fluorescence microscopy to lysosome-like cytoplasmic vacuoles (Feely and Dyer, 1987) that were shown to contain protease activity by sedimentation analysis (Lindmark, 1988). The vacuoles move to the periphery of *Giardia* cells at the time of excystation, and then release their contents between the cyst wall and emerging trophozoites (Coggins and Schaefer, 1986; Feely and Dyer, 1987). At least one of the *Giardia* cysteine proteases appears to be required for the excystation of the parasite. E-64 and a number of fluoromethyl ketone cysteine protease inhibitors blocked the excystation of cultured *G. muris*, a parasite closely related to *G. lamblia* (Ward *et al.*, 1997). One of the *G. lamblia* cysteine proteases, CP2, was purified (Ward *et al.*, 1997). The enzyme cleaved a typical cysteine protease peptide substrate at neutral pH. It also bound one of the fluoromethyl ketone cysteine protease inhibitors that blocked excystation, strongly suggesting that this protease mediates the process. As is the case with other protozoan cysteine proteases, the *G. lamblia* cysteine protease may be an appropriate chemotherapeutic target.

11. *CRYPTOSPORIDIUM PARVUM*

Infection with the coccidian parasite *C. parvum* was considered only a rare cause of human disease until the last few decades. Intestinal infection with this parasite is now known to be a major cause of severe diarrhoea in patients with AIDS (Petersen, 1992) and also a significant cause of waterborne outbreaks of diarrhoea in immunocompetent individuals (Guerrant, 1997). No effective therapy for cryptosporidiosis is currently available.

A number of *C. parvum* protease activities have been identified. A neutral protease with broad specificity was partially purified from *C. parvum* sporozoites (Nesterenko *et al.*, 1995). The protease was inhibited by both cysteine protease inhibitors and a chelator, suggesting that it is a metal-dependent cysteine protease. The protease was immunolocalized to the surface of sporozoites. A *C. parvum* cysteine protease gene, possibly encoding the same protease, has been isolated and sequenced (C. Petersen, personal communication). The gene has a cathepsin L-like sequence and predicts an Mr 45 000 preproenzyme and Mr 25 000 mature protease. Cysteine protease inhibitors block the invasion of epithelial cells by *C. parvum* sporozoites, suggesting that the cysteine protease is required for this process (Petersen, C., personal communication).

Protease activity was also noted in a homogenate of partially excysted *C. parvum* oocysts (Forney et al., 1996b). Protease activity was maximally inhibited by a combination of cysteine and serine protease inhibitors, and serine but not cysteine protease inhibitors blocked excystation. In studies utilizing a bovine fallopian tube epithelial cell culture system, both serine and cysteine protease inhibitors blocked host cell infection by *C. parvum* (Forney et al., 1996a, 1997). Infection was inhibited when oocysts, but not sporozoites, were incubated with the inhibitors, suggesting that it is excystation and not cell invasion that requires proteolytic activity.

An integral membrane aminopeptidase was identified in freshly excysted *C. parvum* sporozoites (Okhuysen et al., 1994). The peptidase cleaved amino-terminal arginines. Aminopeptidase inhibitors and a metal chelator, but not endopeptidase inhibitors, blocked aminopeptidase activity and inhibited excystation (Okhuysen et al., 1996).

12. *TRICHOMONAS VAGINALIS*

T. vaginalis is a flagellated protozoan that is a major cause of sexually transmitted vaginitis and urethritis, and also causes serious gynaecological complications (Heine and McGregor, 1993). It is generally treated with metronidazole, although this drug can be toxic, is contraindicated during pregnancy and is not always effective.

Multiple trichomonal protease activities were identifed using substrate gel electrophoresis (Coombs and North, 1983; Lockwood et al., 1987; Neale and Alderete, 1990). All of the enzymes had general characteristics of papain-family cysteine proteases, but they differed in specificities for substrates and inhibitors (North et al., 1990). Similar protease activities were identified in flagellated forms of the parasite, which have been most studied, and in amoeboid forms, which probably interact with vaginal epithelial cells (Scott et al., 1995b). It was not clear from initial studies how many distinct cysteine protease gene products were present.

More recent molecular analyses have identified trichomonal cysteine protease genes (Mallinson et al., 1994, 1995). Genes encoding four cysteine proteases of *T. vaginalis* (Mallinson *et al.*, 1994) and seven cysteine proteases of the related cattle parasite *Tritrichomonas foetus* (Mallinson *et al.*, 1995) have been identified. The genes predict cathepsin L-like cysteine proteases that are similar but clearly distinct in sequence. Sequence identity among the cysteine protease genes is 53–72% in *T. vaginalis* and 30–78% in *T. foetus*. All but one of the genes is single copy. Results with *T. vaginalis* and *T. foetus* indicate that at least some of the multiplicity of trichomonal proteases identified with substrate gels was due to the expression of multiple enzymes.

The functions of the trichomonal proteases are unknown. Proteases are released by cultured parasites (Lockwood *et al.*, 1988; Garber and Lemchuk-Favel, 1989), possibly via secretion from an endosomal compartment (Scott *et al.*, 1995a). These results suggest that trichomonal proteases may act extracellularly, perhaps to mediate attachment of parasites to vaginal epithelium. Studies of clinical isolates of *T. vaginalis* showed that one of two studied protease activities was present in all isolates, while the other activity was present in less than half (Garber and Lemchuk-Favel, 1994). Studies with cysteine protease inhibitors suggest that not all of the trichomonal proteases are essential, but that the inhibition of certain cysteine proteases blocks parasite growth (Irvine *et al.*, 1997). Thus inhibitors of trichomonal cysteine proteases may be effective chemotherapeutic agents.

13. *TOXOPLASMA GONDII*

Infection with *T. gondii* is extremely common. Infection is usually asymptomatic, but acute infection can cause a febrile illness with lymphadenopathy (Beaman *et al.*, 1995). Toxoplasmosis can be a much more serious illness in the immunocompromised, including AIDS patients, in whom encephalitis, pneumonitis and retinitis commonly develop. Congenital toxoplasmosis is also a significant problem. *T. gondii* infection in normal hosts usually does not require therapy, but in the immunocompromised long courses of combination antimicrobial therapy are required. These therapies are often complicated by toxicity.

Two proteases were partially purified from *T. gondii*. One enzyme appeared to be an acid cysteine protease, the other a neutral ATP-dependent serine protease (Choi *et al.*, 1989). In more recent studies, neutral cathepsin B-like activity has been identified in *T. gondii* lysates (S.L. Reed, personal communication). Two *T. gondii* cysteine protease genes have been identified, one of which appears to encode an enzyme most closely related to cathepsin B (Reed, S.L., personal communication). Additional biochemical and molecular characterization of these proteases is under way.

14. SUMMARY

Many proteases of medically important protozoan parasites have been identified. Great progress has been made in recent years in the biochemical and molecular characterization of these enzymes, and in beginning to understand their biological roles. It appears that most protozoans express multiple proteases. All of the medically important protozoans appear to express cysteine proteases, which act either extracellularly or intracellularly in lysosome-like organelles. In many cases, multiple copies of similar cysteine protease genes are expressed. Serine proteases of protozoans have been less well characterized, but they appear to play key roles in some organisms. Individual protozoan metalloproteases and aspartic proteases have been well characterized, although there is not yet evidence for the expression of these classes of proteases in most protozoans. Important functions of protozoan proteases probably include the invasion of host cells and tissues, the degradation of components of the host immune response and the hydrolysis of host proteins for nutritional purposes. For example, the roles of cysteine and aspartic proteases in haemoglobin degradation by malaria parasites is quite well characterized. The important roles of proteases in the life cycles of medically important protozoan parasites suggest that the inhibition of these enzymes may be a useful new means of chemotherapy. Indeed, intitial results from multiple systems suggest that the inhibition of protozoan proteases may be a powerful new mode of antiparasitic chemotherapy.

ACKNOWLEDGEMENTS

I thank Jacques Bouvier, James McKerrow, Jeremy Mottram, Carolyn Petersen, Sharon Reed, Judy Sakanari and Linda Troeberg for helpful discussions and for critical reviews of the manuscript. Work in the author's laboratory is supported by the National Institutes of Health, the World Health Organization and the American Heart Association. The author is an established investigator of the American Heart Association.

REFERENCES

Aikawa, M. (1988). Fine structure of malaria parasites in the various stages of development. In: *Malaria: Principles and Practice of Malariology* (Wernsdorfer, W.H. and McGregor, I., eds.), pp. 97–129. Edinburgh: Churchill Livingstone.

Aissi, E., Charet, P., Bouquelet, S. and Biguet, J. (1983). Endoprotease in *Plasmodium yoelii nigeriensis*. *Comparative Biochemistry and Physiology* **74B**, 559–566.

Ankri, S., Miron, T., Rabinkov, A., Wilchek, M. and Mirelman, D. (1997). Allicin from garlic strongly inhibits cysteine proteinases and cytopathic effects of *Entamoeba histolytica*. *Antimicrobial Agents and Chemotherapy* **41**, 2286–2288.

Asawamahasakda, W., Ittarat, I., Chang, C.-C., McElroy, P. and Meshnick, S.R. (1994). Effects of antimalarials and protease inhibitors on plasmodial hemozoin production. *Molecular and Biochemical Parasitology* **67**, 183–191.

Ashall, F. (1990). Characterisation of an alkaline peptidase of *Trypanosoma cruzi* and other trypanosomatids. *Molecular and Biochemical Parasitology* **38**, 77–87.

Ashall, F., Angliker, H. and Shaw, E. (1990a). Lysis of trypanosomes by peptidyl fluoromethyl ketones. *Biochemical and Biophysical Research Communications* **170**, 923–929.

Ashall, F., Harris, D., Roberts, H., Healy, N. and Shaw, E. (1990b). Substrate specificity and inhibitor sensitivity of a trypanosomatid alkaline peptidase. *Biochimica et Biophysica Acta* **1035**, 293–299.

Ashford, R.W., Desjeux, P. and deRaadt, P. (1992). Estimation of population at risk of infection and number of cases of leishmaniasis. *Parasitology Today* **8**, 104–105.

Aslund, L., Henriksson, J., Campetella, O., Frasch, A.C., Pettersson, U. and Cazzulo, J.J. (1991). The C-terminal extension of the major cysteine proteinase (cruzipain) from *Trypanosoma cruzi*. *Molecular and Biochemical Parasitology* **45**, 345–347.

Authie, E., Duvallet, G., Robertson, C. and Williams, D.J. (1993). Antibody responses to a 33 kDa cysteine protease of *Trypanosoma congolense*: relationship to 'trypanotolerance' in cattle. *Parasite Immunology* **15**, 465–474.

Avila, E.E. and Calderon, J. (1993). *Entamoeba histolytica* trophozoites: a surface-associated cysteine protease. *Experimental Parasitology* **76**, 232–241.

Bailly, E., Savel, J., Mahouy, G. and Jaureguiberry, G. (1991). *Plasmodium falciparum*: isolation and characterization of a 55-kDa protease with a cathepsin D-like activity from *P. falciparum*. *Experimental Parasitology* **72**, 278–284.

Bailly, E., Jambou, R., Savel, J. and Jaureguiberry, G. (1992). *Plasmodium falciparum*: differential sensitivity in vitro to E-64 (cysteine protease inhibitor) and pepstatin A (aspartyl protease inhibitor). *Journal of Protozoology* **39**, 593–599.

Ball, E.G., McKee, R.W., Anfinsen, C.B., Cruz, W.O. and Geiman, Q.M. (1948). Studies on malarial parasites. IX. Chemical and metabolic changes during growth and multiplication in vivo and in vitro. *Journal of Biological Chemistry* **175**, 547–571.

Bannister, L.H. and Mitchell, G.H. (1989). The fine structure of secretion by *Plasmodium knowlesi* merozoites during red cell invasion. *Journal of Protozoology* **36**, 362–367.

Banyal, H.S., Misra, G.C., Gupta, C.M. and Dutta, G.P. (1981). Involvement of malarial proteases in the interaction between the parasite and host erythrocyte in *Plasmodium knowlesi* infections. *Journal of Parasitology* **67**, 623–626.

Barrett, A.J. (1994). Classification of peptidases. *Methods in Enzymology* **244**, 1–15.

Barry, D.N. (1982). Metabolism of *Babesia* parasites in vitro: amino acid production by *Babesia rodhaini* compared to *Plasmodium berghei*. *Australian Journal of Experimental Biology and Medical Sciences* **60**, 175–180.

Bart, G., Coombs, G.H. and Mottram, J.C. (1995). Isolation of lmcpc, a gene encoding a *Leishmania mexicana* cathepsin-B-like cysteine proteinase. *Molecular and Biochemical Parasitology* **73**, 271–274.

Bart, G., Frame, M.J., Carter, R., Coombs, G.H. and Mottram, J.C. (1997). Cathepsin B-like cysteine proteinase-deficient mutants of *Leishmania mexicana*. *Molecular and Biochemical Parasitology* **88**, 53–61.

Bates, P.A., Robertson, C.D. and Coombs, G.H. (1994). Expression of cysteine proteinases by metacyclic promastigotes of *Leishmania mexicana*. *Journal of Eukaryotic Microbiology* **41**, 199–203.

Beaman, M.H., McCabe, R.E., Wong, S.-Y. and Remington, J.S. (1995). *Toxoplasma gondii*. In: *Principles and Practice of Infectious Diseases* (Mandell, G.L., Bennett, J.E. and Dolin, R., eds.), pp. 2455–2475. New York: Churchill Livingstone.

Berman, J.D. (1997). Human leishmaniasis: clinical, diagnostic, and chemotherapeutic developments in the last 10 years. *Clinical Infectious Diseases* **24**, 684–703.

Bernard, F. and Schrevel, J. (1987). Purification of a *Plasmodium berghei* neutral endopeptidase and its localization in merozoite. *Molecular and Biochemical Parasitology* **26**, 167–174.

Berti, P.J. and Storer, A.C. (1995). Alignment/phylogeny of the papain superfamily of cysteine proteases. *Journal of Molecular Biology* **246**, 273–283.

Bhattacharya, J., Dey, R. and Datta, S.C. (1993). Calcium dependent thiol protease caldonopain and its specific endogenous inhibitor in *Leishmania donovani*. *Molecular and Cellular Biochemistry* **126**, 9–16.

Blackman, M.J. and Holder, A.A. (1992). Secondary processing of the *Plasmodium falciparum* merozoite surface protein-1 (MSP1) by a calcium-dependent membrane-bound serine protease: shedding of $MSP1_{33}$ as a noncovalently associated complex with other fragments of the MSP1. *Molecular and Biochemical Parasitology* **50**, 307–316.

Blackman, M.J., Heidrich, H.-G., Donachie, S., McBride, J.S. and Holder, A.A. (1990). A single fragment of a malaria merozoite surface protein remains on the parasite during red cell invasion and is the target of invasion-inhibiting antibodies. *Journal of Experimental Medicine* **172**, 379–382.

Blackman, M.J., Whittle, H. and Holder, A.A. (1991). Processing of the *Plasmodium falciparum* major merozoite surface protein-1: identification of a 33-kilodalton secondary processing product which is shed prior to erythrocyte invasion. *Molecular and Biochemical Parasitology* **49**, 35–44.

Blackman, M.J., Chappel, J.A., Shai, S. and Holder, A.A. (1993). A conserved parasite serine protease processes the *Plasmodium falciparum* merozoite surface protein-1. *Molecular and Biochemical Parasitology* **62**, 103–114.

Bonaldo, M.C., d'Escoffier, L.N., Salles, J.M. and Goldenberg, S. (1991). Characterization and expression of proteases during *Trypanosoma cruzi* metacyclogenesis. *Experimental Parasitology* **73**, 44–51.

Bond, J.S. and Butler, P.E. (1987). Intracellular proteases. *Annual Review of Biochemistry* **56**, 333–364.

Bontempi, E., Martinez, J. and Cazzulo, J.J. (1989). Subcellular localization of a cysteine proteinase from *Trypanosoma cruzi*. *Molecular and Biochemical Parasitology* **33**, 43–47.

Bordier, C., Etges, R.J., Ward, J., Turner, M.J. and Cardoso de Almeida, M.L. (1986). *Leishmania* and *Trypanosoma* surface glycoproteins have a common glycophospholipid membrane anchor. *Proceedings of the National Academy of Sciences of the USA* **83**, 5988–5991.

Bouvier, J., Etges, R.J. and Bordier, C. (1985). Identification and purification of membrane and soluble forms of the major surface protein of *Leishmania* promastigotes. *Journal of Biological Chemistry* **260**, 15504–15509.

Bouvier, J., Etges, R. and Bordier, C. (1987). Identification of the promastigote surface protease in seven species of *Leishmania*. *Molecular and Biochemical Parasitology* **24**, 73–79.

Bouvier, J., Bordier, C., Vogel, H., Reichelt, R. and Etges, R. (1989). Characterization of the promastigote surface protease of *Leishmania* as a membrane-bound zinc endopeptidase. *Molecular and Biochemical Parasitology* **37**, 235–245.

Bouvier, J., Schneider, P., Etges, R. and Bordier, C. (1990). Peptide substrate specificity of the membrane-bound metalloprotease of *Leishmania*. *Biochemistry* **29**, 10113–10119.

Bouvier, J., Schneider, P. and Etges, R. (1995). Leishmanolysin: surface metalloproteinase of *Leishmania*. *Methods in Enzymology* **248**, 614–633.

Braun-Breton, C., Rosenberry, T.L. and Pereira da Silva, L. (1988). Induction of the proteolytic activity of a membrane protein in *Plasmodium falciparum* by phosphatidyl inisitol-specific phospholipase C. *Nature* **332**, 457–459.

Braun-Breton, C., Blisnick, T., Jouin, H., Barale, J.C., Rabilloud, T., Langsley, G. and Pereira da Silva, L.H. (1992). *Plasmodium chabaudi* p68 serine protease activity required for merozoite entry into mouse erythrocytes. *Proceedings of the National Academy of Sciences of the USA* **89**, 9647–9651.

Brittingham, A., Morrison, C.J., McMaster, W.R., McGwire, B.S., Chang, K.P. and Mosser, D.M. (1995). Role of the *Leishmania* surface protease gp63 in complement fixation, cell adhesion, and resistance to complement-mediated lysis. *Journal of Immunology* **155**, 3102–3111.

Brown, P.D. and Giavazzi, R. (1995). Matrix metalloproteinase inhibition: a review of anti-tumour activity. *Annals of Oncology* **6**, 967–974.

Bruchhaus, I. and Tannich, E. (1996). A gene highly homologous to ACP1 encoding cysteine proteinase 3 in *Entamoeba histolytica* is present and expressed in *E. dispar*. *Parasitology Research* **82**, 189–192.

Bruchhaus, I., Jacobs, T., Leippe, M. and Tannich, E. (1996). *Entamoeba histolytica* and *Entamoeba dispar*: differences in numbers and expression of cysteine proteinase genes. *Molecular Microbiology* **22**, 255–263.

Burleigh, B.A. and Andrews, N.W. (1995). A 120-kDa alkaline peptidase from *Trypanosoma cruzi* is involved in the generation of a novel Ca(2+)-signaling factor for mammalian cells. *Journal of Biological Chemistry* **270**, 5172–5180.

Burleigh, B.A., Caler, E.V., Webster, P. and Andrews, N.W. (1997). A cytosolic serine endopeptidase from *Trypanosoma cruzi* is required for the generation of Ca2+ signaling in mammalian cells. *Journal of Cell Biology* **136**, 609–620.

Button, L.L. and McMaster, W.R. (1988). Molecular cloning of the major surface antigen of *Leishmania*. *Journal of Experimental Medicine* **167**, 724–729.

Button, L.L., Russell, D.G., Klein, H.L., Medina-Acosta, E., Karess, R.E. and McMaster, W.R. (1989). Genes encoding the major surface glycoprotein in *Leishmania* are tandemly linked at a single chromosomal locus and are constitutively transcribed. *Molecular and Biochemical Parasitology* **32**, 271–283.

Button, L.L., Wilson, G., Astell, C.R. and McMaster, W.R. (1993). Recombinant *Leishmania* surface glycoprotein GP63 is secreted in the baculovirus expression system as a latent metalloproteinase. *Gene* **134**, 75–81.

Bzik, D.J., Li, W.-B., Horii, T. and Inselburg, J. (1988). Amino acid sequence of the serine-repeat antigen (SERA) of *Plasmodium falciparum* determined from cloned cDNA. *Molecular and Biochemical Parasitology* **30**, 279–288.

Campetella, O., Martinez, J. and Cazzulo, J.J. (1990). A major cysteine proteinase is developmentally regulated in *Trypanosoma cruzi*. *FEMS Microbiology Letters* **55**, 145–149.

Campetella, O., Henriksson, J., Aslund, L., Frasch, A.C., Pettersson, U. and Cazzulo, J.J. (1992). The major cysteine proteinase (cruzipain) from *Trypanosoma cruzi* is encoded by multiple polymorphic tandemly organized genes located on different chromosomes. *Molecular and Biochemical Parasitology* **50**, 225–234.
Cazzulo, J.J., Couso, R., Raimondi, A., Wernstedt, C. and Hellman, U. (1989). Further characterization and partial amino acid sequence of a cysteine proteinase from *Trypanosoma cruzi*. *Molecular and Biochemical Parasitology* **33**, 33–41.
Cazzulo, J.J., Cazzulo Franke, M.C., Martinez, J. and Franke de Cazzulo, B.M. (1990). Some kinetic properties of a cysteine proteinase (cruzipain) from *Trypanosoma cruzi*. *Biochimica et Biophysica Acta* **1037**, 186–191.
Cenedella, R.J., Rosen, H., Angel, C.R. and Saxe, L.H. (1968). Free amino-acid production in vitro by *Plasmodium berghei*. *American Journal of Tropical Medicine and Hygiene* **17**, 800–803.
Chang, K.P., Chaudhuri, G. and Fong, D. (1990). Molecular determinants of *Leishmania* virulence. *Annual Review of Microbiology* **44**, 499–529.
Chaudhuri, G. and Chang, K.P. (1988). Acid protease activity of a major surface membrane glycoprotein (gp63) from *Leishmania mexicana* promastigotes. *Molecular and Biochemical Parasitology* **27**, 43–52.
Chaudhuri, G., Chaudhuri, M., Pan, A. and Chang, K.P. (1989). Surface acid proteinase (gp63) of *Leishmania mexicana*. A metalloenzyme capable of protecting liposome-encapsulated proteins from phagolysosomal degradation by macrophages. *Journal of Biological Chemistry* **264**, 7483–7489.
Choi, W.Y., Nam, H.W. and Youn, J.H. (1989). Characterization of proteases of *Toxoplasma gondii*. *Kisaengchunghak Chapchi* **27**, 161–170.
Coggins, J.R. and Schaefer, F.W. (1986). *Giardia muris*: ultrastructural analysis of in vitro excystation. *Experimental Parasitology* **61**, 219–228.
Coombs, G.H. (1982). Proteinases of *Leishmania mexicana* and other flagellate protozoa. *Parasitology* **84**, 149–155.
Coombs, G.H. and Baxter, J. (1984). Inhibition of *Leishmania* amastigote growth by antipain and leupeptin. *Annals of Tropical Medicine and Parasitology* **78**, 21–24.
Coombs, G.H. and North, M.J. (1983). An analysis of the proteinases of *Trichomonas vaginalis* by polyacrylamide gel electrophoresis. *Parasitology* **86**, 1–6.
Cooper, J.A. and Bujard, H. (1992). Membrane-associated proteases process *Plasmodium falciparum* merozoite surface antigen-1 (MSA1) to fragment gp41. *Molecular and Biochemical Parasitology* **56**, 151–160.
Curley, G.P., O'Donovan, S.M., McNally, J., Mullally, M., O'Hara, H., Troy, A., O'Callaghan, S.A. and Dalton, J.P. (1994). Aminopeptidases from *Plasmodium falciparum*, *Plasmodium chabaudi* and *Plasmodium berghei*. *Journal of Eukaryotic Microbiology* **41**, 119–123.
Dame, J.B., Reddy, G.R., Yowell, C.A., Dunn, B.M., Kay, J. and Berry, C. (1994). Sequence, expression and modeled structure of an aspartic proteinase from the human malaria parasite *Plasmodium falciparum*. *Molecular and Biochemical Parasitology* **64**, 177–190.
Davies, C.R., Cooper, A.M., Peacock, C., Lane, R.P. and Blackwell, J.M. (1990). Expression of LPG and GP63 by different developmental stages of *Leishmania major* in the sandfly *Phlebotomus papatasi*. *Parasitology* **101**, 337–343.
De Souza Leao, S., Lang, T., Prina, E., Hellio, R. and Antoine, J.C. (1995). Intracellular *Leishmania amazonensis* amastigotes internalize and degrade MHC class II molecules of their host cells. *Journal of Cell Science* **108**, 3219–3231.
Debrabant, A. and Delplace, P. (1989). Leupeptin alters the proteolytic processing of

P126, the major parasitophorous vacuole antigen of *Plasmodium falciparum*. *Molecular and Biochemical Parasitology* **33**, 151–158.

Deguercy, A., Hommel, M. and Schrevel, J. (1990). Purification and characterization of 37-kilodalton proteases from *Plasmodium falciparum* and *Plasmodium berghei* which cleave erythrocyte cytoskeletal components. *Molecular and Biochemical Parasitology* **38**, 233–244.

Dejkriengkraikhul, P. and Wilairat, P. (1983). Requirement of malarial protease in the invasion of human red cells by merozoites of *Plasmodium falciparum*. *Zeitschrift für Parasitenkunde* **69**, 313–317.

Del Nery, E., Juliano, M.A., Lima, A.P., Scharfstein, J. and Juliano, L. (1997a). Kininogenase activity by the major cysteinyl proteinase (cruzipain) from *Trypanosoma cruzi*. *Journal of Biological Chemistry* **272**, 25713–25718.

Del Nery, E., Juliano, M.A., Meldal, M., Svendsen, I., Scharfstein, J., Walmsley, A. and Juliano, L. (1997b). Characterization of the substrate specificity of the major cysteine protease (cruzipain) from *Trypanosoma cruzi* using a portion-mixing combinatorial library and fluorogenic peptides. *Biochemical Journal* **323**, 427–433.

Delplace, P., Fortier, B., Tronchin, G., Dubremetz, J. and Vernes, A. (1987). Localization, biosynthesis, processing and isolation of a major 126 kDa antigen of the parasitophorous vacuole of *Plasmodium falciparum*. *Molecular and Biochemical Parasitology* **23**, 193–201.

Diamond, L.S. and Clark, C.G. (1993). A redescription of *Entamoeba histolytica* Schaudinn, 1903 (Emended Walker, 1911) separating it from *Entamoeba dispar* Brumpt, 1925. *Eukaryotic Microbiology* **40**, 340–344.

Divo, A.A., Geary, T.G., Davis, N.L. and Jensen, J.B. (1985). Nutritional requirements of *Plasmodium falciparum* in culture. I. Exogenously supplied dialyzable components necessary for continuous growth. *Journal of Protozoology* **32**, 59–64.

Dluzewski, A.R., Rangachari, K., Wilson, R.J.M. and Gratzer, W.B. (1986). *Plasmodium falciparum*: protease inhibitors and inhibition of erythrocyte invasion. *Experimental Parasitology* **62**, 416–422.

Domínguez, J.N., López, S., Charris, J., Iarruso, L., Lobo, G., Semenov, A., Olson, J.E. and Rosenthal, P.J. (1997). Synthesis and antimalarial effects of phenothiazine inhibitors of a *Plasmodium falciparum* cysteine protease. *Journal of Medicinal Chemistry* **40**, 2726–2732.

Duboise, S.M., Vannier-Santos, M.A., Costa-Pinto, D., Rivas, L., Pan, A.A., Traub-Cseko, Y., De Souza, W. and McMahon-Pratt, D. (1994). The biosynthesis, processing, and immunolocalization of *Leishmania pifanoi* amastigote cysteine proteinases. *Molecular and Biochemical Parasitology* **68**, 119–132.

Eakin, A.E., Bouvier, J., Sakanari, J.A., Craik, C.S. and McKerrow, J.H. (1990). Amplification and sequencing of genomic DNA fragments encoding cysteine proteases from protozoan parasites. *Molecular and Biochemical Parasitology* **39**, 1–8.

Eakin, A.E., Mills, A.A., Harth, G., McKerrow, J.H. and Craik, C.S. (1992). The sequence, organization and expression of the major cysteine protease (cruzain) from *Trypanosoma cruzi*. *Journal of Biological Chemistry* **267**, 7411–7420.

Eakin, A.E., McGrath, M.E., McKerrow, J.H., Fletterick, R.J. and Craik, C.S. (1993). Production of crystallizable cruzain, the major cysteine protease from *Trypanosoma cruzi*. *Journal of Biological Chemistry* **268**, 6115–6118.

El-Sayed, N.M.A. and Donelson, J.E. (1997). African trypanosomes have differentially expressed genes encoding homologues of the leishmania GP63 surface protease. *Journal of Biological Chemistry* **272**, 26742–26748.

Elford, B.C., Cowan, G.M. and Ferguson, D.J.P. (1995). Parasite-regulated membrane transport processes and metabolic control in malaria-infected erythrocytes. *Biochemical Journal* **308**, 361–374.

Engel, J.C., Doyle, P.S., Palmer, J., Hsieh, I., Bainton, D.F. and McKerrow, J.H. (1998). Cysteine protease inhibitors alter Golgi complex ultrastructure and function in *Trypanosoma cruzi*. *Journal of Cell Science* **111**, 597–606.

Etges, R. (1992). Identification of a surface metalloproteinase on 13 species of *Leishmania* isolated from humans, *Crithidia fasciculata* and *Herpetomonas samuelpessoai*. *Acta Tropica* **50**, 205–217.

Etges, R., Bouvier, J. and Bordier, C. (1986). The major surface protein of *Leishmania* promastigotes is anchored in the membrane by a myristic acid-labeled phospholipid. *EMBO Journal* **5**, 597–601.

Feely, D.E. and Dyer, J.K. (1987). Localization of acid phosphatase activity in *Giardia lamblia* and *Giardia muris* trophozoites. *Journal of Protozoology* **34**, 80–83.

Forney, J.R., Yang, S., Du, C. and Healey, M.C. (1996a). Efficacy of serine protease inhibitors against *Cryptosporidium parvum* infection in a bovine fallopian tube epithelial cell culture system. *Journal of Parasitology* **82**, 638–640.

Forney, J.R., Yang, S. and Healey, M.C. (1996b). Protease activity associated with excystation of *Cryptosporidium parvum* oocysts. *Journal of Parasitology* **82**, 889–892.

Forney, J.R., Yang, S. and Healey, M.C. (1997). Antagonistic effect of human alpha-1-antitrypsin on excystation of *Cryptosporidium parvum* oocysts. *Journal of Parasitology* **83**, 771–774.

Francis, S.E., Gluzman, I.Y., Oksman, A., Knickerbocker, A., Mueller, R., Bryant, M.L., Sherman, D.R., Russell, D.G. and Goldberg, D.E. (1994). Molecular characterization and inhibition of a *Plasmodium falciparum* aspartic haemoglobinase. *EMBO Journal* **13**, 306–317.

Francis, S.E., Gluzman, I.Y., Oksman, A., Banerjee, D. and Goldberg, D.E. (1996). Characterization of native falcipain, an enzyme involved in *Plasmodium falciparum* haemoglobin degradation. *Molecular and Biochemical Parasitology* **83**, 189–200.

Francis, S.E., Banerjee, R. and Goldberg, D.E. (1997a). Biosynthesis and maturation of the malaria aspartic haemoglobinases plasmepsins I and II. *Journal of Biological Chemistry* **272**, 14961–14968.

Francis, S.E., Sullivan, D.J. and Goldberg, D.E. (1997b). Hemoglobin metabolism in the malaria parasite *Plasmodium falciparum*. *Annual Review of Microbiology* **51**, 97–123.

Franke de Cazzulo, B.M., Martinez, J., North, M.J., Coombs, G.H. and Cazzulo, J.J. (1994). Effects of proteinase inhibitors on the growth and differentiation of *Trypanosoma cruzi*. *FEMS Microbiology Letters* **124**, 81–86.

Frommel, T.O., Button, L.L., Fujikura, Y. and McMaster, W.R. (1990). The major surface glycoprotein (GP63) is present in both life stages of *Leishmania*. *Molecular and Biochemical Parasitology* **38**, 25–32.

Fulton, J.D. and Grant, P.T. (1956). The sulphur requirements of the erythrocytic form of *Plasmodium knowlesi*. *Biochemical Journal* **63**, 274–282.

Funk, V.A., Jardim, A. and Olafson, R.W. (1994). An investigation into the significance of the N-linked oligosaccharides of *Leishmania* gp63. *Molecular and Biochemical Parasitology* **63**, 23–35.

Gadasi, H. and Kessler, E. (1983). Correlation of virulence and collagenolytic activity in *Entamoeba histolytica*. *Infection and Immunity* **39**, 528–531.

Gadasi, H. and Kobiler, D. (1983). *Entamoeba histolytica*: correlation between virulence and content of proteolytic enzymes. *Experimental Parasitology* **55**, 105–110.

Gamboa de Domínguez, N.D. and Rosenthal, P.J. (1996). Cysteine proteinase inhibitors block early steps in hemoglobin degradation by cultured malaria parasites. *Blood* **87**, 4448–4454.

Garber, G.E. and Lemchuk-Favel, L.T. (1989). Characterization and purification of extracellular proteases of *Trichomonas vaginalis*. *Canadian Journal of Microbiology* **35**, 903–909.

Garber, G.E. and Lemchuk-Favel, L.T. (1994). Analysis of the extracellular proteases of *Trichomonas vaginalis*. *Parasitology Research* **80**, 361–365.

Gillmor, S.A., Craik, C.S. and Fletterick, R.J. (1997). Structural determinants of specificity in the cysteine protease cruzain. *Protein Science* **6**, 1603–1611.

Gluzman, I.Y., Francis, S.E., Oksman, A., Smith, C.E., Duffin, K.L. and Goldberg, D.E. (1994). Order and specificity of the *Plasmodium falciparum* hemoglobin degradation pathway. *Journal of Clinical Investigation* **93**, 1602–1608.

Goldberg, D.E., Slater, A.F.G., Cerami, A. and Henderson, G.B. (1990). Hemoglobin degradation in the malaria parasite *Plasmodium falciparum*: an ordered process in a unique organelle. *Proceedings of the National Academy of Sciences USA* **87**, 2931–2935.

Goldberg, D.E., Slater, A.F.G., Beavis, R., Chait, B., Cerami, A. and Henderson, G.B. (1991). Hemoglobin degradation in the human malaria pathogen *Plasmodium falciparum*: a catabolic pathway initiated by a specific aspartic protease. *Journal of Experimental Medicine* **173**, 961–969.

Gonzalez, J., Ramalho-Pinto, F.J., Frevert, U., Ghiso, J., Tomlinson, S., Scharfstein, J., Corey, E.J. and Nussenzweig, V. (1996). Proteasome activity is required for the stage-specific transformation of a protozoan parasite. *Journal of Experimental Medicine* **184**, 1909–1918.

Gonzalez-Aseguinolaza, G., Almazan, F., Rodriguez, J.F., Marquet, A. and Larraga, V. (1997). Cloning of the gp63 surface protease of *Leishmania infantum*. Differential post-translational modifications correlated with different infective forms. *Biochimica et Biophysica Acta* **1361**, 92–102.

Grellier, P., Picard, I., Bernard, F., Mayer, R., Heidrich, H.-G., Monsigny, M. and Schrevel, J. (1989). Purification and identification of a neutral endopeptidase in *Plasmodium falciparum* schizonts and merozoites. *Parasitology Research* **75**, 455–460.

Groman, N.B. (1951). Dynamic aspects of the nitrogen metabolism of *Plasmodium gallinaceum* in vivo and in vitro. *Journal of Infectious Diseases* **88**, 126–150.

Guerrant, R.L. (1997). Cryptosporidiosis: an emerging, highly infectious threat. *Emerging Infectious Diseases* **3**, 51–57.

Gyang, F.N., Poole, B. and Trager, W. (1982). Peptidases from *Plasmodium falciparum* cultured in vitro. *Molecular and Biochemical Parasitology* **5**, 263–273.

Hadley, T., Aikawa, M. and Miller, L.H. (1983). *Plasmodium knowlesi*: studies on invasion of rhesus erythrocytes by merozoites in the presence of protease inhibitors. *Experimental Parasitology* **55**, 306–311.

Hare, D.F., Jarroll, E.L. and Lindmark, D.G. (1989). *Giardia lamblia*: characterization of proteinase activity in trophozoites. *Experimental Parasitology* **68**, 168–175.

Harth, G., Andrews, N., Mills, A.A., Engel, J.C., Smith, R. and McKerrow, J.H. (1993). Peptide-fluoromethyl ketones arrest intracellular replication and intercellular transmission of *Trypanosoma cruzi*. *Molecular and Biochemical Parasitology* **58**, 17–24.

Heine, P. and McGregor, J.A. (1993). *Trichomonas vaginalis*: a reemerging pathogen. *Clinical Obstetrics and Gynecology* **36**, 137–144.
Hellman, U., Wernstedt, C. and Cazzulo, J.J. (1991). Self-proteolysis of the cysteine proteinase, cruzipain, from *Trypanosoma cruzi* gives a major fragment corresponding to its carboxy-terminal domain. *Molecular and Biochemical Parasitology* **44**, 15–21.
Hill, J., Tyas, L., Phylip, L.H., Kay, J., Dunn, B.M. and Berry, C. (1994). High level expression and characterisation of plasmepsin II, an aspartic proteinase from *Plasmodium falciparum*. *FEBS Letters* **352**, 155–158.
Hua, S., To, W.Y., Nguyen, T.T., Wong, M.L. and Wang, C.C. (1996). Purification and characterization of proteasomes from *Trypanosoma brucei*. *Molecular and Biochemical Parasitology* **78**, 33–46.
Huet, G., Richet, C., Demeyer, D., Bisiau, H., Soudan, B., Tetaert, D., Han, K.K. and Degand, P. (1992). Characterization of different proteolytic activities in *Trypanosoma brucei brucei*. *Biochimica et Biophysica Acta* **1138**, 213–221.
Ilg, T., Harbecke, D. and Overath, P. (1993). The lysosomal gp63-related protein in *Leishmania mexicana* amastigotes is a soluble metalloproteinase with an acidic pH optimum. *FEBS Letters* **327**, 103–107.
Inverso, J.A., Medina-Acosta, E., O'Connor, J., Russell, D.G. and Cross, G.A. (1993). *Crithidia fasciculata* contains a transcribed leishmanial surface proteinase (gp63) gene homologue. *Molecular and Biochemical Parasitology* **57**, 47–54.
Irvine, J.W., North, M.J. and Coombs, G.H. (1997). Use of inhibitors to identify essential cysteine proteinases of *Trichomonas vaginalis*. *FEMS Microbiology Letters* **149**, 45–50.
Jackson, E.K. and Garrison, J.C. (1996). Renin and angiotensin. In: *The Pharmacological Basis of Therapeutics* (Hardman, J.G., Limbird, L.E., Molinoff, P.B., Ruddon, R.W. and Gilman, A.G., eds.), pp. 733–758. New York: McGraw-Hill.
Jacobs, T., Bruchhaus, I., Dandekar, T., Tannich, E. and Leippe, M. (1998). Isolation and molecular characterization of a surface-bound proteinase of *Entamoeba histolytica*. *Molecular Microbiology* **27**, 269–276.
Joshi, P.B., Sacks, D.L., Modi, G. and McMaster, W.R. (1998). Targeted gene deletion of *Leishmania major* genes encoding developmental stage-specific leishmanolysin (GP63). *Molecular Microbiology* **27**, 519–530.
Kamchonwongpaisan, S., Samoff, E. and Meshnick, S.R. (1997). Identification of hemoglobin degradation products in *Plasmodium falciparum*. *Molecular and Biochemical Parasitology* **86**, 179–186.
Keene, W.E., Petitt, M.G., Allen, S. and McKerrow, J.H. (1986). The major neutral proteinase of *Entamoeba histolytica*. *Journal of Experimental Medicine* **163**, 536–549.
Keene, W.E., Hidalgo, M.E., Orozco, E. and McKerrow, J.H. (1990). *Entamoeba histolytica*: correlation of the cytopathic effect of virulent trophozoites with secretion of a cysteine proteinase. *Experimental Parasitology* **71**, 199–206.
Kelsall, B.L. and Ravdin, J.I. (1993). Degradation of human IgA by *Entamoeba histolytica*. *Journal of Infectious Diseases* **168**, 1319–1322.
Kink, J.A. and Chang, K.P. (1988). N-glycosylation as a biochemical basis for virulence in *Leishmania mexicana amazonensis*. *Molecular and Biochemical Parasitology* **27**, 181–190.
Kirchhoff, L.V. (1993a). American trypanosomiasis (Chagas' disease) — a tropical disease now in the United States. *New England Journal of Medicine* **329**, 639–644.

Kirchhoff, L.V. (1993b). Chagas disease. American trypanosomiasis. *Infectious Disease Clinics of North America* **7**, 487–502.

Kirchhoff, L.V. (1995). Agents of African trypanosomiasis (sleeping sickness). In: *Principles and Practice of Infectious Diseases* (Mandell, G.L., Bennett, J.E. and Dolin, R., eds.), pp. 2450–2455. New York: Churchill Livingstone.

Knapp, B., Hundt, E., Nau, U. and Küpper, H.A. (1989). Molecular cloning, genomic structure and localization in a blood stage antigen of *Plasmodium falciparum* characterized by a serine stretch. *Molecular and Biochemical Parasitology* **32**, 73–84.

Knapp, B., Nau, U., Hundt, E. and Küpper, H.A. (1991). A new blood stage antigen of *Plasmodium falciparum* highly homologous to the serine-stretch protein SERP. *Molecular and Biochemical Parasitology* **44**, 1–14.

Kolakovich, K.A., Gluzman, I.Y., Duffin, K.L. and Goldberg, D.E. (1997). Generation of hemoglobin peptides in the acidic digestive vacuole of *Plasmodium falciparum* implicates peptide transport in amino acid production. *Molecular and Biochemical Parasitology* **87**, 123–135.

Kornblatt, M.J., Mpimbaza, G.W. and Lonsdale-Eccles, J.D. (1992). Characterization of an endopeptidase of *Trypanosoma brucei brucei*. *Archives of Biochemistry and Biophysics* **293**, 25–31.

Krogstad, D.J. and Schlesinger, P.H. (1987). Acid-vesicle function, intracellular pathogens and the action of chloroquine against *Plasmodium falciparum*. *New England Journal of Medicine* **317**, 542–549.

Kuzoe, F.A. (1993). Current situation of African trypanosomiasis. *Acta Tropica* **54**, 153–162.

Kweider, M., Lemesre, J.L., Darcy, F., Kusnierz, J.P., Capron, A. and Santoro, F. (1987). Infectivity of *Leishmania braziliensis* promastigotes is dependent on the increasing expression of a 65,000-dalton surface antigen. *Journal of Immunology* **138**, 299–305.

Lalmanach, G., Mayer, R., Serveau, C., Scharfstein, J. and Gauthier, F. (1996). Biotin-labelled peptidyl diazomethane inhibitors derived from the substrate-like sequence of cystatin: targeting of the active site of cruzipain, the major cysteine proteinase of *Trypanosoma cruzi*. *Biochemical Journal* **318**, 395–399.

Leon, G., Fiori, C., Das, P., Moreno, M., Tovar, R., Sanchez-Salas, J.L. and Munoz, M.L. (1997). Electron probe analysis and biochemical characterization of electron-dense granules secreted by *Entamoeba histolytica*. *Molecular and Biochemical Parasitology* **85**, 233–242.

Li, R., Kenyon, G.L., Cohen, F.E., Chen, X., Gong, B., Dominguez, J.N., Davidson, E., Kurzban, G., Miller, R.E., Nuzum, E.O., Rosenthal, P.J. and McKerrow, J.H. (1995). In vitro antimalarial activity of chalcones and their derivatives. *Journal of Medicinal Chemistry* **38**, 5031–5037.

Li, R., Chen, X., Gong, B., Selzer, P.M., Li, Z., Davidson, E., Kurzban, G., Miller, R.E., Nuzum, E.O., McKerrow, J.H., Fletterick, R.J., Gillmor, S.A., Craik, C.S., Kuntz, I.D., Cohen, F.E. and Kenyon, G.L. (1996). Structure-based design of parasitic protease inhibitors. *Bioorganic and Medicinal Chemistry* **4**, 1421–1427.

Lindmark, D.G. (1988). *Giardia lamblia*: localization of hydrolase activities in lysosome-like organelles of trophozoites. *Experimental Parasitology* **65**, 141–147.

Liu, X. and Chang, K.P. (1992). Extrachromosomal genetic complementation of surface metalloproteinase (gp63)-deficient *Leishmania* increases their binding to macrophages. *Proceedings of the National Academy of Sciences of the USA* **89**, 4991–4995.

Lockwood, B.C., North, M.J., Scott, K.I., Bremner, A.F. and Coombs, G.H. (1987).

The use of a highly sensitive electrophoretic method to compare the proteinases of trichomonads. *Molecular and Biochemical Parasitology* **24**, 89–95.
Lockwood, B.C., North, M.J. and Coombs, G.H. (1988). The release of hydrolases from *Trichomonas vaginalis* and *Tritrichomonas foetus*. *Molecular and Biochemical Parasitology* **30**, 135–142.
Lomo, P.O., Coetzer, T.H. and Lonsdale-Eccles, J.D. (1997). Characterization of a multicatalytic proteinase complex (20S proteasome) from *Trypanosoma brucei brucei*. *Immunopharmacology* **36**, 285–293.
Lonsdale-Eccles, J.D. and Grab, D.J. (1987). Lysosomal and non-lysosomal peptidyl hydrolases of the bloodstream forms of *Trypanosoma brucei brucei*. *European Journal of Biochemistry* **169**, 467–475.
Lonsdale-Eccles, J.D. and Mpimbaza, G.W.N. (1986). Thiol-dependent proteases of African trypanosomes. Analysis by electrophoresis in sodium dodecyl sulphate/polyacrylamide gels co-polymerized with fibrinogen. *European Journal of Biochemistry* **155**, 469–473.
Lowndes, C.M., Bonaldo, M.C., Thomaz, N. and Goldenberg, S. (1996). Heterogeneity of metalloprotease expression in *Trypanosoma cruzi*. *Parasitology* **112**, 393–399.
Luaces, A.L. and Barrett, A.J. (1988). Affinity purification and biochemical characterization of histolysin, the major cysteine proteinase of *Entamoeba histolytica*. *Biochemical Journal* **250**, 903–909.
Luker, K.E., Francis, S.E., Gluzman, I.Y. and Goldberg, D.E. (1996). Kinetic analysis of plasmepsins I and II, aspartic proteases of the *Plasmodium falciparum* digestive vacuole. *Molecular and Biochemical Parasitology* **79**, 71–78.
Lushbaugh, W.B., Hofbauer, A.F. and Pittman, F.E. (1984). Proteinase activities of *Entamoeba histolytica* cytotoxin. *Gastroenterology* **87**, 17–27.
Lushbaugh, W.B., Hofbauer, A.F. and Pittman, F.E. (1985). *Entamoeba histolytica*: purification of cathepsin B. *Experimental Parasitology* **59**, 328–336.
Lyon, J.A. and Haynes, J.D. (1986). *Plasmodium falciparum* antigens synthesized by schizonts and stabilized at the merozoite surface when schizonts mature in the presence of protease inhibitors. *Journal of Immunology* **136**, 2245–2251.
Macdonald, M.H., Morrison, C.J. and McMaster, W.R. (1995). Analysis of the active site and activation mechanism of the *Leishmania* surface metalloproteinase GP63. *Biochimica et Biophysica Acta* **1253**, 199–207.
Magos, M.A., de la Torre, M. and Munoz, M.L. (1992). Collagenase activity in clinical isolates of *Entamoeba histolytica* maintained in xenic cultures. *Archives of Medical Research* **23**, 115–118.
Mallinson, D.J., Lockwood, B.C., Coombs, G.H. and North, M.J. (1994). Identification and molecular cloning of four cysteine proteinase genes from the pathogenic protozoon *Trichomonas vaginalis*. *Microbiology* **140**, 2725–2735.
Mallinson, D.J., Livingstone, J., Appleton, K.M., Lees, S.J., Coombs, G.H. and North, M.J. (1995). Multiple cysteine proteinases of the pathogenic protozoon *Tritrichomonas foetus*: identification of seven diverse and differentially expressed genes. *Microbiology* **141**, 3077–3085.
Mayer, R., Picard, I., Lawton, P., Grellier, P., Barrault, C., Monsigny, M. and Schrevel, J. (1991). Peptide derivatives specific for a *Plasmodium falciparum* proteinase inhibit the human erythrocyte invasion by merozoites. *Journal of Medicinal Chemistry* **34**, 3029–3035.
McDonald, C.K. and Kuritzkes, D.R. (1997). Human immunodeficiency virus type 1 protease inhibitors. *Archives of Internal Medicine* **157**, 951–959.
McGowan, K., Deneke, C.F., Thorne, G.M. and Gorbach, S.L. (1982). *Entamoeba*

histolytica cytotoxin: purification, characterization, strain virulence, and protease activity. *Journal of Infectious Diseases* **146**, 616–625.

McGrath, M.E., Eakin, A.E., Engel, J.C., McKerrow, J.H., Craik, C.S. and Fletterick, R.J. (1995). The crystal structure of cruzain: a therapeutic target for Chagas' disease. *Journal of Molecular Biology* **247**, 251–259.

McGwire, B. and Chang, K.P. (1994). Genetic rescue of surface metalloproteinase (gp63)-deficiency in *Leishmania amazonensis* variants increases their infection of macrophages at the early phase. *Molecular and Biochemical Parasitology* **66**, 345–347.

McGwire, B.S. and Chang, K.P. (1996). Posttranslational regulation of a *Leishmania* HEXXH metalloprotease (gp63). The effects of site-specific mutagenesis of catalytic, zinc binding, N-glycosylation, and glycosyl phosphatidylinositol addition sites on N-terminal end cleavage, intracellular stability, and extracellular exit. *Journal of Biological Chemistry* **271**, 7903–7909.

McKerrow, J.H., Sun, E., Rosenthal, P.J. and Bouvier, J. (1993). The proteases and pathogenicity of parasitic protozoa. *Annual Review of Microbiology* **47**, 821–853.

McKerrow, J.H., McGrath, M.E. and Engel, J.C. (1995). The cysteine protease of *Trypanosoma cruzi* as a model for antiparasite drug design. *Parasitology Today* **11**, 279–282.

McLaughlin, J. and Faubert, G. (1977). Partial purification and some properties of a neutral sulfhydryl and an acid proteinase from *Entamoeba histolytica*. *Canadian Journal of Microbiology* **23**, 420–425.

McPherson, R.A., Donald, D.R., Sawyer, W.H. and Tilley, L. (1993). Proteolytic digestion of band 3 at an external site alters the erythrocyte membrane organisation and may facilitate malarial invasion. *Molecular and Biochemical Parasitology* **62**, 233–242.

Medina-Acosta, E., Karess, R.E., Schwartz, H. and Russell, D.G. (1989). The promastigote surface protease (gp63) of *Leishmania* is expressed but differentially processed and localized in the amastigote stage. *Molecular and Biochemical Parasitology* **37**, 263–273.

Medina-Acosta, E., Karess, R.E. and Russell, D.G. (1993). Structurally distinct genes for the surface protease of *Leishmania mexicana* are developmentally regulated. *Molecular and Biochemical Parasitology* **57**, 31–45.

Meirelles, M.N., Juliano, L., Carmona, E., Silva, S.G., Costa, E.M., Murta, A.C. and Scharfstein, J. (1992). Inhibitors of the major cysteinyl proteinase (GP57/51) impair host cell invasion and arrest the intracellular development of *Trypanosoma cruzi* in vitro. *Molecular and Biochemical Parasitology* **52**, 175–184.

Miller, R.A., Reed, S.G. and Parsons, M. (1990). *Leishmania* gp63 molecule implicated in cellular adhesion lacks an Arg-Gly-Asp sequence. *Molecular and Biochemical Parasitology* **39**, 267–274.

Mills, B.J. and Lang, C.A. (1996). Differential distribution of free and bound glutathione and cyst(e)ine in human blood. *Biochemical Pharmacology* **52**, 401–406.

Mirelman, D., Nuchamowitz, Y., Bohm-Gloning, B. and Walderich, B. (1996). A homologue of the cysteine proteinase gene (ACP1 or Eh-CPp3) of pathogenic *Entamoeba histolytica* is present in non-pathogenic *E. dispar* strains. *Molecular and Biochemical Parasitology* **78**, 47–54.

Montfort, I., Perez-Tamayo, R., Gonzalez Canto, A., Garcia de Leon, M.C., Olivos, A. and Tello, E. (1993). Role of cysteine proteinases of *Entamoeba histolytica* on the cytopathogenicity of axenic trophozoites on rat and hamster hepatocytes in vitro. *Journal of Parasitology* **79**, 98–105.

Moon, R.P., Tyas, L., Certa, U., Rupp, K., Bur, D., Jacquet, C., Matile, H., Loetscher, H., Grueninger-Leitch, F., Kay, J., Dunn, B.M., Berry, C. and Ridley, R.G. (1997). Expression and characterisation of plasmepsin I from *Plasmodium falciparum*. *European Journal of Biochemistry* **244**, 552–560.

Mottram, J.C., North, M.J., Barry, J.D. and Coombs, G.H. (1989). A cysteine proteinase cDNA from *Trypanosoma brucei* predicts an enzyme with an unusual C-terminal extension. *FEBS Letters* **258**, 211–215.

Mottram, J.C., Robertson, C.D., Coombs, G.H. and Barry, J.D. (1992). A developmentally regulated cysteine proteinase gene of *Leishmania mexicana*. *Molecular Microbiology* **6**, 1925–1932.

Mottram, J.C., Souza, A.E., Hutchison, J.E., Carter, R., Frame, M.J. and Coombs, G.H. (1996). Evidence from disruption of the lmcpb gene array of *Leishmania mexicana* that cysteine proteinases are virulence factors. *Proceedings of the National Academy of Sciences of the USA* **93**, 6008–6013.

Mottram, J.C., Frame, M.J., Brooks, D.R., Tetley, L., Hutchison, J.E., Souza, A.E. and Coombs, G.H. (1997). The multiple cpb cysteine proteinase genes of *Leishmania mexicana* encode isoenzymes that differ in their stage regulation and substrate preferences. *Journal of Biological Chemistry* **272**, 14285–14293.

Munoz, M.L., Calderon, J. and Rojkind, M. (1982). The collagenase of *Entamoeba histolytica*. *Journal of Experimental Medicine* **155**, 42–51.

Munoz, M.L., Lamoyi, E., Leon, G., Tovar, R., Perez-Garcia, J., De La Torre, M., Murueta, E. and Bernal, R.M. (1990). Antigens in electron-dense granules from *Entamoeba histolytica* as possible markers for pathogenicity. *Journal of Clinical Microbiology* **28**, 2418–2424.

Murray, P.J., Handman, E., Glaser, T.A. and Spithill, T.W. (1990). *Leishmania major*: expression and gene structure of the glycoprotein 63 molecule in virulent and avirulent clones and strains. *Experimental Parasitology* **71**, 294–304.

Murta, A.C., Persechini, P.M., Padron, T. de S., de Souza, W., Guimaraes, J.A. and Scharfstein, J. (1990). Structural and functional identification of GP57/51 antigen of *Trypanosoma cruzi* as a cysteine proteinase. *Molecular and Biochemical Parasitology* **43**, 27–38.

Neal, R.A. (1960). Enzymic proteolysis by *Entamoeba histolytica*: biochemical characteristics and relationship with invasiveness. *Parasitology* **50**, 531–550.

Neale, K.A. and Alderete, J.F. (1990). Analysis of the proteinases of representative *Trichomonas vaginalis* isolates. *Infection and Immunity* **58**, 157–162.

Nesterenko, M.V., Tilley, M. and Upton, S.J. (1995). A metallo-dependent cysteine proteinase of *Cryptosporidium parvum* associated with the surface of sporozoites. *Microbios* **83**, 77–88.

Neurath, H. (1989). The diversity of proteolytic enzymes. In: Proteolytic enzymes: a practical approach (Beynon, R.J. and Bond, J.S., eds.), pp. 1–13. Oxford: IRL Press.

North, M.J., Coombs, G.H. and Barry, J.D. (1983). A comparative study of the proteolytic enzymes of *Trypanosoma brucei*, *T. equiperdum*, *T. evansi*, *T. vivax*, *Leishmania tarentolae* and *Crithidia fasciculata*. *Molecular and Biochemical Parasitology* **9**, 161–180.

North, M.J., Robertson, C.D. and Coombs, G.H. (1990). The specificity of trichomonad cysteine proteinases analysed using fluorogenic substrates and specific inhibitors. *Molecular and Biochemical Parasitology* **39**, 183–193.

Nwagwu, M., Haynes, J.D., Orlandi, P.A. and Chulay, J.D. (1992). *Plasmodium falciparum*: chymotryptic-like proteolysis associated with a 101-kDa acidic-basic repeat antigen. *Experimental Parasitology* **75**, 399–414.

Oaks, S.C., Mitchell, V.S., Pearson, G.W. and Carpenter, C.C.J., eds. (1991). *Malaria: Obstacles and Opportunities*. Washington, D. C.: National Academy Press.
Okhuysen, P.C., DuPont, H.L., Sterling, C.R. and Chappell, C.L. (1994). Arginine aminopeptidase, an integral membrane protein of the *Cryptosporidium parvum* sporozoite. *Infection and Immunity* **62**, 4667–4670.
Okhuysen, P.C., Chappell, C.L., Kettner, C. and Sterling, C.R. (1996). *Cryptosporidium parvum* metalloaminopeptidase inhibitors prevent in vitro excystation. *Antimicrobial Agents and Chemotherapy* **40**, 2781–2784.
Olafson, R.W., Thomas, J.R., Ferguson, M.A., Dwek, R.A., Chaudhuri, M., Chang, K.P. and Rademacher, T.W. (1990). Structures of the N-linked oligosaccharides of gp63, the major surface glycoprotein, from *Leishmania mexicana amazonensis*. *Journal of Biological Chemistry* **265**, 12240–12247.
Olaya, P. and Wasserman, M. (1991). Effect of calpain inhibitors on the invasion of human erythrocytes by the parasite *Plasmodium falciparum*. *Biochemica et Biophysica Acta* **1096**, 217–221.
Olliaro, P., Cattani, J. and Wirth, D. (1996). Malaria, the submerged disease. *JAMA* **275**, 230–233.
Ortega, Y.R. and Adam, R.D. (1997). *Giardia*: overview and update. *Clinical Infectious Diseases* **25**, 545–549.
Ostoa-Saloma, P., Cabrera, N., Becker, I. and Perez-Montfort, R. (1989). Proteinases of *Entamoeba histolytica* associated with different subcellular fractions. *Molecular and Biochemical Parasitology* **32**, 133–143.
Pamer, E.G., So, M. and Davis, C.E. (1989). Identification of a developmentally regulated cysteine protease of *Trypanosoma brucei*. *Molecular and Biochemical Parasitology* **33**, 27–32.
Pamer, E.G., Davis, C.E., Eakin, A. and So, M. (1990). Cloning and sequencing of the cysteine protease cDNA from *Trypanosoma brucei rhodesiense*. *Nucleic Acids Research* **18**, 6141.
Pamer, E.G., Davis, C.E. and So, M. (1991). Expression and deletion analysis of the *Trypanosoma brucei rhodesiense* cysteine protease in *Escherichia coli*. *Infection and Immunity* **59**, 1074–1078.
Parenti, D.M. (1989). Characterization of a thiol proteinase in *Giardia lamblia*. *Journal of Infectious Disease* **160**, 1076–1080.
Perkins, M.E. (1989). Erythrocyte invasion by the malarial merozoite: recent advances. *Experimental Parasitology* **69**, 94–99.
Perkins, M.E. and Ziefer, A. (1994). Preferential binding of *Plasmodium falciparum* SERA and rhoptry proteins to erythrocyte membrane inner leaflet phospholipids. *Infection and Immunity* **62**, 1207–1212.
Petersen, C. (1992). Cryptosporidiosis in patients infected with the human immunodeficiency virus. *Clinical Infectious Diseases* **15**, 903–909.
Piras, M.M., Henriquez, D. and Piras, R. (1985). The effect of proteolytic enzymes and protease inhibitors on the interaction *Trypanosoma cruzi*-fibroblasts. *Molecular and Biochemical Parasitology* **14**, 151–163.
Pral, E.M., Bijovsky, A.T., Balanco, J.M. and Alfieri, S.C. (1993). *Leishmania mexicana*: proteinase activities and megasomes in axenically cultivated amastigote-like forms. *Experimental Parasitology* **77**, 62–73.
Pupkis, M.F. and Coombs, G.H. (1984). Purification and characterization of proteolytic enzymes of *Leishmania mexicana mexicana* amastigotes and promastigotes. *Journal of General Microbiology* **130**, 2375–2383.
Pupkis, M.F., Tetley, L. and Coombs, G.H. (1986). *Leishmania mexicana*: amastigote hydrolases in unusual lysosomes. *Experimental Parasitology* **62**, 29–39.

Ramamoorthy, R., Donelson, J.E., Paetz, K.E., Maybodi, M., Roberts, S.C. and Wilson, M.E. (1992). Three distinct RNAs for the surface protease gp63 are differentially expressed during development of *Leishmania donovani chagasi* promastigotes to an infectious form. *Journal of Biological Chemistry* **267**, 1888–1895.

Ramamoorthy, R., Swihart, K.G., McCoy, J.J., Wilson, M.E. and Donelson, J.E. (1995). Intergenic regions between tandem gp63 genes influence the differential expression of gp63 RNAs in *Leishmania chagasi* promastigotes. *Journal of Biological Chemistry* **270**, 12133–12139.

Ramos, M.A., Stock, R.P., Sanchez-Lopez, R., Olvera, F., Lizardi, P.M. and Alagon, A. (1997). The *Entamoeba histolytica* proteasome alpha-subunit gene. *Molecular and Biochemical Parasitology* **84**, 131–135.

Ravdin, J.I. (1995). Amebiasis. *Clinical Infectious Diseases* **20**, 1453–1464.

Rawlings, N.D. and Barrett, A.J. (1994a). Families of cysteine peptidases. *Methods in Enzymology* **244**, 461–486.

Rawlings, N.D. and Barrett, A.J. (1994b). Families of serine peptidases. *Methods in Enzymology* **244**, 19–61.

Rawlings, N.D. and Barrett, A.J. (1995a). Evolutionary families of metallopeptidases. *Methods in Enzymology* **248**, 183–228.

Rawlings, N.D. and Barrett, A.J. (1995b). Families of aspartic peptidases, and those of unknown catalytic mechanism. *Methods in Enzymology* **248**, 105–120.

Reed, S.L. and Gigli, I. (1990). Lysis of complement-sensitive *Entamoeba histolytica* by activated terminal complement components. Initiation of complement activation by an extracellular neutral cysteine proteinase. *Journal of Clinical Investigation* **86**, 1815–1822.

Reed, S.L., Keene, W.E. and McKerrow, J.H. (1989a). Thiol proteinase expression and pathogenicity of *Entamoeba histolytica*. *Journal of Clinical Microbiology* **27**, 2772–2777.

Reed, S.L., Keene, W.E., McKerrow, J.H. and Gigli, I. (1989b). Cleavage of C3 by a neutral cysteine proteinase of *Entamoeba histolytica*. *Journal of Immunology* **143**, 189–195.

Reed, S.L., Bouvier, J., Pollack, A.S., Engel, J.C., Brown, M., Hirata, K., Que, X., Eakin, A., Hagblom, P., Gillin, F. and McKerrow, J.H. (1993). Cloning of a virulence factor of *Entamoeba histolytica*. Pathogenic strains possess a unique cysteine proteinase gene. *Journal of Clinical Investigation* **91**, 1532–1540.

Reed, S.L., Ember, J.A., Herdman, D.S., DiScipio, R.G., Hugli, T.E. and Gigli, I. (1995). The extracellular neutral cysteine proteinase of *Entamoeba histolytica* degrades anaphylatoxins C3a and C5a. *Journal of Immunology* **155**, 266–274.

Ring, C.S., Sun, E., McKerrow, J.H., Lee, G.K., Rosenthal, P.J., Kuntz, I.D. and Cohen, F.E. (1993). Structure-based inhibitor design by using protein models for the development of antiparasitic agents. *Proceedings of the National Academy of Sciences of the USA* **90**, 3583–3587.

Rivett, A.J. (1993). Proteasomes: multicatalytic proteinase complexes. *Biochemical Journal* **291**, 1–10.

Roberts, S.C., Swihart, K.G., Agey, M.W., Ramamoorthy, R., Wilson, M.E. and Donelson, J.E. (1993). Sequence diversity and organization of the msp gene family encoding gp63 of *Leishmania chagasi*. *Molecular and Biochemical Parasitology* **62**, 157–171.

Robertson, C.D. and Coombs, G.H. (1990). Characterisation of three groups of cysteine proteinases in the amastigotes of *Leishmania mexicana mexicana*. *Molecular and Biochemical Parasitology* **42**, 269–276.

Robertson, C.D. and Coombs, G.H. (1993). Cathepsin B-like cysteine proteases of *Leishmania mexicana*. *Molecular and Biochemical Parasitology* **62**, 271–279.

Robertson, C.D. and Coombs, G.H. (1994). Multiple high activity cysteine proteases of *Leishmania mexicana* are encoded by the lmcpb gene array. *Microbiology* **140**, 417–424.

Robertson, C.D., North, M.J., Lockwood, B.C. and Coombs, G.H. (1990). Analysis of the proteinases of *Trypanosoma brucei*. *Journal of General Microbiology* **136**, 921–925.

Rockett, K.A., Playfair, J.H.L., Ashall, F., Targett, G.A.T., Angliker, H. and Shaw, E. (1990). Inhibition of intraerythrocytic development of *Plasmodium falciparum* by proteinase inhibitors. *FEBS Letters* **259**, 257–259.

Roggwiller, E., Bétoulle, M.E.M., Blisnick, T. and Braun Breton, C. (1996). A role for erythrocyte band 3 degradation by the parasite gp76 serine protease in the formation of the parasitophorous vacuole during invasion of erythrocytes by *Plasmodium falciparum*. *Molecular and Biochemical Parasitology* **82**, 13–24.

Roggwiller, E., Fricaud, A.-C., Blisnick, T. and Braun-Breton, C. (1997). Host urokinase-type plasminogen activator participates in the release of malaria merozoites from infected erythrocytes. *Molecular and Biochemical Parasitology* **86**, 49–59.

Rosales-Encina, J.L., Campos-Salazar, M.S. and Rojkind Matluk, M. (1992). *Entamoeba histolytica* collagen binding proteins. *Archives of Medical Research* **23**, 109–113.

Rosenthal, P.J. (1993). A *Plasmodium vinckei* cysteine proteinase shares unique features with its *Plasmodium falciparum* analogue. *Biochimica et Biophysica Acta* **1173**, 91–93.

Rosenthal, P.J. (1995). *Plasmodium falciparum*: Effects of proteinase inhibitors on globin hydrolysis by cultured malaria parasites. *Experimental Parasitology* **80**, 272–281.

Rosenthal, P.J. (1996). Conservation of key amino acids among the cysteine proteinases of multiple malarial species. *Molecular and Biochemical Parasitology* **75**, 255–260.

Rosenthal, P.J. and Meshnick, S.R. (1996). Hemoglobin catabolism and iron utilization by malaria parasites. *Molecular and Biochemical Parasitology* **83**, 131–139.

Rosenthal, P.J. and Nelson, R.G. (1992). Isolation and characterization of a cysteine proteinase gene of *Plasmodium falciparum*. *Molecular and Biochemical Parasitology* **51**, 143–152.

Rosenthal, P.J., Kim, K., McKerrow, J.H. and Leech, J.H. (1987). Identification of three stage-specific proteinases of *Plasmodium falciparum*. *Journal of Experimental Medicine* **166**, 816–821.

Rosenthal, P.J., McKerrow, J.H., Aikawa, M., Nagasawa, H. and Leech, J.H. (1988). A malarial cysteine proteinase is necessary for haemoglobin degradation by *Plasmodium falciparum*. *Journal of Clinical Investigation* **82**, 1560–1566.

Rosenthal, P.J., McKerrow, J.H., Rasnick, D. and Leech, J.H. (1989). *Plasmodium falciparum*: inhibitors of lysosomal cysteine proteinases inhibit a trophozoite proteinase and block parasite development. *Molecular and Biochemical Parasitology* **35**, 177–184.

Rosenthal, P.J., Wollish, W.S., Palmer, J.T. and Rasnick, D. (1991). Antimalarial effects of peptide inhibitors of a *Plasmodium falciparum* cysteine proteinase. *Journal of Clinical Investigation* **88**, 1467–1472.

Rosenthal, P.J., Lee, G.K. and Smith, R.E. (1993). Inhibition of a *Plasmodium*

vinckei cysteine proteinase cures murine malaria. *Journal of Clinical Investigation* **91**, 1052–1056.
Rosenthal, P.J., Ring, C.S., Chen, X. and Cohen, F.E. (1994). Characterization of a *Plasmodium vivax* cysteine proteinase gene identifies uniquely conserved amino acids that may mediate the substrate specificity of malarial hemoglobinases. *Journal of Molecular Biology* **241**, 312–316.
Rosenthal, P.J., Olson, J.E., Lee, G.K., Palmer, J.T., Klaus, J.L. and Rasnick, D. (1996). Antimalarial effects of vinyl sulfone cysteine proteinase inhibitors. *Antimicrobial Agents and Chemotherapy* **40**, 1600–1603.
Roth, E.F., Brotman, D.S., Vanderberg, J.P. and Schulman, S. (1986). Malarial pigment-dependent error in the estimation of hemoglobin content in *Plasmodium falciparum*-infected red cells: implications for metabolic and biochemical studies of the erythrocytic phases of malaria. *American Journal of Tropical Medicine and Hygiene* **35**, 906–911.
Russell, D.G. and Wilhelm, H. (1986). The involvement of the major surface glycoprotein (gp63) of *Leishmania* promastigotes in attachment to macrophages. *Journal of Immunology* **136**, 2613–2620.
Russo, D.C., Grab, D.J., Lonsdale-Eccles, J.D., Shaw, M.K. and Williams, D.J. (1993). Directional movement of variable surface glycoprotein-antibody complexes in *Trypanosoma brucei*. *European Journal of Cell Biology* **62**, 432–441.
Sakanari, J.A., Nadler, S.A., Chan, V.J., Engel, J.C., Leptak, C. and Bouvier, J. (1997). *Leishmania major*: comparison of the cathepsin L- and B-like cysteine protease genes with those of other trypanosomatids. *Experimental Parasitology* **85**, 63–76.
Salas, F., Fichmann, J., Lee, G.K., Scott, M.D. and Rosenthal, P.J. (1995). Functional expression of falcipain, a *Plasmodium falciparum* cysteine proteinase, supports its role as a malarial hemoglobinase. *Infection and Immunity* **63**, 2120–2125.
Salvesen, G. and Nagase, H. (1989). Inhibition of proteolytic enzymes. In: *Proteolytic Enzymes* (Beynon, R.J. and Bond, J.S., eds.), pp. 83–104. Oxford: IRL Press.
Santana, J.M., Grellier, P., Rodier, M.H., Schrevel, J. and Teixeira, A. (1992). Purification and characterization of a new 120 kDa alkaline proteinase of *Trypanosoma cruzi*. *Biochemical and Biophysical Research Communications* **187**, 1466–1473.
Santana, J.M., Grellier, P., Schrevel, J. and Teixeira, A.R. (1997). A *Trypanosoma cruzi*-secreted 80 kDa proteinase with specificity for human collagen types I and IV. *Biochemical Journal* **325**, 129–137.
Sato, K., Fukabori, Y. and Suzuki, M. (1987). *Plasmodium berghei*: a study of globinolytic enzyme in erythrocytic parasite. *Zentralblatt für Bakteriologie, Mikrobiologie, und Hygiene, Series A* **264**, 487–495.
Scheibel, L.W. and Sherman, I.W. (1988). Plasmodial metabolism and related organellar function during various stages of the life-cycle: proteins, lipids, nucleic acids and vitamins. In: *Malaria: Principles and Practice of Malariology* (Wernsdorfer, W.H. and McGregor, I., eds.), pp. 219–252. Edinburgh: Churchill Livingstone.
Schlagenhauf, E., Etges, R. and Metcalf, P. (1995). Crystallization and preliminary X-ray diffraction studies of leishmanolysin, the major surface metalloproteinase from *Leishmania major*. *Proteins* **22**, 58–66.
Schneider, P. and Glaser, T.A. (1993). Characterization of a surface metalloprotease from *Herpetomonas samuelpessoai* and comparison with *Leishmania major*

promastigote surface protease. *Molecular and Biochemical Parasitology* **58**, 277–282.

Schneider, P., Ferguson, M.A., McConville, M.J., Mehlert, A., Homans, S.W. and Bordier, C. (1990). Structure of the glycosyl-phosphatidylinositol membrane anchor of the *Leishmania major* promastigote surface protease. *Journal of Biological Chemistry* **265**, 16955–16964.

Schneider, P., Rosat, J.P., Bouvier, J., Louis, J. and Bordier, C. (1992). *Leishmania major*: differential regulation of the surface metalloprotease in amastigote and promastigote stages. *Experimental Parasitology* **75**, 196–206.

Scholze, H. and Schulte, W. (1988). On the specificity of a cysteine proteinase from *Entamoeba histolytica*. *Biomedica et Biochimica Acta* **47**, 115–123.

Scholze, H. and Tannich, E. (1994). Cysteine endopeptidases of *Entamoeba histolytica*. *Methods in Enzymology* **244**, 512–523.

Scholze, H. and Werries, E. (1984). A weakly acidic protease has a powerful proteolytic activity in *Entamoeba histolytica*. *Molecular and Biochemical Parasitology* **11**, 293–300.

Scholze, H., Lohden-Bendinger, U., Muller, G. and Bakker-Grunwald, T. (1992). Subcellular distribution of amebapain, the major cysteine proteinase of *Entamoeba histolytica*. *Archives of Medical Research* **23**, 105–108.

Scholze, H., Frey, S., Cejka, Z. and Bakker-Grunwald, T. (1996). Evidence for the existence of both proteasomes and a novel high molecular weight peptidase in *Entamoeba histolytica*. *Journal of Biological Chemistry* **271**, 6212–6216.

Schrevel, J., Grellier, P., Mayer, R. and Monsigny, M. (1988). Neutral proteases involved in the reinvasion of erythrocytes by *Plasmodium* merozoites. *Biology of the Cell* **64**, 233–244.

Schulte, W. and Scholze, H. (1989). Action of the major protease from *Entamoeba histolytica* on proteins of the extracellular matrix. *Journal of Protozoology* **36**, 538–543.

Scott, D.A., North, M.J. and Coombs, G.H. (1995a). The pathway of secretion of proteinases in *Trichomonas vaginalis*. *International Journal of Parasitology* **25**, 657–666.

Scott, D.A., North, M.J. and Coombs, G.H. (1995b). *Trichomonas vaginalis*: amoeboid and flagellated forms synthesize similar proteinases. *Experimental Parasitology* **80**, 345–348.

Seay, M.B., Heard, P.L. and Chaudhuri, G. (1996). Surface Zn-proteinase as a molecule for defense of *Leishmania mexicana amazonensis* promastigotes against cytolysis inside macrophage phagolysosomes. *Infection and Immunity* **64**, 5129–5137.

Seemüller, E., Lupas, A., Stock, D., Löwe, J., Huber, R. and Baumeister, W. (1995). Proteasome from *Thermoplasma acidophilum*: a threonine protease. *Science* **268**, 579–582.

Selzer, P.M., Chen, X., Chan, V.J., Cheng, M., Kenyon, G.L., Kuntz, I.D., Sakanari, J.A., Cohen, F.E. and McKerrow, J.H. (1997). *Leishmania major*: molecular modeling of cysteine proteases and prediction of new nonpeptide inhibitors. *Experimental Parasitology* **87**, 212–221.

Serveau, C., Lalmanach, G., Juliano, M.A., Scharfstein, J., Juliano, L. and Gauthier, F. (1996). Investigation of the substrate specificity of cruzipain, the major cysteine proteinase of *Trypanosoma cruzi*, through the use of cystatin-derived substrates and inhibitors. *Biochemical Journal* **313**, 951–956.

Sherman, I.W. and Mudd, J.B. (1966). Malaria infection (*Plasmodium lophurae*): changes in free amino acids. *Science* **154**, 287–289.

Sherman, I.W. and Tanigoshi, L. (1970). Incorporation of 14C-amino acids by malaria (*Plasmodium lophurae*) IV. In vivo utilization of host cell hemoglobin. *International Journal of Biochemistry* **1**, 635–637.

Sherman, I.W. and Tanigoshi, L. (1983). Purification of *Plasmodium lophurae* cathepsin D and its effects on erythrocyte membrane proteins. *Molecular and Biochemical Parasitology* **8**, 207–226.

Silva, A.M., Lee, A.Y., Gulnik, S.V., Majer, P., Collins, J., Bhat, T.N., Collins, P.J., Cachau, R.E., Luker, K.E., Gluzman, I.Y., Francis, S.E., Oksman, A., Goldberg, D.E. and Erickson, J.W. (1996). Structure and inhibition of plasmepsin II, a haemoglobin-degrading enzyme from *Plasmodium falciparum*. *Proceedings of the National Academy of Sciences of the USA* **93**, 10034–10039.

Slater, A.F.G. (1992). Malaria pigment. *Experimental Parasitology* **74**, 362–365.

Sogin, M.L., Gunderson, J.H., Elwood, H.J., Alonso, R.A. and Peattie, D.A. (1989). Phylogenetic meaning of the kingdom concept: an unusual ribosomal RNA from *Giardia lamblia*. *Science* **243**, 75–77.

Souto-Padron, T., Campetella, O.E., Cazzulo, J.J. and de Souza, W. (1990). Cysteine proteinase in *Trypanosoma cruzi*: immunocytochemical localization and involvement in parasite-host cell interaction. *Journal of Cell Science* **96**, 485–490.

Souza, A.E., Waugh, S., Coombs, G.H. and Mottram, J.C. (1992). Characterization of a multi-copy gene for a major stage-specific cysteine proteinase of *Leishmania mexicana*. *FEBS Letters* **311**, 124–127.

Souza, A.E., Bates, P.A., Coombs, G.H. and Mottram, J.C. (1994). Null mutants for the lmcpa cysteine proteinase gene in *Leishmania mexicana*. *Molecular and Biochemical Parasitology* **63**, 213–220.

Stanley, S.L.J., Zhang, T., Rubin, D. and Li, E. (1995). Role of the *Entamoeba histolytica* cysteine proteinase in amebic liver abscess formation in severe combined immunodeficient mice. *Infection and Immunity* **63**, 1587–1590.

Steinkraus, H.B. and Langer, P.J. (1992). The protein sequence predicted from a *Leishmania guyanensis* gp63 major surface glycoprotein gene is divergent as compared with other *Leishmania* species. *Molecular and Biochemical Parasitology* **52**, 141–144.

Stoka, V., Nycander, M., Lenarcic, B., Labriola, C., Cazzulo, J.J., Bjork, I. and Turk, V. (1995). Inhibition of cruzipain, the major cysteine proteinase of the protozoan parasite, *Trypanosoma cruzi*, by proteinase inhibitors of the cystatin superfamily. *FEBS Letters* **370**, 101–104.

Tannich, E., Scholze, H., Nickel, R. and Horstmann, R.D. (1991). Homologous cysteine proteinases of pathogenic and nonpathogenic *Entamoeba histolytica*. Differences in structure and expression. *Journal of Biological Chemistry* **266**, 4798–4803.

Tannich, E., Nickel, R., Buss, H. and Horstmann, R.D. (1992). Mapping and partial sequencing of the genes coding for two different cysteine proteinases in pathogenic *Entamoeba histolytica*. *Molecular and Biochemical Parasitology* **54**, 109–111.

Theakston, R.D.G., Fletcher, S.A. and Maegraith, B.G. (1970). The use of electron microscope autoradiography for examining the uptake and degradation of haemoglobin by *Plasmodium berghei*. *Annals of Tropical Medicine & Parasitology* **64**, 63–71.

To, W.Y. and Wang, C.C. (1997). Identification and characterization of an activated 20S proteasome in *Trypanosoma brucei*. *FEBS Letters* **404**, 253–262.

Tomas, A.M. and Kelly, J.M. (1996). Stage-regulated expression of cruzipain, the major cysteine protease of *Trypanosoma cruzi* is independent of the level of RNA1. *Molecular and Biochemical Parasitology* **76**, 91–103.

Tomas, A.M., Miles, M.A. and Kelly, J.M. (1997). Overexpression of cruzipain, the major cysteine proteinase of *Trypanosoma cruzi*, is associated with enhanced metacyclogenesis. *European Journal of Biochemistry* **244**, 596–603.

Tran, V.Q., Herdman, D.S., Torian, B.E. and Reed, S.L. (1998). The neutral cysteine proteinase of *Entamoeba histolytica* degrades IgG and prevents its binding. *Journal of Infectious Diseases* **177**, 508–511.

Traub-Cseko, Y.M., Duboise, M., Boukai, L.K. and McMahon-Pratt, D. (1993). Identification of two distinct cysteine proteinase genes of *Leishmania pifanoi* axenic amastigotes using the polymerase chain reaction. *Molecular and Biochemical Parasitology* **57**, 101–115.

Troeberg, L., Pike, R.N., Morty, R.E., Berry, R.K., Coetzer, T.H. and Lonsdale-Eccles, J.D. (1996). Proteases from *Trypanosoma brucei brucei*. Purification, characterisation and interactions with host regulatory molecules. *European Journal of Biochemistry* **238**, 728–736.

Turk, B., Turk, V. and Turk, D. (1997). Structural and functional aspects of papain-like cysteine proteinases and their protein inhibitors. *Biological Chemistry* **378**, 141–150.

Tzinia, A.K. and Soteriadou, K.P. (1991). Substrate-dependent pH optima of gp63 purified from seven strains of *Leishmania*. *Molecular and Biochemical Parasitology* **47**, 83–89.

Vander Jagt, D.L., Baack, B.R. and Hunsaker, L.A. (1984). Purification and characterization of an aminopeptidase from *Plasmodium falciparum*. *Molecular and Biochemical Parasitology* **10**, 45–54.

Vander Jagt, D.L., Caughey, W.S., Campos, N.M., Hunsaker, L.A. and Zanner, M.A. (1989). Parasite proteases and antimalarial activities of protease inhibitors. *Progress in Clinical and Biological Research* **313**, 105–118.

Vander Jagt, D.L., Hunsaker, L.A., Campos, N.M. and Scaletti, J.V. (1992). Localization and characterization of hemoglobin-degrading aspartic proteinases from the malarial parasite *Plasmodium falciparum*. *Biochemica et Biophysica Acta* **1122**, 256–264.

Walsh, J.A. (1989). Disease problems in the Third World. *Annals of the New York Academy of Sciences* **569**, 1–16.

Ward, W., Alvarado, L., Rawlings, N.D., Engel, J.C., Franklin, C. and McKerrow, J.H. (1997). A primitive enzyme for a primitive cell: the protease required for excystation of *Giardia*. *Cell* **89**, 437–444.

Webb, J.R., Button, L.L. and McMaster, W.R. (1991). Heterogeneity of the genes encoding the major surface glycoprotein of *Leishmania donovani*. *Molecular and Biochemical Parasitology* **48**, 173–184.

Weber, J.L., Lyon, J.A., Wolff, R.H., Hall, T., Lowell, G.H. and Chulay, J.D. (1988). Primary structure of a *Plasmodium falciparum* malaria antigen located at the merozoite surface and within the parasitophorous vacuole. *Journal of Biological Chemistry* **263**, 11421–11425.

Werries, E., Franz, A., Hippe, H. and Acil, Y. (1991). Purification and substrate specificity of two cysteine proteinases of *Giardia lamblia*. *Journal of Protozoology* **38**, 378–383.

Westling, J., Yowell, C.A., Majer, P., Erickson, J.W., Dame, J.B. and Dunn, B.M. (1997). *Plasmodium falciparum*, *P. vivax*, and *P. malariae*: a comparison of the active site properties of plasmepsins cloned and expressed from three different species of the malaria parasite. *Experimental Parasitology* **87**, 185–193.

Wilson, M.E., Hardin, K.K. and Donelson, J.E. (1989). Expression of the major surface glycoprotein of *Leishmania donovani chagasi* in virulent and attenuated promastigotes. *Journal of Immunology* **143**, 678–684.

Wilson, M.E., Paetz, K.E., Ramamoorthy, R. and Donelson, J.E. (1993). The effect of ongoing protein synthesis on the steady state levels of Gp63 RNAs in *Leishmania chagasi*. *Journal of Biological Chemistry* **268**, 15731–15736.

Wiser, M.F. (1995). Proteolysis of a 34 kDa phosphoprotein coincident with a decrease in protein kinase activity during the erythrocytic schizont stage of the malaria parasite. *Journal of Eukaryotic Microbiology* **42**, 659–664.

Yokoyama-Yasunaka, J.K., Pral, E.M., Oliveira Junior, O.C., Alfieri, S.C. and Stolf, A.M. (1994). *Trypanosoma cruzi*: identification of proteinases in shed components of trypomastigote forms. *Acta Tropica* **57**, 307–315.

Zarchin, S., Krugliak, M. and Ginsburg, H. (1986). Digestion of the host erythrocyte by malaria parasites is the primary target for quinoline-containing antimalarials. *Biochemical Pharmacology* **35**, 2435–2442.

Proteinases and Associated Genes of Parasitic Helminths

Jose Tort[1], Paul J. Brindley[2], Dave Knox[3], Kenneth H. Wolfe[4] and John P. Dalton[1]

[1] *School of Biological Sciences, Dublin City University, Dublin 9, Republic of Ireland;*
[2] *Molecular Parasitology Unit, and Australian Centre for International & Tropical Health & Nutrition, Queensland Institute of Medical Research, Post Office, Royal Brisbane Hospital, Queensland 4029, Australia;*
[3] *Moredun Research Institute, 408 Gilmerton Road, Edinburgh EH17 7JH;*
[4] *Department of Genetics, University of Dublin, Trinity College, Dublin 2, Republic of Ireland*

Abstract 162
1. Introduction 163
2. Digenean Trematodes 164
3. Proteinases of Schistosomes 165
   3.1. Proteinases of eggs and miracidia 165
   3.2. Proteinases of cercariae 169
   3.3. Proteinases of adults and schistosomules 174
4. Proteinases of *Fasciola hepatica* and Other Fasciolidae 190
   4.1. Cathepsin L 190
   4.2. Cathepsin B 192
   4.3. Dipeptidylpeptidase (DPP) and leucine aminopeptidase (LAP) 193
   4.4. Other proteinases of *F. hepatica* 193
   4.5. Proteinases of other Fasciolidae (*F. gigantica*) 194
5. Proteinases of Other Flukes 194
   5.1. *Paragonimus westermani* 194
   5.2. *Haplometra cylindracea* 195
   5.3. *Clonorchis sinensis* 196
   5.4. *Diplostomum pseudopathaceum* 196
6. Proteinases of Cestodes 196
   6.1. *Taenia* 196
   6.2. *Echinococcus* 198

        6.3. *Proteocephalus*.. 199
        6.4. *Schistocephalus*.. 199
        6.5. *Spirometra*... 199
    7. Proteinases of Parasitic Nematodes.. 200
        7.1. Rhabditoidea.. 201
        7.2. Ascaridoidea 201
        7.3. Filaroidea 209
        7.4. Strongyloidea 211
        7.5. Trichostrongyloidea... 214
        7.6. Trichuroidea... 221
        7.7. Parasitic nematodes of plants and *Caenorhabditis elegans* 223
    8. Phylogenetic Analysis of Cysteine Proteinases of the Papain Superfamily..... 225
        8.1. Introduction 225
        8.2. The papain superfamily 226
        8.3. The ERFNIN motif 230
        8.4. Branch A of the papain superfamily... 237
        8.5. Branch B of the papain superfamily.. 240
    9. Concluding Remarks.. 244
    References.. 247

ABSTRACT

Many parasites have deployed proteinases to accomplish some of the tasks imposed by a parasitic life style, including tissue penetration, digestion of host tissue for nutrition and evasion of host immune responses. Information on proteinases from trematodes, cestodes and nematode parasites is reviewed, concentrating on those worms of major medical and economical importance. Their biochemical characterization is discussed, along with their putative biological roles and, where available, their associated genes. For example, proteinases expressed by the various stages of the schistosome life-cycle, in particular the well-characterized cercarial elastase which is involved in the penetration of the host skin and the variety of proteinases, such as cathepsin B (Sm31), cathepsin L1, cathepsin L2, cathepsin D, cathepsin C and legumain (Sm32), which are believed to be involved in the catabolism of host haemoglobin. The various endo- and exoproteinases of *Fasciola hepatica*, the causative agent of liver fluke disease, are reviewed, and recent reports of how these enzymes have been successfully employed in cocktail vaccines are discussed. The various proteinases of cestodes and of the diverse superfamilies of parasitic nematodes are detailed, with special attention being given to those parasites for which most is known, including species of *Taenia, Echinococcus, Spirometra, Necator, Acylostoma* and *Haemonchus*.

By far the largest number of papers in the literature and entries to the sequence data bases dealing with proteinases of parasitic helminths report on enzymes belonging to the papain superfamily of cysteine proteinases. Accordingly, the final section of the review is devoted to a phylogenetic analysis of this superfamily using over 150 published sequences. This analysis shows that the papain superfamily can be divided into two major branches. Branch A contains the cathepin Bs, the cathepsin Cs and a novel family termed cathepsin Xs, while Branch B contains the cruzipains, cathepsin Ls, papain-like and aleurain/cathepsin H-like proteinases. The relationships of the helminth proteinases, and similar proteinases from protozoan parasites and other organisms, within these groups are discussed.

1. INTRODUCTION

Parasitism has arisen many times in the evolution of diverse species, but the challenges encountered by all parasitic organisms are in certain ways similar. The first challenge for a parasite is to invade its host. Once on or inside the host, it has to find its final destination — a specific cell type, organ or vessel. This process often involves traversing host tissue, extracellular matrix, basement membranes and blood or lymph vessel walls. Throughout its life the parasite utilizes resources produced by the hosts. At the same time, it must overcome the physico-chemical, immunological and behavioural barriers exhibited by the host.

The methods different parasites have evolved to achieve these objectives are diverse, and provide fascinating examples of the plasticity of essential physiological mechanisms. On the other hand, the range of biochemical and molecular mechanisms available for adaption to parasitism is finite, and so we observe the evolution of analogous strategies in taxonomically unrelated parasites to deal with the same problem. In this respect it is not surprising to find that peptidases, the major enzymes involved in cellular catabolic processes, have been deployed to accomplish some of the tasks imposed by a parasitic life style, including tissue penetration, digestion of host tissue for nutrition and evasion of host immune responses.

Proteolytic enzymes must have originated very early in evolution, since all organisms require them for digestion and metabolism of their own proteins. In the early days of life, a limited set of peptidases, probably with a broad substrate specificity, could have accounted for this activity (Jensen, 1976). It is reasonable to assume that from this limited set of ancestral enzymes a complex series of proteinases evolved by a process of gene duplication and divergence. Subsequent to this divergence, the enzymes would have acquired a higher degree of specificity by tailoring their catalysis to a discrete suite of peptide bonds located at specific sites in protein substrates. At the same time, the expression and distribution of these new proteinases could also be restricted, generating specialized proteinases for specific purposes (Neurath, 1984; Creighton and Darby, 1989).

The diversification of proteinases has undoubtedly contributed to the success of the parasitic helminths. Here we review the nature and functions so far ascribed to these enzymes characterized from parasitic helminths. The main selection pressure that initiated the parasite–host association might have been the ease by which nutrients could be acquired from the host by the parasite (Halton, 1997); hence we find that much of the predominant proteolytic activity in helminth parasites is involved in this function. However, the development of mechanisms to facilitate the parasites' migration through host tissue, defend against host immunological attack and otherwise ensure the completion of their life cycles would obviously have presented other selection pressures. Energy saved by the proximity and abundance of food was re-directed and employed in the development and diversification of structures and processes to deal with these new obstacles. It is not unlikely that the diversification and specialization of proteinases has played a major role in the evolution of these processes.

In this paper we review information on proteinases and their associated genes from parasitic helminths. We have concentrated on proteinases from helminths of medical and economical importance, because more information is available about them compared to other species. More specifically, we deal with proteinases of digenean trematodes, in particular of schistosomes and *Fasciola hepatica*, from cestodes and from nematodes parasitizing both animals and plants. Interestingly, the scientific literature and the sequence databases contains a surprisingly large number of reports dealing with proteinases of parasitic helminths. It turns out, however, that although proteinases of all the four major groups of peptidases — serine, aspartic, cysteine and metalloproteinases — have been characterized in parasitic helminths, by far the largest number reported up to now belong to the papain superfamily of cysteine proteinases. Accordingly, we devote the final section of our review to a phylogenetic analysis of this group of proteinases from parasitic worms. These are compared and contrasted with each other and with similar proteinases from protozoan parasites and from other organisms, and the phylogenetic relationships of the papain-like enzymes are interpreted and discussed.

2. DIGENEAN TREMATODES

Most of the information available about the molecular biology and biochemistry of proteinases is known from the human schistosomes and from the liver fluke *Fasciola hepatica*; hence, we have concentrated on them. However, in the past few years reports have appeared about genes encoding proteinases from several other genera, including *Clonorchis* and *Paragonimus*.

3. PROTEINASES OF SCHISTOSOMES

Schistosomiasis is the most important of the human helminthiases in terms of morbidity and mortality. It is endemic in more than 70 tropical and subtropical countries; about one in every 30 humans are infected and more than 600 million others are at risk of infection (World Health Organization, 1996). Although targeted chemotherapy and other public health measures are employed to control schistosomiasis and its spread, there is a need for a vaccine and for the development of new anti-schistosome drugs. Because of the key roles they play in key aspects of the schistosome biology, including the hatching of the egg, the penetration of the skin of the mammalian host, immunoresistance and nutrition (see McKerrow and Doenhoff, 1988; Dalton *et al.*, 1995a), proteolytic enzymes of these and other parasites are considered potential targets at which to develop and direct anti-schistosomal therapies (Ring *et al.*, 1993; Li *et al.*, 1996; Wasilewski *et al.*, 1996; Brindley *et al.*, 1997).

A decade has passed since McKerrow and Doenhoff (1988) wrote an informative review of schistosome proteinases. Since that time, because of the widespread application of gene cloning technologies, including the availability of the polymerase chain reaction (PCR) technique, genes encoding a number of the enzyme activities referred to by McKerrow and Doenhoff (1988) have been characterized, and recombinant enzymes are now available. Accordingly, it appears timely to review the current status of knowledge on the proteinases of schistosomes, including aspects of their structure, function and utility in respect of diagnosis and as targets for anti-schistosomal interventions. More recently, we have written reviews of schistosome proteinases that participate in nutrition, particularly in the digestion of haemoglobin obtained from ingested red blood cells (RBC) (Dalton *et al.*, 1995a; Brindley *et al.*, 1997). In this section of our review we focus on a number of proteinases from the different life cycle stages of schistosomes for which a gene has been cloned and/or for which the recombinant form of the proteinase has been characterized.

3.1. Proteinases of Eggs and Miracidia

An array of proteolytic enzyme activities belonging to the cysteine, metallo- and serine classes of proteinases have been reported from schistosome eggs, although the genes encoding these activities have yet to be characterized. Asch and Dresden (1979) reported the presence of a 25–27 kDa proteolytic activity in eggs of *Schistosoma mansoni* that required thiols and EDTA for maximum activity. Soon after, Dresden and co-workers reported the isolation and characterization of an IgG1 anti-cysteine proteinase monoclonal Ab (mAb 1.1A7A) that inhibited the activity of the enzyme against the cathepsin B-specific substrate CBZ-Arg-Arg-AFC (Dresden *et al.*, 1983). Subsequently, Sung and Dresden (1986) partially purified three cysteine proteinase activities from *S. mansoni* eggs, all of which had a pH optimum for activity against CBZ-Arg-Arg-AFC at pH 5.5 and which were inhibited by the cysteine proteinase inhibitor L-*trans*-epoxysuccinyl-leucylamido(4-guanidino) butane (E-64) (Barrett *et al.*, 1982). Based on size exclusion chromatography, the molecular sizes of the two most prominent of these three cysteine proteinases were reported as 25 300 daltons and 30 500 daltons. The 25 300 dalton activity was apparently the proteinase reported previously by Asch and Dresden (1979) and the cognate antigen of mAb1.1A7A.

Using native polyacrylamide gels which were probed with the synthetic substrate Z-Phe-Arg-NHMec (-amino methyl coumarin), Day *et al.* (1995) showed that soluble extracts of eggs contained cathepsin L activity with physical characteristics that were the same or very similar to those of adult *S. japonicum* (see Section 3.3.3, p. 180). They suggested that this activity might contribute to the liver pathogenesis associated with schistosomiasis, either directly through its proteolytic activity or through immunological responses directed against the proteinase(s). Sung and Dresden (1986) suggested a similar role in pathology for cathepsin B-like cysteine proteinase activities secreted by eggs of *S. mansoni*. Pino-Heiss *et al.* (1985) also detected a cysteine proteinase activity in eggs and miracidia of *S. mansoni* that was capable of digesting glycoproteins but not collagen or elastin. They suggested that eggs and miracidia may employ these enzymes to facilitate passage through the wall of the intestines or bladder, and that the miracidium may use the activity to penetrate the smooth muscle, extracellular matrix of the snail.

A novel cysteine proteinase activity, which specifically hydrolyses Z-Phe-Arg-NHMec, has been reported from miracidia and sporocysts of *S. mansoni* by Yoshino *et al.* (1993). Although the activity was not ascribed to a cathepsin L, and although the sizes of the partially purified enzymes (19 kDa and 36 kDa) do not conform with that predicted for the processed cathepsin L of adult *S. mansoni* ($\sim$ 24 kDa), it is not unlikely that schistosome miracidia and sporocysts may also employ cathepsins L. Yoshino *et al.* (1993) suggested that the miracidial cysteine proteinase(s) might play a role in the penetration and infection of the intermediate snail host.

Leucine aminopeptidase activity (LAP; EC 3.4.11.1), a metalloproteinase, has been reported from schistosomes, predominantly in secretions of *S. mansoni* eggs and hatching fluid, but also in other life cycle stages of *S. mansoni*, *S. haematobium* and *S. rodhani* (Fripp, 1967; Coles, 1970; Auriault et al., 1982; Bogitsh, 1983; Xu and Dresden, 1986). Bogitsh (1983) demonstrated the presence of LAP activity in the germ cell of *S. mansoni* eggs, in the epidermis of the miracidium within the egg and also within the vitelline membrane surrounding the miracidium. Xu and Dresden (1986) showed that this egg LAP activity, measured using H-Leu-NHMec, possessed a pH optimum of 7.2 and was inhibited with bestatin. They suggested that LAP plays a role in the hatching of the schistosome egg, possibly through the degradation of the eggshell to allow the release of the miracidium. The LAP of adult *S. mansoni* is also optimally active at neutral pH and shows a dependence on metal ions.

Doenhoff and co-workers have recently reported a schistosome antigen named Sm480, in reference to its apparent molecular size as determined by size exclusion chromatography, that has serine proteinase activity (Curtis et al., 1996; Doenhoff, 1998). The Sm480 antigen was immunoprecipitated from soluble homogenates of *S. mansoni* eggs using a mouse infection serum and, subsequently, with rabbit antiserum. The immunoprecipitated Sm480 exhibited peptidolytic activity against the chromogenic substrate N-acetyl-DL-phenylalanine beta-naphthyl ester (NAPBNE; Doenhoff et al., 1988) at pH 7.2. The activity against NAPBNE was inhibited by pre-incubation with chymotrypsin-like, but not by trypsin-like, inhibitors. Sm480 could be radiolabelled with tritiated di-isopropyl fluorophosphate, a general inhibitor of serine proteinases. Antigens cross-reactive with Sm480 were detected in miracidia, cercariae and adult schistosomes, although the peptidolytic activity of Sm480 was not present in the extracts of adult schistosomes. Binding to concanavalin A-Sepharose indicated that Sm480 was glycosylated (Curtis et al., 1996).

Although the genes encoding these various proteinase activities in eggs have yet to be characterized, it is not unlikely that the genes encoding egg proteinases may also be expressed in other life cycle stages, based on similarities of enzyme activities. Thus, for example, cathepsin L-like activity with similar electrophoretic mobility and substrate profile is present both in soluble extracts of adults and of eggs of *S. japonicum* (Day et al., 1995) and in cercariae, schistosomules and adults of *S. mansoni* (Dalton et al., 1997). Indeed, whereas complete cDNA sequences encoding cathepsin L and cathepsin D have been characterized from adult schistosomes (Table 1), partial cDNA sequences have been isolated for both of these from egg stage cDNA libraries as part of the continuing World Health Organization-sponsored *Schistosoma* Genome Initiative (Johnston, 1997).

Table 1 Schistosome proteinases for which a gene and/or cDNA has been reported in the literature.

Proteinase	Synonym	Protein class (Family)	Stage/site of expression	Function	Genbank Accession Number	Key references
Cercarial elastase	m28	Serine	Sporocyst; acetabular glands of cercaria; somule surface	Skin penetration. Removal of glycocalyx. Immunoresistance.	J03946 U31768 U31769	Newport et al., 1988 Pierrot et al., 1995 Price et al., 1997
Calpain EC 3.4.22.17	Calcium-activated neutral proteinase (CANP)	Cysteine	Adults, sporocyst cytosol	Cytoplasmic proteinase. May be involved in membrane synthesis.	M74233 AFO44 407	Andresen et al., 1991 Karcz et al., 1991
Cathepsin B EC 3.4.22.1	Sm31/Sj31	Cysteine (papain)	Gastrodermis	Haemoglobin proteolysis?	M21309 X70968	Klinkert et al., 1987 Merckelbach et al., 1994 Lipps 1996
Legumain EC 3.4.22.34	Sm32/Sj32, asparaginyl endopepidase	Cysteine (legumain)	Gastrodermis cecum	Processing of proteinases?	M17423 M21308 X70967	Davis et al., 1987 El Menawey et al., 1990 Merckelbach et al., 1994 Dalton and Brindley, 1998
Cathepsin L1 EC 3.4.22.15	SmCL1/SjCL1	Cysteine (papain)	Egg, cercaria, adult	Haemoglobin proteolysis?	U07345 U38475	Smith et al., 1994 Day et al., 1995
Cathepsin L2 EC 3.4.22.15	SmCL2/SjCL2	Cysteine (papain)	Ovaries, testis	Egg shell synthesis; seminal fluid component.	Z32529 U38476	Michel et al., 1995 Day et al., 1995
Cathepsin D EC 3.4.23.5	Sjasp/Smasp	Aspartic	Gastrodermis of adults, egg	Haemoglobin proteolysis	L41346 U60995 U90750	Becker et al., 1995 Wong et al., 1997 Bogitsh et al., 1986
Cathepsin C EC 3.4.14.1	Dipeptidylpeptidase I (DPP I)	Cysteine (papain)	Gastrodermis	Haemoglobin proteolysis	Z32531 U77932	Butler et al., 1996 Hola-Jamriska et al., 1998 Bogitsh et al., 1983
ER 60 proteinase	N11 protein	Cysteine	Lumen of ER, protonephidia, gastrodermis	Pre-Golgi protein degradation	Z22934	Finken-Eigen et al., 1996
Kallikrein-like EC 3.4.21.35?	SmSP1	Serine	Cercaria, adult	Vasodilation?		Cocude et al., 1997

3.2. Proteinases of Cercariae

Schistosome cercariae infect the mammalian host by penetration of the skin, which involves exploration of the outer squamous layers and then entry and migration through the epidermal and dermal layers (Stirewalt, 1974). Concurrently, the cercariae lose their glycocalyx and transform into schistosomules. Although the glycocalyx plays a vital role in protecting the cercariae against the hypo-osmotic environment of fresh water, it activates the complement cascade system making cercariae and newly transformed schistosomula susceptible to lysis by the alternative pathway, and hence must be shed rapidly after skin penetration (Stirewalt, 1974; Minard et al., 1977; Fishelson, 1989; Marikovsky et al., 1988b; McKerrow et al., 1991; Dalton et al., 1997).

The cercarial acetabular glands, the contents of which are released via ducts to the exterior following stimulation with skin lipids, contained multiple proteinases (Stirewalt, 1974; Minard et al., 1997). These activities were described as gelatinase-like or serine-like, but dissimilar to trypsin or chymotrypsin, particularly in their activity towards elastin. Since these findings, a number of serine proteinases have been characterized, including those migrating in polyacrylamide gels at 25 kDa (Landsperger et al., 1982), 27 kDa (Darani et al., 1997), 28 kDa (Marikovsky et al., 1988b), 30 kDa (McKerrow et al., 1985), 47 kDa (Chavez-Olortegui et al., 1992) and 60 kDa (Marikovsky et al., 1988b). Although the 28 and 30 kDa proteinases were initially thought to be separate enzymes because they differed in glycosylation, molecular size and dependence on metal ions for activity (Marikovsky et al., 1988b), they may be differentially processed forms of the same gene product (McKerrow et al., 1991; Fishelson et al., 1992). Moreover, the 25 kDa proteinase was suggested be a post-translational derivative of the 28/30 kDa enzyme (McKerrow et al., 1985, 1991). Although the 28/30 kDa and 47 kDa enzymes have similar high-pH optima for activity (pH 9–10) and cleave native elastin and collagen, monoclonal antibodies prepared against the 47 kDa enzyme did not react with the 28/30 kDa proteinase (Chavez-Olortegui et al., 1992). The 28/30 kDa and 60 kDa proteinases were separately isolated by Marikovsky et al. (1988b) from cercarial secretions using ion exchange followed by gel filtration chromatography; the 28/30 kDa proteinase represented 90% of the total caseinolytic/gelatinlytic activity in the cercarial secretions and showed distinct biochemical properties, including discrete susceptibility to inhibitors and detergents, activation by calcium and isoelectric point, to the 60 kDa proteinase.

In contrast to serine-like proteinases, other classes of proteolytic enzymes of schistosome cercariae have not been located or studied comprehensively. However, Dalton et al. (1997) recently reported cathepsin L-like and B-like cysteine proteinase activities in extracts of *S. mansoni* cercariae. As in adult schistosomes (Smith *et al.*, 1994; Day *et al.*, 1995), cathepsin L activity predominated compared to cathepsin B in soluble, acidic extracts of cercariae. Immunolocalization studies showed that both cathepsin L and B proteinases were present in the post-acetabular glands. Unlike cercarial elastase and other serine proteinases (see below), which are expressed only or predominantly in the sporocyst stage of the schistosome, expression of the cathepsin L and B proteinases continues after the cercariae transform into schistosomules within the definitive host (Dalton *et al.*, 1997).

Schistosomatium douthitti, a schistosome that infects small mammals and causes 'swimmers' itch' dermatitis in humans appears to use a proteolytic enzyme to facilitate infection of its definitive hosts, in a fashion similar to the human schistosomes of the genus *Schistosoma*. However, unlike the cercarial elastase from *S. mansoni*, which is a serine proteinase (see below — cercarial elastase), the *Schistosomatium douthitti* enzyme is a metallo-proteinase. This enzyme is about 50 kDa in size and has a pH optimum of 9 and a dependence on calcium ions for activity (Amiri *et al.*, 1988).

3.2.1. *Cercarial Elastase*

Of all the schistosome proteinases reported to date, the 28/30 kDa serine proteinase from cercariae is perhaps the best characterized, and has become known as cercarial elastase. Both skin penetration and glycocalyx shedding are believed to be facilitated by the activities of cercarial elastase (Marikovsky *et al.*, 1990a,b; McKerrow *et al.*, 1991; Fishelson *et al.*, 1992). Furthermore, a surface-localized, schistosomular serine proteinase with characteristics similar to cercarial elastase has been implicated in immunomodulatory roles, including the cleavage of host complement components (Marikovsky *et al.*, 1988a,b, 1990a,b) and of host IgG and IgE bound via Fc receptors to the their surface (Auriault *et al.*, 1981; Verwaerde *et al.*, 1988).

Cercarial elastase was originally purified from excretions of cercariae, with the excretions stimulated by skin lipid or linoleic acid (McKerrow *et al.*, 1983, 1985). The native enzyme has a molecular weight of $\sim$ 30 000, a pI of 8, a pH optimum of 9 against elastin, and an optimal calcium dependence of 2 mM. The cercarial elastase is also active against azocoll, gelatin, laminin, fibronectin, keratin and type IV collagen (McKerrow *et al.*, 1985). The cDNA encoding the cercarial elastase was reported by Newport *et al.* (1988). To isolate the cDNA, a sporocyst stage cDNA library was constructed using schistosome-infected hepatopancreas of the snail *Biomphalaria glabrata*, which was screened with a radiolabelled oligonucleotide (3'-CTT AAG GGA/T AAG TAG/T AAG GA-5') encoding NH_2-terminal residues determined from the purified native elastase. The elastase cDNA included a 5'-untranslated region (UTR) of 254 bp, an open reading frame of 810 bp and a short 3'-UTR. Comparison of the deduced amino acid sequence of the cDNA indicated that the elastase was synthesized as a preproenzyme of 267 residues, including a signal peptide of 27 amino acids. The catalytic triad of residues characteristic of serine proteinase occur here as Ser 218, His 68 and Asp 126 (Newport *et al.*, 1988; Price *et al.*, 1997). The mRNA transcript encoding the cercarial elastase was determined to be 1350 nt in length by Northern blot analysis using RNA from infected snails (Newport *et al.*, 1988). Subsequent analysis of the gene showed that it includes three exons and is probably present as a single copy in the genome of *S. mansoni* (Price *et al.*, 1997). Two introns, of 141 bp and 604 bp in length, separate amino acids 78 and 79 and 184 and 185, respectively (McKerrow *et al.*, 1991; Pierrot *et al.*, 1995; Price *et al.*, 1997). Several minor allelic differences have been reported in the cDNA encoding the cercarial elastase (Newport *et al.*, 1988; Price *et al.*, 1997). Analysis of the proximal sequences upstream of the initiation codon of the cercarial elastase gene have indicated the presence of several potential regulatory sequences, including NF-IL6 and AP1 consensus recognition sites (Pierrot *et al.*, 1995). Comparisons with sequences of other proteinases indicate that the cercarial elastase is more closely related to eukaryotic serine proteinases of the trypsin group, such as rat pancreatic elastases I and II, than to prokaryotic serine proteinases of the subtilisin family (Newport *et al.*, 1988; Price *et al.*, 1997).

In situ hybridization studies have demonstrated that the cercarial elastase is expressed in large cells that lie in the posterior portion of the developing cercarial 'head', just anterior to the site of attachment of the tail. Expression of the elastase is completed by the time cercariae have matured, at which point the cercarial elastase is stored as secretory material in vesicles in the lumen of the acetabular cells. During invasion of human skin, groups of intact vesicles are released through the cytoplasmic processes of the acetabular glands and rupture within the host skin. Ruptured proteinase vesicles can be located adjacent to degraded epidermal cells and the dermal-epidermal basement membrane and along the surface of the penetrating cercariae. The location of the ruptured vesicles in the skin and on the surface of the schistosome larvae is consistent with the cercarial elastase playing a dual role in the facilitation of the invasion of the skin by the larva and in facilitating the release of the glycocalyx from the surface of the schistosome during metamorphosis from cercaria to schistosomulum (Newport *et al.*, 1988; McKerrow *et al.*, 1991; Fishelson *et al.*, 1992). Analysis by reverse-transcription-PCR (RT-PCR) and by immunolocalization suggests that the gene encoding cercarial elastase may also be transcribed in eggs and adults, as well as in sporocysts. However, whereas the proteinase is detectable in eggs by immunoassays, it is not detectable in adults, indicating the presence of some form of translational regulation in the adult stage of the schistosome (Pierrot et al., 1996).

Pierrot et al. (1995) reported the presence of a second gene that might encode a closely related cercarial elastase. The gene, named EL2, was isolated from genomic DNA, and the deduced amino acid sequence was found to be 79% identical to the cercarial elastase (Newport et al., 1988). Like the cercarial elastase gene, EL2 contained three exons separated by two introns. These introns were located at the same sites as the two introns in the cercarial elastase, but the introns of EL2 of 131 bp and 254 bp in length differed in size from those of cercarial elastase gene. Interestingly, the aspartic acid residue at position 125, which forms part of the catalytic triad of residues of serine proteinases, is replaced in EL2 with an alanine. This indicates that the EL2 proteinase has a discrete substrate specificity to the cercarial elastase. Alternatively, it may represent a pseudogene, as the stage specificity of expression of the EL2 protein has yet to be reported (Pierrot et al., 1995).

Using the crystallographic co-ordinates of alphalytic proteinase of *Streptococcus griseus* and/or mammalian pancreatic elastase, Cohen et al. (1991) created a three-dimensional model of the 28/30 kDa proteinase. A deep hydrophobic substrate binding cleft was visualized, which correlated with the chymotrypsin-like specificity of cercarial elastase for large hydrophobic amino acids in the P1 site of peptide substrates. In contrast to chymotrypsin, there was not much steric hindrance for large hydrophobic amino acids at the P4 site. Based on the results of substrate specificity studies, which indicated that cercarial elastase proteinase preferred aromatic residues at P1 (McKerrow et al., 1985; Cohen et al., 1991), these workers tested the predictions of their model with a number of synthetic peptide inhibitors by *in vitro* enzyme kinetic and cercarial skin penetration studies. Although some peptides inhibited enzyme activity and cercarial invasion, they were not very potent; the most promising substrate, Z-Ala-Ala-Pro-Phe-CMK and its boronic acid derivative, blocked cercarial invasion by 80% at 50 µM (Cohen et al., 1991). More recently, Dalton et al. (1997) showed that substrates such as Z-Gly-Pro-Lys-NHMec and Z-Gly-Pro-Arg-NHMec, which contain basic residues at the P1 position, are more efficiently cleaved by the cercarial serine proteinases than substrates containing aromatic P1 residues, which indicates that the binding cleft of cercarial elastase is more acidic than hydrophilic.

A membrane-bound 28 kDa serine proteinase (m28) expressed on the surface of schistosomules, lung-stage and adult worms has recently been reported by Ghendler et al. (1996). The cercarial elastase of 28/30 kDa and m28 are antigenically related and possess similar substrate and inhibitor profiles, and both enzymes are reactive with antibodies to the cross-reactive determinant of glycophosphatidylinositol-anchored proteins. Schistosomular surface m28 appears to be a processed variant of cercarial elastase. Since m28 can be removed from the schistosomule surface membrane by phosphatidyl inositol-specific phospholipase C, it may be held in the membrane by a glycosyl phosphatidylinositol anchor, and the 28/30 kDa proteinase (cercarial elastase) might be the soluble form of m28 (Ghendler et al., 1996). Marikovsky et al. (1988b) suggested that cercarial elastase facilitates the removal of the glycocalyx from transforming cercarial upon penetration of host skin, and thereby facilitates the acquisition of resistance to complement (Fishelson, 1989). Support for the suggestion was obtained from experiments that demonstrated that the cercarial elastase, when added exogenously to transforming cercariae, accelerated release of the glycocalyx and conversion to complement resistance. The fragment patterns of the naturally released glycocalyx and the glycocalyx released after addition of exogenous elastase were identical, but addition of trypsin, by contrast, had no effect on the process. Both the m28 serine proteinase from the schistosomulum surface and cercarial elastase can degrade the complement proteins C3, C3b, iC3b and C9 to yield the same pattern of degradation products (Marikovsky et al., 1988a; Parizade et al., 1990, cited in McKerrow et al., 1991). These results suggest that the soluble and membrane-bound serine proteinases may further protect the schistosomulum against immune attack by cleaving off complement components that associate with the surface of the developing schistosome.

Cercarial elastase has been considered a worthy target for immunological intervention, as blocking its action may block infection by cercariae and developing schistosomules (Pierrot et al., 1996; Darani et al., 1997; Doenhoff, 1998). Indeed, infected humans appear to raise serum antibodies to cercarial elastase (Toy et al., 1987). On the other hand, most rabbit antibodies raised against the cercarial elastase did not appear to inhibit its proteolytic activity (Doenhoff et al., 1988). The cercarial elastase may be a poor immunogen, perhaps because it can cleave immunoglobulins and/or because of interaction with serpins (Modha et al., 1988; Ghendler et al., 1994; Mohda et al., 1996). Nevertheless, recombinant fragments encoded by the three exons of the cercarial elastase gene all elicited strong antibody responses when used to immunize rabbits (Price et al., 1997). By contrast, cercarial elastase appears to be capable of enhancing larval infectivity of challenge infections with S. mansoni in mice, perhaps by its direct hydrolysis of host tissue (Teixiera et al., 1993; Fallon et al., 1996). Recent reports by Doenhoff and colleagues indicate that although the cercarial elastase is poorly immunogenic in mice and rabbits, and probably humans as well, the small percentage of experimental mice in which immunization with alum-adjuvantated cercarial elastase elicited specific antibody responses was subsequently partially protected against challenge infection with S. mansoni cercariae (Darani et al., 1997; Doenhoff, 1998).

3.3. Proteinases of Adults and Schistosomules

More is known about genes encoding proteinases expressed in the adult stage of the schistosomes than in other stages. This reflects two facts: first, most extant cDNA libraries have been constructed using messenger RNA isolated from adult schistosomes, principally from *S. mansoni*; and second, the adults are parasites of humans and laboratory mice and thus are easily obtained. It is pertinent to note that a genome project examining both *S. mansoni* and *S. japonicum* is under way, and has already provided information on gene sequences encoding proteinases (see Johnston, 1997). Of the proteinase genes from adults, a number encode genes that appear to play a role in the nutrition of the parasites. Maturing and adult schistosomes obtain amino acids for growth, development and reproduction by catabolizing haemoglobin from ingested host erythrocytes. Although the biochemical pathway involved has not been determined definitively, a number of endoproteinases and exoproteinases have been implicated in the progressive degradation of haemoglobin (Hb) to diffusible peptides (Figure 1). Since Timms and Bueding (1959) reported this phenomenon about four decades ago, a sizeable literature has accumulated dealing with candidate proteinases and their genes that have been thought to participate in this proteolysis of Hb (see Brindley *et al.*, 1997). A number of these genes are discussed below, including cathepsin B and cathepsin D. In addition to nutrition, recent studies on the immune responses to schistosomes and on the cell biology of the adult parasite have resulted in information of a number of other proteinases expressed by adults, including calpain, ER60 proteinase and kallikrein (Table 1). As indicated above, the schistosome proteinases, including the cercarial elastase, appear capable of modulating the host immune response. Moreover, this and/or other serine proteinases have been shown to be capable of regulating the synthesis of specific and other IgE antibodies *in vitro* and *in vivo* in rats (Verwaerde *et al.*, 1986, 1988).

3.3.1. *Schistosome Calpain*

In mammalian tissues, calpains (EC 3.4.22.17) are ubiquitous, intracellular, cytosolic, calcium-activated cysteine proteinases. They appear to be involved with regulatory functions that are mediated by calcium, but act by modifying rather than degrading proteins. Although firm evidence is lacking concerning the function of calpains, they are implicated in the regulation of cytoskeletal proteins, receptor proteins and protein kinases. In addition, calpains may be involved in the intracellular protein degradation that takes place outside the lysosomes and in membrane biogenesis (Moncrief *et al.*,

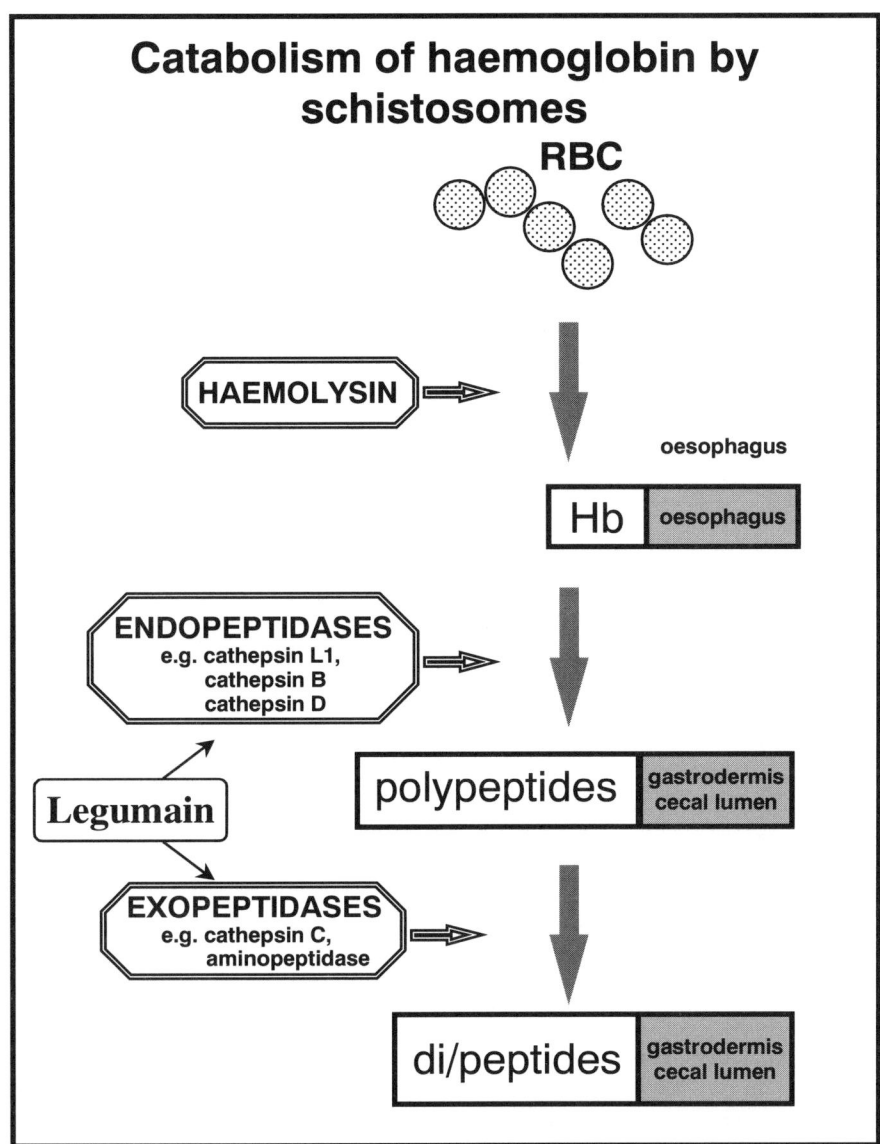

Figure 1 Schematic representation of the schistosome proteolytic pathway(s) and proteinases hypothesized to accomplish the catabolism of host haemoglobin (Hb) from ingested red blood cells (RBC). Sites of action and substrates of haemolysin and proteinases are shown. Schistosome legumain may process other proteinases from inactive to active forms. (Adapted and reproduced with permission from Brindley et al., 1997.)

1990; Croall and Demartino, 1991). Two types of calpains have been identified in mammalian cells, the µ-calpain and m-calpains, which have similar catalytic functions but differ in the molarity of calcium required for activation (see Ohja, 1996). From the differential expression of µ- and m-calpain in response to Ca^{2+}, µ-calpain appears to be in housekeeping roles, whereas m-calpain may function in response to cellular signals (Sorimachi et al., 1995). Both types of calpains have a discrete large subunit of approximately 80 kDa and a small subunit of approximately 30 kDa (Croall and Demartino, 1991). The µ- and m-calpain subunits share 50% sequence identity. The large subunit appears to bear the proteolytic activity of calpain, whereas the 30 kDa subunit plays a regulatory role in activating the large subunit (Goll et al., 1992). Calpain function is regulated in vivo by the endogenous proteinaceous inhibitor calpastatin, a polypeptide of 713–718 residues (Croall and McGrody, 1994).

Both µ- and m-calpain activities have been purified from soluble extracts of adult S. mansoni (Siddiqui et al., 1993) but a schistosome calpastatin has not been reported. Furthermore, cDNAs encoding the large subunit of an adult S. mansoni calpain have been isolated and shown to have high similarity with human µ-calpain and chicken calpain (Andresen et al., 1991; Karcz et al., 1991). Northern blotting analysis revealed that calpain is also expressed by the intramolluscan sporocyst (Andresen et al., 1991), indicating that calpain is important in the metabolism of all the life-cycle stages of the schistosome. The nucleotide and deduced amino sequence of the S. japonicum homologue, which shows 82% identity with the S. mansoni calpain large subunit, has been lodged in GenBank (accession no. AFO44407).

The calpain large subunit consists of four domains, I to IV. Domain II contains the papain-like cysteine proteinase activity, while Domain IV is the calmodulin-like, calcium-binding domain containing EF hand motifs (Ohno et al., 1984). Domain I in S. mansoni calpain (residues 1 to 125) is the least conserved domain compared to calpains of other species. It is 40 amino acid residues longer than human µ-calpain, and has homology with the receptor binding sequence of colicin 1a and 1b. Andresen et al. (1991) suggested that domain I might play a role in binding schistosome calpain to the plasma membrane. Domain II (residues 126 to 369) is the most highly conserved region in the S. mansoni calpain, with the proteinase catalytic Cys and His residues at positions 154 and 313 respectively (Croall and Demartino, 1991; Karcz et al., 1991). Domain III (residues 370 to 613) is also well conserved between S. mansoni and chicken and human forms of calpain. Schistosome calpain is unusual in that domain IV (residues 614 to 758) appears to contain only three of the four EF hand motifs typical of chicken and mammalian calpain (although the fourth domain may be present), and because it has an additional EF hand motif between domains II and III (Figure 2). The functional significance of these novel structural features of the schistosome calpain is not yet understood.

Immunolocalization studies showed that schistosome calpain is located throughout the surface epithelial layer and underlying muscle fibres. Within the syncytium they are free in the cytoplasm, and are associated with membranes of inclusion bodies and the apical surface (Siddiqui et al., 1993). When adult worms were incubated in the presence of inhibitors of μ- and m-calpain, the C3-stimulated, calcium-mediated incorporation of phosphatidylcholine and methionine into the overlying envelope was inhibited. Thus it appears that the schistosome calpains may be involved in the biogenesis of the schistosome surface membrane (Siddiqui et al., 1993).

Except for the inhibition studies of schistosome calpain reported by Siddiqui et al. (1993), little information is available on the enzymology or substrates of schistosome calpain. By contrast, increasing evidence about the immunological reactivity of schistosome calpain indicates that it might be a useful target for development of an anti-schistosome vaccine. Jankovic et al. (1996) have implicated calpain as a target of protective immunity in schistosomes, because calpain was found to be the target antigen of a murine CD4+ T lymphocyte clone named clone B obtained by vaccination of mice with a preparation of soluble worm antigens adjuvanted with the bacillus Calmette Guérin, that transfers protective immunity against challenge

Large subunit of *Schistosoma mansoni* calpain

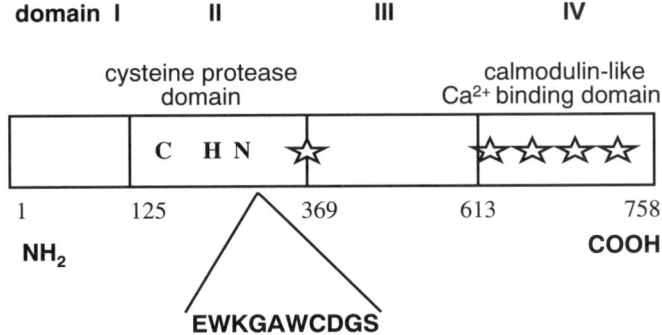

Figure 2 Schematic representation of the large sub-unit of calpain of *Schistosoma mansoni*. Calpain domains I to IV are indicated. The active site Cys, His, and Asn residues of the cysteine proteinase of domain II are indicated, as is the EWKGAWCDGS motif identified by Jankovic et al. (1996) as a protective T cell epitope. Stars in domains III and IV indicate the positions of EF hand, potential calcium-binding motifs.

infection with *S. mansoni*. Mouse recipients of clone B displayed significant protection against cercarial challenge; moreover, clone B could kill schistosomula *in vitro*. Clone B exhibits a characteristic Th1 phenotype of lymphokine secretion. By screening an adult *S. mansoni* cDNA library with clone B, with screening based on capacity to stimulate synthesis of interferon and interleukin 2, cDNAs encoding part of the large sub-unit of *S. mansoni* calpain were isolated. Analysis using truncated-length cDNA clones followed by studies with overlapping peptides determined that residues 344 to 353, EWKGAWCDGS (within Domain II, the cysteine proteinase domain of calpain), represented the minimal linear core T cell epitope of clone B, and thereby represented the immunologically reactive epitope (Figure 2). Clone B was not stimulated by a peptide spanning the corresponding sequence of human calpain. Since clone B responds to supernatants from cultured schistosomula, Jankovic *et al.* (1996) postulated that recognition of calpain released by invading larval schistosomes and resulting induction of Th1 cytokines accounted for the protection mediated by the adoptively transferred clone.

Hota-Mitchell *et al.* (1997) expressed the large subunit of *S. mansoni* calpain in a baculovirus-insect cell expression system. They observed that this recombinant subunit of calpain (termed Sm-p80) induced modest levels of protection in mice against challenge infection with *S. mansoni* (29% to 39%) and that Sm-p80 was recognized by IgA, IgM, IgG1 and IgG3 isotype antibodies in all or some of eight of 12 *S. mansoni*-infected people. Other workers have also observed anti-calpain antibodies in sera of infected humans (Andresen *et al.*, 1991; Karcz *et al.*, 1991).

3.3.2. *Cathepsin B*

Whereas cercarial elastase may be the best characterized of any schistosome proteinase, the cathepsin B cysteine proteinase (EC 3.4.22.1) known as Sm31 has been examined more fully than other proteinases expressed by adult schistosomes. (Cathepsin B is a category of papain-like cysteine proteinases located in lysosomes in mammalian cells; see Section 8.4). Sm31 was first reported by Ruppel and co-workers while investigating the diagnostic potential of schistosome antigens (Ruppel *et al.*, 1985a,b, 1987). During the 1980s, immunoblots revealed that proteins of 31 kDa and 32 kDa (see Section 3.3.7) were highly immunogenic in infected humans and mice. Whereas the biological functions of these antigens were not known, they were localized by histochemical techniques to the schistosome gut. This localization and earlier reports that cathepsin B-like cysteine proteinase activity was responsible for haemoglobin proteolysis, the major source of amino acids for the adult schistosome, resulted in Sm31 gaining prominence as an important or even *the* schistosome haemoglobinase (McKerrow and Doenhoff, 1988). About that time, a cDNA encoding Sm31 was isolated by screening an expression library with infection sera. Sequence analysis revealed that the Sm31 gene encoded a cathepsin B-like cysteine proteinase (Klinkert *et al.*, 1987, 1989). Schistosome cathepsin B is synthesized as a preproenzyme of 340 amino acid residues, with the processed, mature enzyme comprising 250 residues. The apparent molecular mass of 31 kDa is larger than the $\sim$ 28 kDa predicted from its amino acid sequence and results from asparagine-linked glycosylation of the mature cathepsin B (Felleisen and Klinkert, 1990). The mechanism of activation of the mature enzyme is not fully clear, although auto-processing and processing by another cysteine proteinase are possibilities (Gotz and Klinkert, 1993; Lipps *et al.*, 1996).

Gotz and Klinkert (1993) expressed recombinant Sm31 in insect cells using baculovirus, and showed that the recombinant enzyme could digest the cathepsin B-specific substrate Z-Arg-Arg-NHMec. Using the coordinates of the crystal structure of human liver cathepsin B, Klinkert *et al.* (1994) reported a three-dimensional model for *S. mansoni* cathepsin B complexed with synthetic inhibitors, and predicted likely differences in the inhibition profiles between the two enzymes. Biochemical studies confirmed their predictions: Z-Trp-MetCHN$_2$ is a more potent inhibitor of schistosome than of human cathepsin B, whereas the reverse is true with CA-074 (*N*-(L-3-trans-carboxyoxirane-2-carbonyl)-L-isoleucyl-L-proline), a derivative of the epoxide E-64.

Lipps *et al.* (1996) expressed recombinant procathepsin B from *S. mansoni* in *Saccharomyces cerevisiae*. Because they could not activate the enzyme *in vitro*, they concluded that the enzyme might not undergo autoprocessing *in vivo*, but that it would be processed by another proteinase. Indeed, exposure of the recombinant procathepsin B to porcine pepsin produced enzymatically active Sm31. The active, recombinant schistosome cathepsin B preferentially cleaved Z-Arg-Arg-NHMec over Z-Phe-Arg-NHMec under reducing conditions, with a pH optimum of 6.0. It was inactive below pH 4. These characteristics are similar to those of mammalian cathepsin B and similar to the pH profile for the cysteine proteinase activity against Z-Arg-Arg-NHMec in crude extracts of adult *S. mansoni* and ES (Dalton *et al.*, 1996a). Recombinant Sm31 expressed in yeast cells does not show a marked substrate preference for Hb, although it cleaves this substrate at several sites. Based on NH$_2$-terminus analysis of digestion products of Hb, Lipps *et al.* (1996) calculated a consensus cleavage sequence of 6X1*181 (1, hydrophilic/small aliphatic; 6, aliphatic; 8, hydrophobic) (P3 P2 P1*P1'P2'P3').

As noted, it has been widely suggested that the degradation of Hb within the gut of *S. mansoni* was accomplished by the action of Sm31 (Sj31 in *S. japonicum*) (Merckelbach *et al.*, 1994) (Dresden and Deelder, 1979; Chappell and Dresden, 1986; Lindquist *et al.*, 1986). Indeed, based on studies with peptidyl fluoromethyl ketones and the peptidyl (acyclo) methyl ketones, broad-spectrum inhibitors of papain-like cysteine proteinases, Wasilewski *et al.* (1996) recently ascribed a central role for Sm31 in the development and fecundity of schistosomes. However, the catabolism of Hb appears to require more than just the action of Sm31. Indeed, it seems to involve an aspartic proteinase in addition to other cysteine proteinases including cathepsin L (below) (see Brindley *et al.*, 1997), and it may have intrinsic redundancies in terms of which proteinases can accomplish the catabolism of Hb (Figure 1).

3.3.3. *Cathepsin L1*

As discussed below, cathepsin L (like cathepsin B) is a cysteine proteinase of the papain superfamily (see Section 8). However, the prosegment of cathepsin Ls exhibits the ERFNIN motif of Karrer *et al.* (1993), which together which other characteristics distinguishes cathepsin L and other cathepsins from cathepsin B (see Section 8.3). Activity ascribable to cathepsin L has been characterized from adult and other stages of schistosomes. Indeed, using soluble extracts and ES products of *S. mansoni* and *S. japonicum*, and class-specific, synthetic peptidyl substrates, including several that could discriminate between classes of papain-like, cysteine endoproteinases, namely cathepsin L (Z-Phe-Arg-NHMec, Z-Phe-Val-Arg-NHMec), cathepsin B (Z-Arg-Arg-NHMec; Z-Arg-NHMec) and cathepsin H (Z-Arg-NHMec), Dalton and colleagues concluded that the dominant proteinase activity between pH 4 and pH 6 was cathepsin L-like (Smith *et al.*, 1994; Day *et al.*, 1995; Dalton *et al.*, 1996a), in contrast to cathepsin B as reported by other workers (Chappell and Dresden, 1986; Lindquist *et al.*, 1986; Wasilewski *et al.*, 1996). This activity against cathepsin L (EC 3.4.22.15) substrates at acidic pH is many times greater than that for cathepsin B substrates, although cathepsin B activity with a pH optimum of ~6 against Z-Arg-Arg-NHMec is apparent (Dalton *et al.*, 1996). The activity of schistosome cathepsin L in tissue extracts of adult *S. mansoni* against the substrate Z-Phe-Arg-NHMec can be totally blocked by nanomolar concentrations of the diazomethanes Z-Phe-AlaCHN$_2$ and Z-Phe-PheCHN$_2$. Z-Phe-AlaCHN$_2$ (K_i 50% = 50 nM at pH 5) is a more potent inhibitor than Z-Phe-PheCHN$_2$ (K_i 50% = 0.4 µm), which contrasts with mammalian cathepsin L where Z-Phe-PheCHN$_2$ is the more potent inhibitor (Smith *et al.*, 1994; Day *et al.*, 1995a, 1996). Similar cathepsin L activity has been described from schistosome eggs (Day *et al.*, 1995).

cDNAs encoding two cysteine proteinases with similarity to vertebrate cathepsin Ls have been characterized from *S. mansoni* (Smith *et al.*, 1994; Michel *et al.*, 1995) and homologues of both genes reported from *S.*

japonicum (Day *et al.*, 1995). Using PCR with degenerate primers designed to anneal to regions of cysteine proteinase genes encoding the active site residues Cys25 and Asn179, Smith *et al.* (1994) isolated the first cathepsin L from adult *S. mansoni* cDNA. The full-length transcript encoded a preproenzyme of 319 amino acid residues, including the signal peptide of eight residues, a prosegment of 96 residues and the mature enzyme of 215 residues. Its apparent molecular mass is 24.3 kDa, with pI of 4.95. This proteinase has been termed SmCL1 (*S. mansoni* Cathepsin L1) (Day *et al.*, 1995) since a second cathepsin L gene has been reported from adult *S. mansoni* (Michel *et al.*, 1995). A cDNA encoding the apparent homologue, SjCL1, from *S. japonicum* has also been reported. Although the cDNA encoding SjCL1 did not encode the full-length preproenzyme, its deduced amino acid sequence revealed that the mature proteinase would include 215 amino acid residues of similar length to SmCL1, with a predicted molecular mass of 24.1 kDa and pI of 5.63. The deduced amino acid sequences of SjCL1 and SmCL1 are 92% identical, and their eight S2 subsite residues are identical, indicative of identical substrate preferences (Day *et al.*, 1995).

The tissue or cellular localization of SmCL1 has yet to be reported. However, it may be gut-associated, based on the presence of cathepsin L activity in the gut and in ES of schistosomes and because the only other cathepsin L known from schistosomes, SmCL2/SjCL2 (see Section 3.3.4), has been localized to the reproductive organs and other sites distinct from the gut and gastrodermis of the schistosome (Michel *et al.*, 1995).

3.3.4. *Cathepsin L2*

The cDNA encoding the second cathepsin L-like proteinase has been reported from *S. mansoni* (Michel *et al.*, 1995) and *S. japonicum* (Day *et al.*, 1995). We have termed the proteinase SmCL2 and SjCL2 in each species (Day *et al.*, 1995). The cDNA encoding the *S. mansoni* proteinase was isolated by subtractive hybridization undertaken to locate female specific transcripts in adult *S. mansoni*. The *S. japonicum* homologue was isolated by heterologous DNA screening of a *S. japonicum* cDNA library, using the radiolabelled cDNA sequence from *S. mansoni* as the probe. The preprocathepsin SmCL2 contains 330 amino acids, including a signal peptide of 18, a pro-peptide of 97 and a mature enzyme of 215 amino acid residues. Its apparent molecular mass is predicted to be 24.3 kDa, with a pI of 9.25. Active site labelling and immunoblotting of schistosome extracts demonstrated that native SmCL2 is $\sim$ 31 kDa rather than 24.3 kDa, as predicted from its deduced amino acid sequence, indicating that it might be glycosylated (Michel *et al.*, 1995; Wasilewski *et al.*, 1996). The amino acid sequence deduced from the cDNA encoding SjCL2 is comprised of 332 amino acid residues, with the signal peptide comprised of 16 residues, the prosegment of 99 residues and the mature enzyme of 217 residues. The apparent molecular mass of SjCL2 is predicted to be 24.5 kDa, with a pI of 8.98. The deduced amino acid sequences of SjCL2 and SmCL2 are 78% identical, and share five of eight S2 subsite residues (Day et al., 1995). SmCL2 is overexpressed in females compared to male schistosomes and, in particular, is expressed in the reproductive system of the female parasite and in the gynecophoric canal of the male. Based on this localization, Michel et al. (1995) postulated that its role might be associated with the activation of phenyloxidase, an enzyme involved with the cross-linking of egg-shell proteins. They posulated further that SmCL2 might play a role in altering the viscosity of the seminal and uterine fluids.

Recent phylogenetic analyses based on both propeptide and mature enzyme sequences, as reported here (see Section 8.5), show that the schistosome cathepsin L2, i.e. SmCL2 and SjCL2, are close relatives of mammalian cathepsin L and indeed of cathepsins L1 and L2 of *Fasciola hepatica*. By contrast, schistosome cathepsin L1, i.e. SmCL1 and SjCL1, is not as closely related to the mammalian cathepsin L as is SmCL2 and SjCL2. Rather, they are more closely related to cruzipain from *Trypanosoma cruzi* and the neutral proteinase from *Paragonimus westermani* (see Figure 7, see Section 8).

3.3.5. *Cathepsin C*

Cathepsin C, also known as dipeptidylpeptidase I (DPP I) (EC 3.4.14.1), is a member of the papain family of cysteine proteinases. Unlike papain and cathepsin L, however, cathepsin C or DPP I is an exopeptidase, which hydrolyses the removal of dipeptides from the amino-terminus of the substrate. Several other dipeptidylpeptidases other than cathepsin C have been reported, but these are not cysteine proteinases (see Hui, 1988). In mammalian tissues, cathepsin C exhibits an array of biochemical properties distinct from those of other papain-like enzymes, including a dependence on halide ions for enzymatic activity, slow inhibition by the specific cysteine proteinase inhibitor E-64, and formation of oligomeric structures. Furthermore, cathepsin C enzymes possess prosegments comprised of $\sim$ 200 amino acid residues, which is unusually long compared to the 60 to 100 residues of other papain-like proteinases. Interest in these enzymes has been stimulated by reports which suggest they are central to the activation of serine proteinases in inflammatory cells, and that elevated serum cathepsin C activity is associated with pathogenesis of myocardial infarctions, diabetes mellitus and hepatitis (Dolenc et al., 1995). Few genes or cDNAs encoding cathepsin C have been described; these include those encoding cathepsin C from human, rat and mouse tissues (see McGuire *et al.*, 1997) and cathepsin C from *S. mansoni* and *S. japonicum* (Butler *et al.*, 1995; Hola-Jamriska, Tort, Dalton, Day, Fan, Aaskov and Brindley, unpublished).

Bogitsh and Dresden (1983) detected a cathepsin C-like activity of *S. mansoni* and *S. japonicum* using Pro-Arg-4-methoxy-β-naphthylamide (Pro-Arg-4MNA) as the substrate. They probed thin sections of adult worms with the substrate at pH 5.5 under reducing conditions, and detected activity against the substrate in the gastrodermis of both schistosomes and in the ceca close to the gut luminal surface. More activity was present in female than in male worms. The vitelline cells in females and the testes in males also showed activity. They also reported, in this same study, similar activities and localization against the cathepsin B substrate CBZ-Ala-Arg-Arg-4MNA and against the dipeptidylpeptidase II-substrate Lys-Ala-4MNA. Based on their location in the gastrodermis and caecum, they proposed that all three of these proteinases, including cathepsin C, were involved in the digestion of Hb. Kramer and Bogitsh (1985) subsequently partially characterized a Lys-Ala-4MNA-degrading activity in extracts of adult *S. japonicum* as a serine proteinase with a pH optimum for activity of 6.3.

Butler *et al.* (1995) have reported the sequence of a cathepsin C from adult *S. mansoni*. The deduced amino acid sequence of *S. mansoni* preprocathepsin C is 454 amino acids in length, comprising a signal peptide of 24, a long pro-peptide of 193 and a mature enzyme of 237 residues. The coding region of the *S. mansoni* cathepsin C cDNA is 43% identical and 52% similar to rat cathepsin C. Furthermore, *S. mansoni* cathepsin C is 30% identical to both Sm31 and SmCL2, and shows structural similarities with papain. Hola-Jamriska *et al.* (1998) reported the sequence of a cathepsin C from adult *S. japonicum*. The deduced amino acid sequence of *S. japonicum* cathepsin C comprised 458 amino acid residues: 22 NH_2-terminal residues corresponding to the signal peptide, 199 residues corresponding to the propeptide and 237 COOH-terminal residues corresponding to the mature enzyme region. The amino acid sequence of this preprocathepsin showed 43% and 50% identity to preprocathepsin C of human and rat, respectively. The preproenzyme shared only 59% identity with the cathepsin C from *S. mansoni*, differing from it in active site residues and in its potential N-glycosylation sites. The differences indicate that these two schistosome cathepsins C may not be species homologues. The function of the long propeptide in these schistosome cathepsins C is not clear but, as in other papain-like proteinases, it probably plays a role in the correct folding of the nascent enzyme and in inactivation of the zymogen. Northern blot analysis showed that *S. japonicum* cathepsin C was expressed in greater quantities in female than in male parasites. Southern hybridization analysis indicated that *S. japonicum* cathepsin C may be encoded by a few genes rather than a single one (Hola-Jamriska, Tort, Dalton, Day, Fan, Aaskov and Brindley, unpublished). The phylogenetic relationships of the two schistosome cathepsins C to each other and to other papain-like proteinases is detailed in Section 8 (see Figures 5 and 6).

Exoproteinases such as dipeptidylpeptidases and aminopeptidases probably play a downstream role in the degradation of host-derived Hb to readily absorbable peptides, acting on cleavage fragments released after the activities of endopeptidases such as cathepsin D (Figure 1). In support of the Northern blot analysis which demonstrated that females expressed more cathepsin C than did male parasites, Brindley and colleagues (unpublished) detected a cathepsin C activity capable of cleaving the diagnostic substrate H-Gly-Arg-NHMec with a pH optimum for activity of 5.5 in extracts of adult *S. japonicum*, with extracts of females exhibiting several times more activity than males.

3.3.6. *Cathepsin D*

Aspartic proteinases are considered the most conserved group of the four classes (cysteine, serine, aspartic and metallo-) of proteinases. A primordial gene duplication event is considered to have resulted in the presence of a catalytic dyad of aspartic acid residues, from which the class derives its name (Tang *et al.*, 1978). Five types of aspartic proteinase have been reported from human and other mammalian tissues. These are pepsin, renin, chymosin, cathepsin D and cathepsin E. Using an homology PCR-based strategy which employed a primer (5′-TTGAYACNGGNTCATCAAAYCTNTGG, where Y = C or T) designed to anneal to the nucleotide sequence region encoding one of aspartic acid residues of the catalytic dyad and adjacent residues (DTGSSNLW) (Harrop *et al.*, 1996), we isolated a cDNA from adult *S. japonicum* that encodes an aspartic proteinase. The *S. japonicum* cDNA encodes a zymogen of 424 amino acid residues, including a signal peptide of 14 residues, a pro-peptide of 37 residues and the mature enzyme of 373 residues (Becker *et al.*, 1995). The homologous *S. mansoni* gene was isolated by screening an adult cDNA library with the radiolabelled cDNA from *S. japonicum* as the probe. The *S. mansoni* cDNA encodes a zymogen of 428 residues, with a signal peptide of 14, a pro-peptide of 37 and a mature enzyme of 377 residues (Wong *et al.*, 1997). The predicted molecular mass of the mature aspartic proteinase from both species of schistosomes is about 41 kDa. Comparison of the deduced amino acids sequences with other enzymes revealed that the schistosome aspartic proteinase was more like mammalian cathepsin D (EC 3.4.23) than other types of aspartic proteinases. The schistosome proteinase is 48% to 55% identical to cathepsin D from species of mammals. However, there are differences, including the presence of a carboxyl extension of about 43 residues in the schistosome cathepsin D-like aspartic proteinase, which is absent from mammalian cathepsin D. Other differences include the absence of the β-hairpin loop 3, which is cleaved during maturation of vertebrate cathepsins D to yield light and heavy chain subunits, differences in residues involved in substrate binding and specificity, and in phosphorylation (Becker *et al.*, 1995; Wong *et al.*, 1997) (Figure 3).

Bogitsh and colleagues localized an activity ascribable to a cathepsin D-like aspartic proteinase to the caecum and gastrodermis of both *S. japonicum* and *S. mansoni* (Bogitsh and Kirschner, 1986, 1987; Bogitsh *et al.*, 1992). These studies involved electron microscopical localization using mercury-

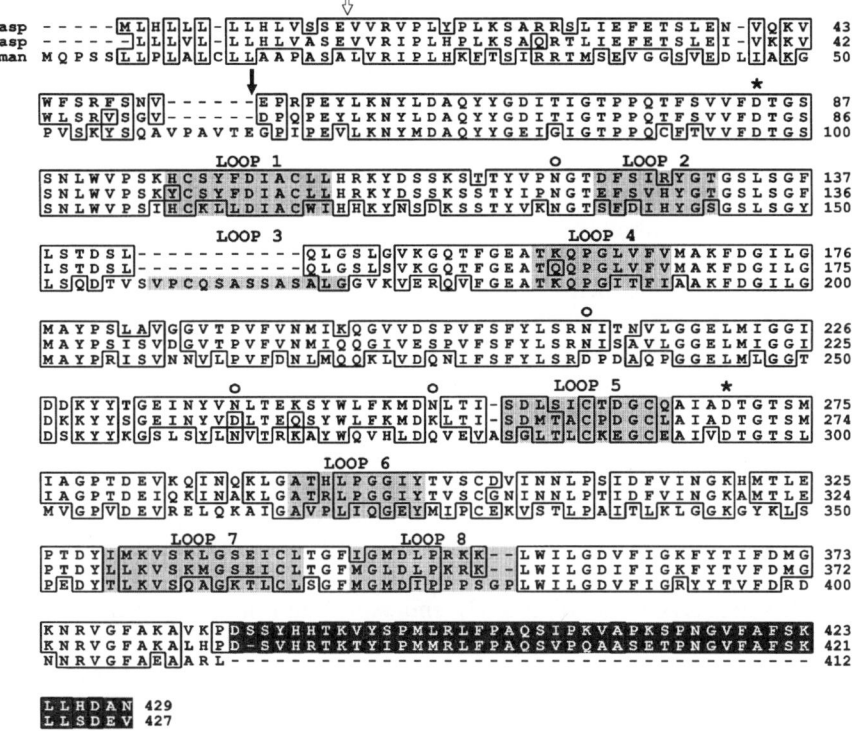

Figure 3 Alignment of deduced amino acid sequences of schistosome and human cathepsin D aspartic proteinases. Sjasp, *S. japonicum* cathepsin D (GenBank) accession number (U90750); Smasp, *S. mansoni* cathepsin D (U60995); and human kidney cathepsin D (P07339). Blocks denote conserved residues. Functional and structural motifs are designated as follows: cleavage point of signal peptide, open arrow; cleavage point of pro- and mature enzyme, solid arrow; catalytic dyad of aspartic acid residues, asterisks; loop regions, numbered and shaded; and the COOH-extension that characterizes the schistosome cathepsins D, dark shaded–light letters.

labelled pepstatin, immunohistochemical localization using anti-bovine cathepsin D antibodies and immunoinhibitory studies on Hb degradation by cultured schistosomula and schistosome extracts using anti-cathepsin D sera. Recent results obtained with a rabbit antiserum raised against recombinant *S. japonicum* aspartic proteinase produced in *E. coli* indicate that schistosome cathepsin D is expressed in the gastrodermal cells of adult *S. japonicum*, with much more in female compared to male schistosomes (Bogitsh and Brindley, unpublished). Dalton and co-workers assayed aspartic proteinase activity in soluble extracts of adult *S. japonicum* and *S. mansoni* and in ES. Using the cathepsin D-specific substrate Butyloxycarbonyl (Boc)-Phe-Ala-Ala-Phe (p-NO$_2$)-Phe-Val-Leu-OM4P, activity was present and was optimally active at pH 3 to 4, and could be blocked by pepstatin, a general inhibitor of aspartic proteinases. Further, the extracts and ES digested human haemoglobin with a similar pH optimum for activity, activity which was also inhibited by pepstain (Becker *et al.*, 1995). Ghonheim and Klinkert (1995) reported that an aspartic proteinase activity in extracts of adult *S. mansoni* appeared to be more efficient than schistosome cysteine proteinases including cathepsin B Sm31 in degradation of Hb. Together these results strongly suggest that the schistosome cathepsin D-like aspartic proteinase plays a central role in haemoglobin proteolysis, and might be the schistosome haemoglobin-degrading proteinase first described by Timms and Bueding (1959).

3.3.7. *Schistosome Legumain*

Like the cathepsin B proteinase Sm31, a second *S. mansoni* antigen — Sm32 — was initially identified as an immunogenic component of soluble extracts of schistosomes, and consequently attracted interest as a diagnostic antigen for schistosomiasis (Ruppel *et al.*, 1985a, 1987). Davis *et al.* (1987) isolated the cDNA encoding Sm32 and expressed it in *Esherichia coli* as a β-galactosidase fusion protein. Because the fusion protein was reported to degrade haemoglobin, Sm32 (like Sm31) has been touted as *the S. mansoni* haemoglobinase. This was later challenged on the basis that the sequence of Sm32 displayed no obvious similarity to other proteinases, and because recombinant Sm32 expressed in insect cells failed to exhibit a similar activity against Hb, as had been reported for the fusion protein (Davis *et al.*, 1987). The sequence of Sj32, the *S. japonicum* homologue of Sm32, has also been reported (Mercklebach *et al.*, 1994).

Recently the identity of Sm32/Sj32 was clarified when sequences encoding asparaginyl endopeptidases from seeds of legumes were shown to be similar to Sm32 (see Dalton *et al.*, 1995a and Ischii, 1994). Following this, Dalton *et al.* (1995b) demonstrated the presence of asparaginyl endopeptidase activity in soluble extracts of adult *S. mansoni*, the first report of this kind of proteolytic activity in any animal tissue, using diagnostic substrates introduced by Kembhavi *et al.* (1993). The schistosome activity has a similar substrate specificity (Z-Ala-Ala-Asn-NHMec, specific activity 0.013 nmol/min/mg protein, preferred over Z-Ala-Pro-Asn-NHMec; little activity against Z-Asn-NHMec) and sensitivity to inhibitors of cysteine proteinases (potent inhibition by iodoacetamide or *N*-ethylmaleimide, poor inhibition by the epoxide E-64 or the diazomethane Z-Phe-Ala-CHN$_2$) as an asparaginyl endopeptidase from the seed of the jack bean *Canavalia ensiformis*, but differs in having a near-neutral pH optimum (pH 6.8) and a higher temperature optimum for activity of 37–45°C.

Asparaginyl endopeptidases are members of a novel family of cysteine proteinases termed legumains (EC 3.4.22.34), since they were first characterized in seeds of leguminous plants. These enzymes cleave peptide bonds on the carboxyl side of Asn residues, except where the Asn occurs at the NH$_2$-terminus or at the second position from the NH$_2$-terminus of the polypeptide, or where the Asn is glycosylated. Legumains function in the post-translational modification of storage proteins in legume seeds by cleaving asparaginyl peptide bonds between pro-peptides and mature proteins (Ischii, 1994). Based on biochemical and sequence similarities (Dalton *et al.*, 1995a,b), the Sm32/Sj32 proteinase has recently been renamed schistosome legumain (Dalton and Brindley, in press).

The mRNA of *S. mansoni* legumain encodes a 50 kDa single-chain protein of 429 amino acid residues (El Meanawy *et al.*, 1990). The *S. japonicum* and *S. mansoni* legumains are 73% identical, and 30–40% identical to plant legumains (Dalton and Brindley, in press). The 50 kDa proenzyme appears to be processed at both its NH$_2$- and COOH-termini; sequencing of the mature enzyme determined that it is comprised of 260 amino acids, beginning at Val-32 and ending at Arg-292 (El Meanawy *et al.*, 1990). Post-translational processing of the schistosome legumains appears to involve the cleavage of a signal peptide from the NH$_2$-terminus and the removal of the pro-peptide from the COOH-terminus to yield the active proteinase. The block of amino acids -D-V-C-H-A-Y- (residues 56 to 61, *S. mansoni* legumain numbering (El Meanawy *et al.*, 1990)) which is conserved between legume (Dalton and Brindley, 1997, in press; Hara-Nishimura *et al.*, 1993; Takeda *et al.*, 1994) and schistosome legumains may contain the active site Cys residue. The conserved block between residues 144 and 155, -I(V)-F-I(V)-F(Y)-Y(F)-(S)-D-H-G-G-(A)-P-G-, may contain the active site His. (The -D-H-G- motif is also conserved in many papain-like cysteine proteinases.) A glutamic acid residue within a block of conserved residues from position 190 to 196, -E-A-N(C)-E-S-E-G-, may complete the catalytic triad of active site residues in the legumains (Dalton and Brindley, 1997).

Monoclonal antibodies have been used to localize *S. mansoni* legumain to the epithelium of the gut in schistosomula and adults, to the ventral surface of adult males, and to the cephalic glands of cercariae (Zerda *et al.*, 1988; El Meanawy *et al.*, 1990; Zhong *et al.*, 1995). Although it has long been considered by other workers to be one of the cysteine proteinases centrally involved in the degradation of Hb, the function of schistosome legumain remains unclear. Its location in the gut indicates that it might play a role in digestion. However, its low specific activity (at least against Z-Ala-Ala-Asn-NHMec) (Dalton *et al.*, 1995b) argues against a central role in Hb degradation, given the enormous numbers of RBC that schistosomes ingest (Lawrence, 1973). Further, the Hb-degrading activity of the cysteine proteinase(s) reputed to be involved in Hb digestion can be blocked by E-64 and diazomethanes (Wasilewski *et al.*, 1996; Becker *et al.*, 1995), neither of which are potent inhibitors of schistosome legumain (Dalton *et al.*, 1995b). Therefore, if schistosome legumain plays a role in the digestion of Hb, that role may be indirect. As in plant legumains (Ischii, 1994), schistosome legumain may process other proteins (Dalton and Brindley, 1996). Indeed, it may be involved in activation of the proteolytic enzymes which play key roles in schistosome nutrition, such as cathepsins D, C and L (Figure 1). If so, drugs targeted at schistosome legumain may block the activation of other proteolytic enzymes directly involved in Hb digestion, and thereby deliver a profound anti-schistosomal effect. Anecdotal reports from Chinese workers suggest that vaccination using purified native Sj32 might deliver a potent anti-fecundity effect as well as reducing worm loads in *S. japonicum*-infected mice/water buffaloes (see Brindley *et al.*, 1995). Intervention targeting schistosome legumain would have to be specific, however, as a legumain has recently been reported from mammalian tissues (Chen *et al.*, 1997).

3.3.8. *ER60 Proteinase*

ER60 is one of four members of the protein disulfide isomerase (PDI) family of enzymes, which are characterized as a group by the presence of two or three thioredoxin-like domains, each with the central CGHC motif. In addition to ER60 and PDI, the PDI family in mammals includes the ER72 and P5 proteins (Freedman *et al.*, 1994; Finken-Eigen and Kunz, 1997). Finken-Eigen and Kunz (1997) isolated and characterized a cDNA encoding the complete sequence of the homologue of the ER luminal proteinase ER60 in *Schistosoma mansoni*. (They had previously reported the *S. mansoni* gene encoding PDI and determined the tissue sites of expression in these parasites (Finken *et al.*, 1994).) The deduced protein of 484 amino acids of *S. mansoni* ER60 contains two thioredoxin domains, each with the CGHC motifs that characterize the PDI family. The sequence ends in the putative ER retention signal KSEL, which indicates that the protein is resident in the lumen of the ER and indicates, as with mammalian ER60 proteinases, that the schistosome ER60 functions in pre-Golgi protein degradation. Although the primary amino acid sequences indicates that *S. mansoni* ER60 is a cysteine proteinase, proteinase activity ascribable to native or recombinant schistosome ER60 proteinase has yet to be characterized. The corresponding ER60 gene exists as a single copy in the genome of S. *mansoni* as six exons, interrupted by five very small introns of 40, 32, 34, 34 and 34 nucleotides respectively. Immunohistology and *in situ* hybridization using adult schistosomes identified the sustentacular cells of the testes, the lining of the protonephridia and the gastrodermis of the caecum to be the major tissues of ER60 gene expression. These are the same major sites of expression of schistosome PDI, and indicate that both ER60 and PDI (Finken *et al.*, 1994), are located in the lumen of the ER in secretory cells. The native molecular mass of ER60 protein in extracts of adult schistosomes was determined as 53 kDa by Western blotting, and this protein was also detected in extracts of miracidia and cercariae. Similar levels of expression were recorded in female and male adult schistosomes (Finken-Eigen *et al.*, 1997).

3.3.9. *Kallikrein-like Proteinase*

In mammalian species, kallikrein is a serine proteinase involved in an array of functions, including processing of bioactive peptides, haemostasis and the enhancement of glycosylation of IgE binding factors (Iwata *et al.*, 1983). Recently, Cocude *et al.* (1997) reported a gene sequence encoding part of a kallikrein-like proteinase from *S. mansoni*. The sequence, termed SmSP1, showed 35% identity with mouse plasma kallikrein, and was dissimilar in sequence to cercarial elastase (or EL2), the only other serine proteinase from schistosomes for which a gene sequence has been reported. A recombinant form of SmSP1 was produced in bacteria, purified, and used to immunize rats to produce a monospecific anti-SmSP1 antiserum. This antiserum immunoprecipitated a schistosome protein of 30 kDa from the tissues of adult schistosomes metabolically radiolabelled with ^{35}S-methionine. Further, antibodies in the rat antiserum reacted with porcine kallikrein in Western blots, but not recombinant *S. mansoni* cercarial elastase. The full-length SmSP1 cDNA has yet to be reported, as has a function for this enzyme. However, as a serine proteinase of adult schistosomes, SmSP1 may be the cognate proteinase of the endogenous serpin Smpi56 of *S. mansoni* (and its homologues from *S. haematobium* and from *S. japonicum*) or other serpins that have been described from these parasites (Ghendler *et al.*, 1994; Modha *et al.*, 1996). Interestingly, Teixeira *et al.* (1993) have suggested that schistosome cercariae possess a kallikrein-like proteinase, based on the ability of cercarial extracts to cause the formation of proinflammatory kinins after injection of the extracts into guinea pig skin.

4. PROTEINASES OF *FASCIOLA HEPATICA* AND OTHER FASCIOLIDAE

4.1. Cathepsin L

As early as 30 years ago it was suggested that proteolytic enzymes secreted from the gastrodermal cells of the liver fluke *F. hepatica* were involved in facilitating tissue penetration and in the acquisition of nutrients by the parasite (Howell, 1966; Halton, 1967). More recently, Chapman and Mitchell (1982) and Dalton and Heffernan (1989) showed that these secreted enzymes were predominantly cysteine proteinases that could cleave immunoglobulin at the hinge region and thereby yield Fab and Fc fragments. Hence an additional role, one of protecting the parasite against immune attack, was attributed to the enzymes.

Biochemical studies from a number of laboratories have shown that proteinases secreted by the liver fluke are related to the mammalian cathepsin Ls (Rege *et al.*, 1989a; Yamasaki *et al.*, 1989; Smith *et al.*, 1993a,b; Dowd *et al.*, 1994b; Wijffels *et al.*, 1994a,b). Although two-dimensional electrophoresis of ES products demonstrated that these proteinases are highly heterogeneous (Wijffels *et al.*, 1994a), two major forms of 27 kDa and 29.5 kDa have been purified and characterized, and were termed cathepsin L1 (Smith *et al.*, 1993a) and cathepsin L2 (Dowd *et al.*, 1994) respectively. The two enzymes have distinct substrate specificities against fluorogenic peptide substrates; for example, cathepsin L2 will accept substrates with proline in the P2 position, but cathepsin L1 will not (Dowd *et al.*, 1994b). *In vivo* experiments have shown that cathepsin L1 can cleave immunoglobulin at the hinge region and prevent the antibody-mediated attachment of immune effector cells to newly excysted juveniles (Carmona *et al.*, 1993). Cathepsin L2 cleaved fibrinogen in a manner that caused the formation of a novel type of clot; however, the relevance of this property to the aetiology of disease is unclear, although it has been suggested that fibrin clots that may form around the parasite could prevent access of immune effector cells to the parasite surface. Both cathepsin Ls can degrade intracellular matrix proteins such as collagen, laminin and fibronectin (Berasain *et al.*, 1997).

Early histochemical studies revealed that proteinase activity was associated with cells of the gut epithelium (Howell, 1966, 1973; Halton,

1967, 1997). More recently, immunolocalization studies by Yamasaki *et al.* (1992) and Smith *et al.* (1993a) have shown that the cathepsin L proteinases are synthesized and packaged in vesicles within these cells. Following the ingestion of a blood meal, the cells lining the digestive tract go through regular phases of secretion and adsorption. During the secretory phase, the vesicles are released into the gut lumen, where the proteinases are exuded and perform their digestive function (Halton, 1997). In addition, since the gut contents are regurgitated at frequent intervals, the proteinases are regularly released to the exterior, where they may be involved in other functions such as tissue degradation and immune defence. The high stability of the cathepsin L at neutral pH may be ideal for enabling these enzymes to prepare a migratory path for the parasite, but this property might also contribute to the liver pathology characteristic of fasciolosis (Dowd and Dalton, unpublished). The discovery of Wijffels *et al.* (1994b) which showed that certain proline residues of cathepsin Ls are modified to 3-hydroxy-prolines, may explain why the enzymes are highly stable at physiological pH and temperature. Histochemical studies also demonstrated the presence of cathepsin L proteinases in the adult *F. hepatica* Mehlis glands, which led to the suggestion that these proteinases are somehow involved in eggshell formation (Wijffels *et al.*, 1994b; Spithill and Morrison, 1997).

By employing consensus primers in PCR, Heussler and Dobbelaere (1994) isolated cDNA clones and gene fragments encoding five different cathepsin L proteinases that varied in their similarity from 48 to 84%, and thus demonstrated the presence of a multigenic family of cathepsin Ls in *F. hepatica*. The cathepsin L genes isolated by Yamasaki and Aoki (1993), Wijffels *et al.* (1994b) and Roche *et al.* (1997) show highest amino acid identity ($> 94\%$) to the Fcp6 gene of the Heussler and Dobbelaere family of genes. Since cathepsin Ls are ubiquitous in cells and function in normal cellular processes, and since PCR was used in their isolation, some of these genes would undoubtedly encode the cellular cathepsin proteinases. However, the cDNA described Roche *et al.* (1997) was isolated using anti-cathepsin L1 antiserum, and was shown by functional expression in *Saccharomyces cerevisiae* to encode a proteinase with similar substrate specificity to native *F. hepatica*-secreted cathepsin L1. Another gene isolated using anti-cathepsin L2 serum was also functionally expressed in yeast and shown to have a similar substrate specificity to the secreted *F. hepatica* cathepsin L2 (Dowd *et al.*, 1997; Tort, 1997). Cathepsin L2 shows 78% similarity to the cathepsin L1 proteinases, and is 97% similar to the Fcp1 member of the gene family identified by Heussler and Dobbelaere (1994) (see Section 8.5).

Because the cathepsin L proteinases are involved in pivotal functions such as tissue penetration, nutrition, immune evasion and possibly eggshell formation, they are considered important targets to which immunoprophylaxis could be directed. Wijffels *et al.* (1994a) performed vaccination studies in sheep using liver fluke cysteine proteinases (which probably contained both cathepsin L1 and cathepsin L2), but did not elicit significant protection against a challenge fluke infection. However, the faecal egg counts in the vaccinated animals were significantly reduced (52–70%) compared to control sheep. Furthermore, the eggs produced by the parasite in the vaccinated sheep had a > 80% reduced viability. Vaccine studies by Dalton *et al.* (1996b) induced high levels of protection (50 to 73%) in cattle against a challenge of *F. hepatica* metacercariae by vaccination with purified cathepsin L1 and cathepsin L2 in combination with liver fluke haemoglobin. In addition, the combination vaccine induced a > 98% anti-embryonation effect. The anti-embyonation/fecundity effects of these vaccines may be elicited by a direct effect of antibodies on enzymes present in the parasites' Mehlis glands and/or an indirect effect through the blocking of the parasites' ability to acquire nutrients from host tissues. More recently, the studies of Mulcahy *et al.* (1997) indicate that protection in cattle is related to the induction of Th1-type immune responses and to the induction of high titres of high-affinity IgG2 antibodies against the fluke proteinases. These and other vaccine studies were reviewed recently by Spithill and Dalton (1998).

4.2. Cathepsin B

The presence of cathepsin B-like activity in *F. hepatica* was first indicated by McGinty *et al.* (1993), who used novel biotinylated active-site affinity labels to probe extracts of juvenile and adult worms. They identified major cysteine proteinases in the 25–26 kDa range, and several minor proteinases of higher molecular size. More recently, Wilson *et al.* (in press) have shown that, although the major secreted proteinases of adult worms are cathepsin L-like, the major proteolytic activity of the newly excysted juvenile is a 29 kDa cathepsin B, with a pH optimum for activity at 7.5. The enzyme was localized to the gut epithelial cells of the newly excysted juvenile (NEJ) (Creaney *et al.*, 1996). Similar to the cathepsin Ls, the cathepsin B cleaves immunoglobulin and therefore may be involved in immune evasion by these invasive stages of liver fluke (Wilson *et al.*, in press). Although anti-cathepsin B monoclonal antibodies could not passively protect rats against a challenge infection with liver fluke metacercariae, Tkalevic and Meeusen (1998, unpublished) showed that the treatment induced the proliferation of immune effector cells (monocytes, neutrophils and eosinophils) and suggested that NEJ cathepsin B might downregulate the recruitment of these immune effector cells during infection.

Heussler and Dobbelaere (1994) isolated two adult *F. hepatica* cDNA fragments that encoded proteins with high similarity to the *S. mansoni* cathepsin B gene. A cDNA was also isolated from a newly excysted juvenile library which encoded a mature cathepsin B sequence of 254 amino acids that has 48–51% identity to various mammalian and *S. mansoni* cathepsin B (Wilson *et al.*, 1997). In addition, a clone was isolated from a genomic library, which also showed high similarity to mammalian and schistosome cathepsin B, but was considered to be a pseudogene because it lacked consensus intron–exon boundaries and termination codons within the regions homologous to other cathepsin Bs (Tkalcevic and Meeusen, unpublished).

4.3. Dipeptidylpeptidase (DPP) and Leucine Aminopeptidase (LAP)

Exopeptidases that cleave dipeptides (dipeptidylpeptidases) and single amino acids (aminopeptidases) from the NH_2-terminal end of synthetic substrates were isolated by Carmona *et al.* (1994) and Acosta *et al.* (1998), respectively. These enzymes may function in the latter stages of proteolysis, which would aid the parasite in the degradation of host tissue, providing dipeptides/amino acids that could then be absorbed through the gut wall as nutrients.

The liver fluke dipeptidylpeptidase (DPP), which was detected in all stages of the parasite that exist in the mammalian host, was characterized as a serine proteinase with a pH optimum for activity at 6.8 and a molecular size > 200 kDa. Whereas the enzyme could cleave the substrates H-Lys-Ala-NHMec and H-Gly-Pro-NHMec, it was inactive against H-Gly-Arg-NHMec and H-Arg-Arg-NHMec (Carmona *et al.*, 1994). The aminopeptidase, a metalloproteinase with an optimal activity between pH 8.0 and pH 8.5, cleaves many amino acids from fluorogenic peptide substrates, but preferentially cleaves leucine residues. The aminopeptidase was localized by histochemistry to the cells of the gut epithelium of adult flukes (Acosta *et al.*, 1998). Recent vaccine trials have shown that immunization of sheep with aminopeptidase alone or in combination with cathepsin L proteinases can induce high levels (80%) of protection in sheep against a liver fluke challenge infection (Piacenza, Acosta, Basmadjian, Dalton and Carmona, unpublished). Genes encoding these exopeptidases have not yet been isolated.

4.4. Other Proteinases of *F. hepatica*

There is evidence that *F. hepatica* express asparaginyl endopeptidases (legumain) analogous to the schistosome enzyme. NH_2-terminal sequencing of proteins extracted from liver fluke NEJ identified a legumain-like proteinase (Tkalcevic *et al.*, 1995). Moreover, enzymatic studies detected the presence of an asparaginyl endopeptidase activity in adult liver fluke which has similar physicochemical characteristics, including pH optimum for activity, substrate specificity and inhibitor profile, to schistosome legumain (Dowd and Dalton, unpublished). The presence of asparaginyl residues close to the site where cathepsin L1 and cathepsin L2 are cleaved to produce mature active enzymes supports the hypothesis of Dalton and Brindley (1996) that legumains play a role in the processing/activation of helminth proteinases.

Simpkin *et al.* (1980) described two proteinase activities secreted by adult liver flukes, a collagen-degrading activity with a neutral pH for activity and an acidic proteinase of 12 kDa which has yet to be characterized.

4.5. Proteinases of other Fasciolidae (*F. gigantica*)

Fagbemi and Hillyer (1991) characterized the proteolytic enzymes from extracts of adult worms of the tropical liver fluke, *F. gigantica*. These enzymes were cysteine proteinases that ranged in molecular sizes from 26 to 193 kDa, as revealed by gelatin-substrate PAGE. The most predominant proteinase, of 26–28 kDa, was partially purfied and shown to be capable of degrading immunoglobulin (Fagbemi and Hillyer, 1992). These enzymes are most likely the homologues of the *F. hepatica* cathepsin L proteinases.

5. PROTEINASES OF OTHER FLUKES

5.1. *Paragonimus westermani*

Several cysteine proteinases have been described in *Paragonimus* spp. The metacercariae of *P. westermani* secrete a neutral thiol proteinase (NTP) from the cells lining their digestive tract (Yamakami and Hamajima, 1988). The enzyme has a pH optimum of 7.5 and a substrate preference and inhibitor profile similar to members of the papain family of cysteine proteinases. In addition, the proteinase is active against collagen and haemoglobin, but not against elastin (Yamakami and Hamajima, 1988, 1990). Song and Dresden (1990) described a 20 kDa metacercarial acidic proteinase, while Chung *et al.* (1995) identified two acidic cysteine proteinases, both of which were capable of cleaving collagen, fibronectin and myosin. The 27 kDa cysteine proteinase was purified and shown to exhibit optimum activity at pH 4.0 and to be capable of degrading albumin, immunoglobulin and complement components (Yamakami *et al.*, 1995). Acidic proteinases of 20 kDa and 27 kDa have also been characterized in extracts of adult parasites of *P.*

westermani (Song and Kim, 1994) and *P. ohirai* (Yamakami *et al.*, 1986) respectively. In addition, a 35 kDa cysteine proteinase with optimal activity at pH 6.0 was isolated from eggs of *P. westermani* (Kang *et al.*, 1995).

Infection with *P. westermani* appears to induce immune tolerance, since people are susceptible to repeated infections. Following infection, eosinophils accumulate around the parasite, a typical response observed in other helminth infections (Hamajima *et al.*, 1986, 1991). Intraperitoneal injections of the NTP into guinea pigs and mice resulted in the accumulation of granulocytes into the peritoneal cavity, but a disappearance of macrophages and suppression of delayed-type hypersensitivity reactions against sheep red blood cells in mice (Hamajima *et al.*, 1991, 1994). Furthermore, intraperitoneal injections of NTP into mice suppressed the delayed footpad reaction and the production of haemagglutinin Ab, reduced expression of the major histocompatibility complex and interleukin 2 receptor on lymphocytes, induced suppressor cells in the spleen and suppressed rejections of skin grafts (Hamajima *et al.*, 1994). Collectively, these data suggest that the NTP (and possibly other cathepsin L-like proteinases) can induce immune supression and tolerance to parasite antigens and thereby prevent parasite elimination through its modulation of several immunological effector mechanisms.

A cDNA encoding the metacercarials NTP has been isolated. Sequence alignments shows that the protein exhibits a low similarity to papain, cathepsin H and cathepsin L (Yamamoto *et al.*, 1994) but is more related to the cruzipain-like members of the papain superfamily (see Section 8.5). Southern blot analysis suggests the presence of a cysteine proteinase gene family in *P. westermani* (Yamamoto *et al.*, 1994; Hamajima *et al.*, 1994).

5.2. *Haplometra cylindracea*

Histochemical studies showed the presence of hydrolytic enzymes in the ceacal epithelium of *Haplometra cylindra*, a parasite of the lungs of the European frog *Rana temporaria* (Halton, 1967). The enzymes, which are released following a blood meal, are capable of degrading haemoglobin. Both serine and cysteine proteinase activities have been identified in parasite ES products (Hawthorne *et al.*, 1994). The trypsin-like serine proteinase activities hydrolysed the substrate Z-Gly-Gly-NHMec, and have molecular sizes of 20 and 24 kDa, as determined by reducing SDS-PAGE. The cysteine proteinase activities consists of three cathepsin B-like proteinases of 14, 22/23 and 48 kDa and a cathepsin L-like proteinase of 55 kDa. Inhibition of both enzyme classes prevented the degradation of haemoglobin by ES products (Hawthorne *et al.*, 1994).

5.3. Clonorchis sinensis

The Chinese liver fluke, *C. sinensis*, resides in the bile ducts of humans. Two different proteinases have been described in adult worms, an immunogenic 18.5 kDa cysteine proteinase with a substrate specificity and inhibitor sensitivity similar to cathepsin B-like proteinases (Song *et al.*, 1990) and a secreted 24 kDa neutral cysteine proteinase that is associated with the cytotoxity observed in *C. sinensis* infections (Park *et al.*, 1995). Cysteine proteinase activity is also expressed in the immature flukes and in metacercariae from which a 32 kDa enzyme was purified (Song and Rege, 1991).

5.4. Diplostomum pseudospathaceum

The cercariae of *D. pseudospathaceum*, a parasite of aquatic birds, penetrate the skin of their intermediate fish host. Histochemical studies showed that cercariae express two distinct endoproteinase activities, a serine proteinase associated with the cercarial caecum (Moczon, 1994a) and a 40 kDa cysteine localized in the penetration glands and possibly involved in skin penetration (Moczon, 1994b).

6. PROTEINASES OF CESTODES

Compared to the trematodes and nematodes, there is a paucity of information on proteinases of cestodes and their possible roles in the infection process (for overview of cestode proteinases, see Table 2). The cestodes are obligate parasites with complex life-cycles involving at least two hosts. Infection of the definitive host is usually by ingestion, and adult stages reside in the gut. Before they finally take up residence in the host's gut, the larval stages (metacestodes) migrate through tissues such as muscle and the nervous system. The most important cestode parasites belong to the order Cyclophyllidea and include the *Taenia* spp., *Echinococcus* spp. and *Moniezia* spp. However, the order Pseudophyllidea includes *Spirometra*, which causes a significant zoonosis termed sparganosis.

6.1. Taenia

Metacestodes of *Taenia solium*, the causative agent of human neurocysticercosis, migrate from the gut to the muscles and central nervous system.

Table 2 The proteinases of cestodes

Genus & species	Stage of development	Proteinases described	Proposed function(s)	References
Taenia solium	Metacestode	Metallo, cysteine, aspartic (H)	Digestion, immune evasion	White et al., 1992
Taenia crassiceps	Cyst fluid	Cysteine (H)	Immune evasion	White et al., 1997
Taenia saginata	Onchospheres	Serine, cysteine, aminopeptidase (ES)	Tissue penetration	White et al., 1996
Echinococcus granulosus	Cysts and fluid cysts	Metallo Serine	Undefined Undefined immunogen	Marco and Nieto, 1991 Willis et al., 1997
Proteocephalus ambloplitis	Plerocercoid	Metallo (H)	Tissue penetration	Polzer et al., 1994
Schistocephalus solidus	Procercoid Plerocercoid, adult	Serine (H) Leucine aminopeptidase (H)	Tissue penetration Digestion	Polzer and Conradt, 1994 Polzer and Conradt, 1994
Spirometra erinacei	Plerocercoid	Cysteine (H)	Immunomodulation	Song et al., 1992; Fukase et al., 1985; Lie et al., 1996
Spirometra mansoni	Plerocercoid	Cysteine, serine (H)	Immunomodulation	Song and Chappel, 1993; Kong et al., 1997; Kong et al., 1994

They can survive for long periods, and proteolytic enzymes may be required to facilitate migration or to evade host immune responses (White et al., 1992). Proteinase activity in extracts of T. solium metacestodes was characterized using synthetic peptide and protein (haemoglobin, albumin and immunoglobulin) substrates and included

1. metallo-dependent aminopeptidase and neutral endopeptidase activities which co-eluted at 104 kDa on gel filtration chromatography;
2. an acidic 32 kDa cysteine proteinase activity which cleaved Z-Phe-Arg-AFC and haemoglobin;
3. a 90 kDa aspartic proteinase activity capable of digesting haemoglobin and IgG heavy chain at acid pH (White et al., 1992).

All three proteinase types were also present in cyst fluid, cyst walls and adult parasite homogenates (White et al., 1992). The parasite may use the enzymes to degrade haemoglobin and IgG for nutrition. The degradation of IgG heavy chain may protect the parasite by removing the complement fixing and opsonic portions of the molecule, and the remaining binding portion of antibody may block parasite epitopes which might otherwise be susceptible to destruction (White et al., 1992).

A cysteine proteinase which might mediate IgG degradation was purified to homogeneity from T. crassiceps cyst fluid (White et al., 1997). The proteinase cleaved Z-Phe-Arg-AFC, but not substrates with neutral or positively charged amino acids in the P2 position. NH2-terminal sequencing revealed a sequence similar to the conserved region surrounding the active site cysteine of cysteine proteinases.

T. saginata onchospheres excreted/secreted serine and cysteine proteinases and aminopeptidase activity during in vitro maintenance (White et al., 1996). It was suggested that the serine proteinase (18 kDa) and aminopeptidase (30 kDa) played a role in the invasion of the intestinal mucosa of the intermediate host (White et al., 1996).

6.2. *Echinococcus*

Marco and Nieto (1991) characterized proteinase activity in the hydatid cyst fluid, cyst membranes and protoscoleces of E. granulosus. The cysts were obtained from the liver and lungs of sheep and cattle. A number of proteinase bands ranging in size from 60 to 100 kDa were resolved by gelatin SDS-PAGE of bovine cyst fluid and cyst membranes. The proteinases were detected at pH 7.0, and all were inhibited by the metalloproteinase inhibitors EDTA or 1,10 o-phenanthroline, but not by inhibitors of other proteinase classes. Sonicated protoscoleces, ES from protoscoleces and detergent extracts of protoscoleces exhibited fewer and weaker bands. Since cyst membranes are the boundary between host and parasite, these proteinase may have specific functions at this interface (Marco and Nieto, 1991).

An N-terminal sequence of a major immunogenic cyst antigen from *E. granulosus* (termed antigen 5) showed homology to functional motifs associated with serine proteinase activity (Willis *et al.*, 1997). The posssible proteinase properties of this antigen have yet to be confirmed biochemically, but a cyst proteinase is suggested to digest the host collagenous capsule and provide space for cyst growth (Willis *et al.*, 1997). Aminopeptidase activity has also been identified in protoscoleces by McManus and Barrett (1985).

6.3. *Proteocephalus*

The plerocercoid of the pseudophyllidean cestode *Proteocephalus ambloplitis* can migrate to the visceral organs or, in the case of bass, from these into the intestinal tract of the same fish (Polzer *et al.*, 1994). Migration is rapid, and is facilitated by apical gland secretions. Plerocercoid homogenates contained a 31.5 kDa metalloproteinase which hydrolysed collagen, haemoglobin and elastin at a pH optimum of 9.0. Accordingly, the enzyme was suggested to be secreted by the apical glands to facilitate parasite migration through the tissues (Polzer *et al.*, 1994).

6.4. *Schistocephalus*

Procercoids of another pseudophyllidean cestode, *Schistocephalus solidus*, penetrate the intestinal wall of the second intermediate host, a stickleback, and then transform to plereocercoids in the body cavity. Procercoid extracts contained a 23.5 kDa chymotrypsin-like proteinase with collagenolytic activity, with an optimum at pH 8.0. This activity was not present in the plerocercoid or adult stages and may, therefore, aid penetration of the intestinal wall (Polzer and Conradt, 1994). Leucine aminopeptidase activity with optimal activity at pH 8.5 was detected in both isolated syncytial teguments from plerocercoids (93.5 kDa) and adults (89 kDa) and were suggested to have a nutritional function by degrading oligopeptides at the tegumental surface (Polzer and Conradt, 1994).

6.5. *Spirometra*

Adult *Spirometra* infect dogs, cats and wild carnivores in the Americas, Australia and the Far East. The procercoids are found in crustaceans, and the plerocercoids infect amphibia, birds and mammals. Man can become infected by drinking water contaminated with crustaceans containing procercoids or by eating plerocercoid-infected hosts such as pigs, the resulting zoonosis being termed sparganosis. In sparganum at least five different proteinases have been characterized and implicated in tissue invasion, nutrition and immune evasion. Casein hydolysing enzymes in the scolex of *S. erinacei* were described initially by Kwa (1972). Subsequently, plerocercoid cysteine proteinases of 19/21 kDa (pH optimum 7.0) and 28 kDa (pH optimum 5.7) were purified by gel filtration and ion-exchange chromatography using N-α-benzoyl-DL-arginine-p-nitroanilide or CBC-phe-arg-AFC as enzyme substrates (Fukase *et al.*, 1985 and Song *et al.*, 1992, respectively). Song and Chappel (1993) also purified a cysteine proteinase with optimal activity at pH 5.5 from extracts *S. mansoni* plerocercoids. This enzyme was present in ES, and exhibited collagenolytic and haemoglobinase activity. Since the enzyme was specifically recognized by antibody from infected patients, it was suggested that it had serodiagnostic potential (Song and Chappel, 1993).

More recently, Kong *et al.* (1997) showed that a cysteine proteinases of 53, 27 and 21 kDa induces a specific IgE response, and they suggested that the enzymes might somehow be involved in eliciting the release of cytokines that lead to Th2 type responses. Kong *et al.* (1994) identified a further three proteinases, all of which had optimal activity at pH 7.5 and were glycosylated. Inhibitor studies showed that two of the proteinases, of 198 and 104 kDa, were trypsin-like serine proteinases, while the third, of 36 kDa, was a chymotrysin-like serine proteinase with enhanced activity in the presence of Ca^{2+}. Both of the higher molecular weight enzymes degraded collagen, whereas the 36 kDa enzyme was shown to degrade human rIFNγ and myelin basic protein. All three proteinases were strongly immunogenic in infected individuals. Although the two larger proteinases may be secreted from the worm, the 36 kDa proteinase was located to the tegument (Kong *et al.*, 1994).

A cDNA encoding the 28 kDa *S. erinacei* cysteine proteinase has been isolated, and has shown significant homology to human and mouse cathepsin L (Liu *et al.*, 1996). The cDNA was expressed in *E. coli*, and was demonstrated to be an immunodominant antigen recognized by sparganosis patient sera but not by sera from patients infected with other parasites.

7. PROTEINASES OF PARASITIC NEMATODES

Nematode infections in man and animals are mostly associated with chronic, debilitating conditions and, relatively infrequently, with acute disease. Nematode parasites usually develop from eggs in the environment to an infective free-living larvae. Infection may occur by ingestion or skin penetration, or be transmitted by an insect vector. Because of the life-cycle diversity, these parasites will be discussed in a framework of superfamilies. Our review focuses on those areas where most is known, but the summary table indicates further background references for each species discussed (Table 3).

7.1. Rhabditoidea

This superfamily includes *Strongyloides* spp., which infect humans and domestic animals as well as birds, reptiles and amphibians. Infection can occur by ingestion or skin penetration by L3, which then migrate via the venous system, the lungs and trachea to develop into adult female worms in the intestine. Proteinases have long been implicated in skin penetration by *Strongyloides* spp. infective larvae (Lewert and Lee, 1954, 1956), and both cysteine and metallo proteinases are believed to facilitate this (Dresden *et al.*, 1985; Rege and Dresden, 1987).

Strongyloides stercoralis can migrate rapidly through the dermal extracellular matrix, a process which is mediated by a neutral metalloproteinase secreted by L3 larvae (McKerrow *et al.*, 1990). Carboxymethyl and hydroxamic acid dipeptide are very effective inhibitors of skin invasion, but have no effect on larval viability and motility (McKerrow *et al.*, 1990). This proteinase migrated as a 50 kDa band on non-reducing substrate gel electrophoresis, was inhibited by EDTA and 1,10 phenanthroline and hydrolysed elastin and glycoprotein, but not the collagen, components of an extracellular matrix prepared as outlined by McKerrow *et al.* (1983). The proteinase migrates at 40 kDa (the proteinase is now termed Ss40) under reducing conditions, does not contain any N-linked sugars, is immunogenic in infected humans and may be allergenic because it can stimulate the *in vitro* release of histamine from peripheral blood leukocytes of *S. stercoralis*-infected patients (Brindley *et al.*, 1995).

7.2. Ascaridoidea

This superfamily includes the major human parasites of species of *Ascaris* and *Toxocara*, as well as the sushi worm *Anisakis*. Histolytic and proteolytic activity have been suggested to be important for tissue penetration and maintenance within the final host since the 1930s and 1940s (Hsu, 1933; Rogers, 1941). These suggestions were later supported by ultrastructural observations (Lee *et al.*, 1973) and demonstrations of proteolytic activity in

Table 3 The proteinases of parasitic nematodes

Superfamily, genus and species	Stage of development	Proteinases described	Proposed function(s)	References
RHABDITOIDEA				
Strongyloides ratti	L3	Neutral collagenase (H/ES)	Skin penetration	Lewert and Lee, 1956
	Adult	Type 1 collagenase (H)	–	Wertheim et al., 1983
Strongyloides simiae	Adult	Neutral collagenase (H)	Skin penetration	Lewert and Lee, 1956
Strongyloides ransomi	L3	Cysteine (H/ES)	Skin penetration	Dresden et al., 1985; Rege and Dresden, 1987
		Metallo (H)	–	Dresden et al., 1985
Strongyloides stercoralis	L3	Metallo (ES)	Skin penetration	McKerrow et al., 1990; Brindley et al., 1995
ASCARIDOIDEA				
Ascaris suum	Adult	Aspartate (H)	Digestion	Maki et al., 1982; Rupova et al., 1984
	L2	Cysteine (H)	Digestion	Maki and Yanagisawa, 1986
		Aspartate, cysteine (ES)	Tissue migration	Knox and Kennedy, 1988
	L3/L4	Serine & metallo (ES)	Tissue migration	Knox and Kennedy, 1988
Toxocara canis	L2	Serine (ES)	Tissue migration	Robertson et al., 1989
		Cysteine (H)	Undefined	Loukas et al., 1998

Species	Stage	Protease (source)	Function	Reference
Anisakis simplex	L2	Trypsin-like (ES)	Tissue migration	Matthews, 1982a, 1984
		Serine (ES)	Tissue migration	Morris and Sakanari, 1994; Sakanari et al., 1989
FILAROIDEA				
Onchocerca lienalis	L3	Serine-elastase (ES)	Tissue migration	Lackey et al., 1989
	Microfilariae	Undefined (ES)	Extracellular matrix	Lackey et al., 1989
Onchocerca cervicalis	L3	Undefined (ES)	Extracellular matrix	Lackey et al., 1989
	Microfilariae	Serine and metallo (ES)	Tissue migration	Lackey et al., 1989
Onchocerca volvulus	L3	Cysteine (EF)	Moulting	Lustigman et al., 1996
Brugia pahangi	L3	Aminopeptidase (EF)	Moulting	Hong et al., 1993
	L3, adult female and males	Collagenase (metallo L3; ES-L3/female; H, male)	Undefined	Petralanda et al., 1986
Dirofilaria immitis	L3, L4	Metallo, cysteine (ES, EF)	Tissue migration, moulting	Richer et al., 1993
	Microfilariae	Serine (H)	Immune evasion	Tamashiro et al., 1987
	Adult	Aspartyl, cysteine (H)	Digestion and reproduction	Maki and Yanagisawa, 1986; Sato et al., 1993
STRONGYLOIDEA				
Necator americanus	L3	Undefined (ES)	Skin penetration	Matthews, 1977, 1982b
	L3	Serine, cysteine (ES)	Immune evasion	Pritchard et al., 1990
	L3	Cysteine (EF)	Moulting	Kumar and Pritchard, 1992
	Adult	Cysteine, aspartic, serine (ES)	Digestion	Kumar and Pritchard 1992; Pritchard (1990)

(continued)

Table 3 Continued

Superfamily, genus and species	Stage of development	Proteinases described	Proposed function(s)	References
STRONGYLOIDEA (*continued*)				
Ancylostoma caninum	L3	Metallo (ES, H)	Moulting, digestion	Hotez *et al.*, 1990; Hawdon *et al.*, 1995
	L3, adult	Metallo (ES)	Skin penetration, anticoagulant, digestion	Hotez and Cerami, 1983
	Adult	Cysteine (ES)	Anticoagulant	Dowd *et al.*, 1994; Harrop *et al.*, 1995
	Adult	Metallo	Digestion	Hotez and Cerami, 1983; Harrop *et al.*, 1995
Strongylus vulgaris	Adult	Cysteine (ES)	Digestion	Caffrey and Ryan, 1994
Syngamus trachea	Adult	Undefined	Immunomodulation	Riga *et al.*, 1995
TRICHOSTRONGYLOIDEA				
Dictyocaulus viviparus	L3	Metallo, serine (H)	Tissue migration	Britton *et al.*, 1992
	Adult	Cysteine (H)	Digestion	Rege *et al.*, 1989b
	Adult	Metallo, serine, cysteine (H & ES)	Digestion	Britton *et al.*, 1992
Haemonchus contortus	L3	Serine (elastase, H)	Anticoagulant	Knox and Jones, 1990
	L3	Metallo (EF)	Moulting, anticoagulant	Gamble *et al.*, 1989, 1996
	L4	Cysteine (ES)	Undefined	Gambe and Mansfield (1996)
	Adult	Cysteine (H, ES)	Digestion, immunomodulation, tissue penetration	Karnanu *et al.*, 1993; Knox *et al.*, 1993; Rhoads and Fetterer, 1995; Fetterer and Rhoads, 1997a,b

Species	Stage	Protease (location)	Function	Reference
	Adult	Aspartic (gut)	Digestion	Longbottom et al., 1997
		Metallo (gut)	Digestion	Redmond et al., 1997
		Cysteine (gut)	Digestion	Cox et al., 1990; Knox et al., 1995
				Knox et al., 1995
Ostertagia ostertagi	L3	Serine (gut)	Digestion	DeCock et al., 1993
	L4	Serine, cysteine (H)	Undefined	DeCock et al., 1993
	Adult	Serine (ES)	Undefined	DeCock et al., 1993;
		Serine, cysteine (H)	Digestion	Pratt et al., 1992b
	Adult	Cysteine, aspartyl, metallo (gut)	Digestion	Smith et al., 1993; Knox et al., 1995
Teladorsagia (Ostertagia) circumcincta	L3	Serine, cysteine, metallo (ES)	Undefined	Knox and Jones, 1990.
	L3	Cysteine (ES)	Tissue penetration, digestion	Young et al., 1995
	L4	Serine (ES) + undefined	Undefined	Young et al., 1995
	Adult	Metallo (ES)	Digestion, anticoagulant	Young et al., 1995
	Adult	Aspartic, metallo, cysteine serine (gut)	Digestion	Smith et al., 1993; Knox et al., 1995
Trichostrongylus vitrinus	L4, adult	Serine, metallo (ES)	Tissue penetration, feeding	MacLennan et al., 1997
Trichostrongylus colubriformis	L3, adult	Undefined (H, ES)	Tissue penetration, feeding	Knox and Jones, 1990
Nippostrongylus brasiliensis	L3	Serine (elastase, H)	Skin penetration	Knox and Jones, 1990
	L3	Metallo (ES)	Skin penetration	Healer et al., 1991
	Adult	Aspartic (H)	Feeding, morphogenesis	Bolla and Weinstein, 1980
	Adult	Cysteine (H, ES)	Immunomodulation	Kamata et al., 1995
	Adult	Metallo (ES)	Undefined	Healer et al., 1991

(continued)

Table 3 Continued

Superfamily, genus and species	Stage of development	Proteinases described	Proposed function(s)	References
TRICHOSTRONGYLOIDEA (continued)				
Heligmosomoides polygyrus	Adult	Serine (ES)	Digestion	Monroy et al., 1989
Trichuris suis	Adult	Metallo (ES)	Tissue penetration, digestion	Hill et al., 1993
Trichuris muris	Adult	Serine, metallo, cysteine (H)	Digestion	Drake, et al., 1994
	Adult	Serine (ES)	Tissue degradation	Drake, et al., 1994
Trichinella spiralis	L3	Serine, metallo (H)	Tissue degradation	Criado-Fornelio et al., 1992
	Adult	Serine (ES)	Digestion, anticoagulant	Todorova et al., 1995
NEMATODE PARASITES OF PLANTS				
Meloidogyne incognita	L2	Aminopeptidase, undefined (ES)	Hatching	Perry et al., 1992
	L2	Serine	Hatching	Dasgupta and Ganguy, 1975
Globodera pallida	Adult	Cysteine (H)	Digestion	Koritsas and Atkinson, 1994; Lilley et al., 1996
Heterodera glycines	Adult	Cysteine (H)	Digestion	Lilley et al., 1996; Unwin et al., 1997
FREE-LIVING				
Caenorhabditis elegans	Adult	Aspartic (H)	Digestion	Sarkis et al., 1988a; Jacobson et al., 1988
	Adult	Cysteine (H)	Digestion	Sarkis et al., 1988 a,b; Ray and McKerrow, 1992; Larminie and Johnstone, 1996

worm extracts (Rogers, 1941) and secretions (Ruitenberg and Loendersloot, 1971).

7.2.1. *Ascaris*

Proteinase activity against a variety of protein substrates has been described in intestinal extracts of adult *Ascaris* by a number of authors (Rogers, 1941; Maki *et al.*, 1982; Rupova *et al.*, 1984). These included an adult parasite haemoglobinase activity with a pH optimum of 3.8 that was totally inhibited by pepstatin, the general inhibitor of aspartic proteases (Rupova *et al.*, 1984) and a cysteine proteinase with optimal activity at pH 5.6 against azocoll (Maki and Yanagisawa, 1986).

More recently, attention has focused on the *in vitro* released products (IVR) of the larval stages of *A. suum*, because they possess a number of immunologically active molecules (Kennedy and Qureshi, 1986), some of which may be proteinases required for the penetration of host tissue barriers (Knox and Kennedy, 1988). L2 and L3/L4 larval stages release all four classes of proteinases *in vitro* in a stage-specific manner. Antibody isolated from serum of infected rabbits inhibited these proteinases, suggesting that these responses may impair parasite migration *in vivo* (Knox and Kennedy, 1988).

7.2.2. *Toxocara*

The parasitic nematode *Toxocara canis* infects a large proportion of dogs around the world (Glickman and Schantz, 1981). Because of poor hygiene, humans, mainly young children, also become infected, and this can cause a variety of symptoms, commonly referred to as *visceral larval migrans*, including liver enlargement, lung inflammation and ocular defects. Infective second-stage larvae of *Toxocara* released proteinases *in vitro* which degrade components of extracellular matrix and basement membrane, including glycoproteins (mainly fibronectin), elastin and collagen types I and III, and may, therefore, facilitate tissue migration. Gelatin-based proteinase assays showed that ES contained serine proteinase activity, with maximal activity at pH 9.0 and less so at pHs 5.0 and 7.0; these activities may be due to several proteinases or to a single subunit enzyme with multiple peaks of activity (Robertson *et al.*, 1989).

A cysteine proteinase activity was identified in somatic extracts of *T. canis* larvae. A cDNA with closest sequence homology to cathepsin L-like proteinases from *C. elegans*, a cathepsin L-like enzyme from *Brugia pahangi* and a range of parasite and plant papain-like proteinases was isolated recently (Loukas *et al.*, 1998; see Section 8.5.5). Several charged residues in the S_1 and S_2 subsites which determine substrate specificity were shared with human cathepsin B, including a Glu known to permit cleavage of Arg-Arg peptide bonds. Antiserum to the bacterially expressed gene product recognized a 30 kDa proteinase in larval extracts. The proteinase was specifically recognized by infected patient sera, indicating that it might be a useful diagnostic antigen for human toxocariasis.

7.2.3. *Anisakis*

Anisakis larvae penetrate the gut wall of both invertebrate and vertebrate intermediate hosts, and invade the stomach mucosa of their final hosts. Humans are aberrant hosts, and become infected by eating seafood dishes and undercooked fish. The larvae penetrate the mucosa, submucosa and muscularis of the stomach and small intestine, and may reach the liver, pancreas and gall bladder. Accordingly, extracellular proteinases were suggested to be involved in this process (Hoeppli, 1933; Hsu, 1933). *Anisakis* larvae release a trypsin-like proteinase activity *in vitro* from their oesophageal glands which is optimal at pH 7.5 and at 37°C, the body temperature of the final host (Matthews, 1982a, 1984). Sakanari and McKerrow (1990) also noted that the secretions of tissue invasive larvae contained a 25 kDa serine proteinase, which reacted with rat anti-trypsin sera and exhibited a substrate specifity and inhibitor sensitivity similar to the trypsin-like serine proteinase family.

Recently, Morris and Sakanari (1994) purified a serine proteinase from extracts of *Anisakis* infective larvae, using a combination of anion exchange chromatography and affinity selection based on proteinase-inhibitor interactions. The 30 kDa proteinase was trypsin-like, and preferentially cleaved substrates with the basic amino acid arginine at the P_1 position. The presence of a bulky hydrophobic residue at the P_2 position greatly decreased the rate of substrate hydrolysis. After endo Lys-C digestion, internal amino acid sequence was obtained from three fragments, and the sequences obtained showed identity in 33 out of 37 residues to porcine trypsin. During this study a second serine proteinase with identity to a secreted tissue-destructive serine proteinase from the pathogenic bacterium *Dichelobacter nodosus* was identified. Moreover, Morris and Sakanari (1994) isolated a Kunitz-type serine proteinase inhibitor, which inhibited the *Anisakis* proteinase activity *in vitro*. Histochemical staining suggested that the inhibitor may be complexed with a proteinase in the upper digestive tract of the larvae preventing auto-proteolysis.

Serine proteinase gene fragments were isolated from genomic DNA of *Anisakis* larvae using degenerate oligonucleotide primers (designed to anneal around the active site-encoding regions) in PCR (Sakanari *et al.*, 1989). Of four gene fragments isolated, one was 67% identical to the rat trypsin II gene, and intron–exon junctions were conserved, suggesting structural and functional similarities to rat trypsin.

7.3. Filaroidea

The Filaroidea inhabit the blood or lymphatic system, connective tissue and body cavities of the definitive host, and the majority produce embryos that live in the blood or skin, from where they are sucked out by the intermediate host, a blood-feeding arthropod. They are then introduced to a new host through the skin when the arthropods bite. The L3 and microfilariae of some filariids such as *Onchocerca* and *Dirofilaria* undergo considerable tissue migration, while others, such as *Brugia malayi* and *Wucheria bancrofti*, are located in the lymphatics.

7.3.1. *Onchocerca*

Extracellular matrix was degraded by both L3 and microfilariae of *Onchocerca cervicalis* and *O. lienalis*, but degradation was markedly lower for the latter, suggesting that secreted proteinases may not be crucial for the migration of the smaller microfilariae through the skin (Lackey *et al.*, 1989). In another study, Petralanda *et al.* (1986) demonstrated collagenase activity in microfilariae of *O. volvulus*, but levels were 100-fold less that those detected in adult females. A 43 kDa proteinase secreted by *O. linealis* L3 larvae was defined as a serine-elastase by substrate gel electrophoresis and other proteinase assays. Microfilariae of *O. cervicalis* and *O. cervipedis* produced both a serine and a metallo proteinase, as judged by inhibitor analyses in substrate gels. Microfilariae still present in the uterus of *O. cervicalis* females did not contain the two major proteinases found in the skin microfilariae (60 kDa metallo, 30 kDa serine) but did express proteinases at 50 and 100 kDa (Lackey *et al.*, 1989). The authors suggested that the changes in proteinase expression could be correlated to important morphological and immunologic changes, and are qualitatively and quantitatively distinct from that of lymphatic or blood-dwelling filarial nematode parasites, which remain intravascular. Moreover, stage-specific expression of proteinases may be under the control of a critical stimulatory factor. Petralanda and Piessens (1994) suggested that filarial proteinases may contribute to the pathogenesis of chronic onchocercal dermatitis directly by enzymatically destroying connective tissue in the skin, and indirectly by triggering auto-immune responses to self-determinants on connective tissue proteins normally hidden within the structure of the extracellular matrix complex.

Fluoromethylketone (FMK) inhibitors of cysteine proteinases did not affect the viability of *O. volvulus* fourth-stage larvae, but did reduce the number of L3 that moulted to L4 in a time- and dose-dependent manner (Lustigman *et al.*, 1996). Ultrastructural studies indicated that the new L4 cuticle was synthesized, but was not separated from the L3 cuticle. Using labelled inhibitors, the cysteine proteinases were located to cuticle regions where the separation between the cuticles occurs in moulting larvae. The amino acid sequence deduced from a cysteine proteinase gene, cloned using a PCR-based strategy, showed homology to a variety of cysteine proteinases, but was most similar to cathepsin C and cathepsin L. The *Onchocerca* proteinase, encoded by a single copy gene, contains a unique five-amino-acid repeat, CGSCW, which is part of the cysteine active site motif CGSCWAF, and two additional repeats of GSC (Lustigman *et al.*, 1996; see Section 8.4). Jolodar and Miller (1997) recently isolated a cDNA fragment encoding an aspartic proteinase homologue from *O. volvulus* microfilariae, and suggested that this proteinase might be required for tissue migration.

7.3.2. *Dirofilaria*

Dirofilaria immitis larvae secrete cysteine and metallo proteinases, 49–54 kDa and 34–39 kDa in size respectively, which can degrade components of extracellular matrix at physiological pH. Matrix degradation was greater by L3 and increased twofold at the onset of moulting (Richer *et al.*, 1992). An involvement of the *Dirofilaria* larval proteinases in the moulting process was suggested by a recent *in vitro* study by Richer *et al.* (1993), which showed that in the presence of FMK cysteine proteinase inhibitors L3 larvae remained viable but moulting to the L4 was impaired. Although the L4 cuticle was synthesized, the L3 cuticle was not shed. Proteinase activity was examined *in situ* through the moulting process by culturing L3 in the presence of a synthetic peptide substrate Z-Val-Leu-Arg-4-methoxy-B-naphthylamide. Activity during days 0–4 of the moulting process was first observed on the anterior tip of the larvae, then in the pharynx and progressively down the L4 as it shed its cuticle.

Extracts of *D. immitis* microfilariae contained multiple proteinases active between pHs 7.0 and 9.0, the most prominent of which appeared to be a serine proteinase (Tamashiro *et al.*, 1987). The extracts contained SDS-resistant proteinases of 22 and 76 kDa with the ability to degrade IgG, fibrinogen and, less readily, haemoglobin. IgG degradation occurred in the presence of live parasites and was therefore suggested to be involved in immune evasion.

Carboxyl (aspartyl) and thiol (cysteine) proteinase activity were detected in extracts of adult *D. immitis*. The former had optimal activity against haemoglobin at pH 2.5, whereas the latter was optimal at pH 4.6 (in the presence of DTT) against the substrate azocoll (Maki and Yanagisawa, 1986). A 42 kDa aspartyl proteinase with a broad tissue distribution was purified from adult *D. immitis*. The proteinase activity was highest at pH 2.8 to 3.4, and could degrade haemoglobin and myoglobin to acid-soluble peptides, but not free amino acids (Sato *et al.*, 1993, 1995).

7.3.3. *Brugia*

Petralanda *et al.* (1986) showed that ES of infective larval and female adult worms, and extracts of male worms of *Brugia malayi* contained collagenolytic activity. The larval activity was ascribed to a metallo-proteinase. Antibodies to the microfilarial proteinase were present in patient sera, and immunoprecipitated the collagenase but did not inhibit enzyme activity. Metallo-aminopeptidase activity was detected in ES from L3 *Brugia pahangi*, and was associated with larval moulting (Hong *et al.*, 1993).

7.4. Strongyloidea

The Strongyloidea include the hookworm genera *Ancylostoma* and *Necator*, which together are the causal agents of one of the most important and widespread helminthic diseases in man. Hookworm infection usually occurs by skin penetration, the larvae burrowing until they reach a lymph or blood vessel which carries them to the lungs. They emerge into the airways, thence to the throat, and are swallowed. They then develop into adults, which feed on the gut mucosa and blood of the host. Infected individuals may have an allergic skin reaction to invading larvae and respiratory symptoms including pneumonia, and zoonotic infections may cause visceral larval migrans. In addition, the blood-feeding habits of the adult parasites induce anaemia. The strongyles, exemplified by species infecting horses such as *Strongylus vulgaris*, *S. edentatus* and *S. equinus*, inhabit the caecum and colon, and the infective larvae penetrate the gut wall, from where, depending on the species, they migrate to arterial endothelia, the liver or under the peritoneum to the flanks and hepatic ligaments.

7.4.1. *Necator*

Necator americanus actively secretes proteinases during skin penetration (Matthews, 1977, 1982b). Exsheathing fluid from *N. americanus* L3 contained a single 116 kDa cysteine proteinase which may facilitate exsheathment and initial access to the dermal capillary network (Kumar and Pritchard, 1992). The ES proteinases of *N. americanus* L3 were heterogeneous, ranging in molecular weight from 62 to 219 kDa, and comprised serine, cysteine and metallo proteinases (Kumar and Pritchard, 1992). The proteinases cleaved the Fc portion of IgM, IgG and IgA in decreasing order of preference, but not that of IgD or IgE; this activity was inhibited by the serine proteinase inhibitor PmsF. The authors proposed a correlation between this pattern of immunoglobulin cleavage and the evasion of Ig responses in the skin of human hosts, IgM being an early and IgG a late response. Proteinases in ES of adults preferentially cleaved IgA, the principle anti-parasite immunoglobulin in the gastrointestinal tract (Pritchard, 1990). Pritchard *et al.* (1990) also showed that adult *N. americanus* release a mixture of aspartic, cysteine and serine proteinases *in vitro* with haemoglobinase (pH 5.0 to 7.0) and fibrinogenolytic activities (pH 3.5), and suggested that they may facilitate blood feeding.

7.4.2. *Ancylostoma*

Thorson (1953a,b) was the first to describe proteinase activity in the oesophageal glands of adult *A. caninum* and to show that these were inhibited by sera from dogs immune to infection, and that extracts of the glands could be used to vaccinate against infection. More recently, Hotez and Cerami (1983) described the secretion of an anticoagulant by adult *Ancylostoma caninum* which preferentially cleaved fibrinogen. This 37 kDa metalloproteinase with elastinolytic-like activity in the pH range 9–11 was purified by a series of chromatographic steps. Hotez *et al.* (1985) proposed that the proteinase had a histolytic function degrading the intestinal mucosal bolus lodged in the parasite buccal capsule. Immunoblot analyses showed that the proteinase was present in extracts of infective larvae, an observation which suggested that the proteinase played a role in skin invasion.

A latter study by Hotez *et al.* (1990) showed that larval extracts and adult ES of *Ancylostoma caninum and A. duodenale* contained metalloproteinase activity with pH optima between pH 9.0 and 10.0. The proteinase of both species resolved at 68 and 38 kDa in casein or gelatin substrate gels, and they degraded fibronectin to a 60 kDa intermediate fragment. Unlike the obligate skin-penetrating nematode *S. stercoralis*, *Ancylostoma* sp. larval proteinases could not degrade the connective tissue macromolecules elastin or laminin, suggesting that skin penetration may not be their primary function. Hotez *et al.* (1990) noted that the 38 kDa proteinase had a similar molecular weight to a zinc-metalloproteinase in exsheathing fluid from infective larvae of the trichostrongyle nematode *Haemonchus contortus*, which mediates breakdown of the cuticle (sheath) and release of the anterior cap, allowing the larvae to escape (Gamble et al., 1989). Polyclonal antibody to a metalloproteinase in adult worms cross-reacted with larval homogenates (Hotez et al., 1985).

A metalloproteinase may also be involved in feeding, since release of these enzymes by *A. caninum* infective larvae coincided with the onset of feeding *in vitro*. Moreover, the metalloproteinase inhibitor, 1,10 phenanthroline, completely inhibited *in vitro* parasite feeding, an effect that could be reversed by the addition of zinc chloride to the culture fluid. Candidate proteinases of 50 and 90 kDa, which were inhibited by 1,10 phenanthroline, were detected by gelatin substrate gels in the worm ES (Hawdon et al., 1995).

Dowd et al. (1994a) described the secretion of a cathepsin L-like cysteine proteinase by adult *A. caninum* capable of degrading the plasmin substrate, Boc-Val-Leu-Lys-AMC, a specificity which supports an anticoagulant function. *A. caninum* can develop in the human gut and can cause an eosinophilic enteritis; the latter may be caused in part by cysteine and metalloproteinases in the parasite secretions (Dowd et al., 1994a; Harrop et al., 1995b). cDNAs encoding two distinct cysteine proteinase genes were isolated from an adult *A. caninum* cDNA library using a cloned fragment generated by consensus PCR (Harrop et al., 1995; see Section 8.4). The first gene, AcCP-1, encoded a cathepsin B-like zymogen of 343 amino acids, which was predicted to be processed to a mature protein of 255 amino acids and showed closest homology to a *H. contortus* cysteine proteinase (61%) and to human cathepsin B (60%). The second gene, AcCP-2, encoded a mature enzyme of 254 amino acids that had 86% identity to AcCP-1 and 58% and 47% identity to bovine and human cathepsin Bs respectively. Rabbit antisera to recombinant AcCP-1 reacted with the oesophageal, amphidial and excretory glands in male and female worms, and with a 40 kDa peptide in Western blots of adult hookworm ES, suggesting that the proteinase encoded by AcCP-1 is secreted by the parasite (Harrop et al., 1995b). Recent computer-based structural predictions showed that the substrate binding and specificity of AcCP-1 differed from cathepsin B in that it would preferentially cleave Phe-Arg, a cathepsin L-like substrate preference, over Arg-Arg (Brinkworth et al., 1996).

7.4.3. *Strongylus*

To date, proteinases have been characterized only in adult *Strongylus vulgaris*, the most pathogenic and prevalent of the large strongyles which infect the horse (Caffrey and Ryan, 1994). ES proteinase activity was optimal at pH 6.0, enhanced by DTT and the proteinases hydrolysed cathepsin B and L-type cysteine proteinase substrates. Proteinase activity were resolved in two peaks (18 and 24 kDa) by molecular sieve FPLC and as eight bands (87–29 kDa) by SDS gelatin substrate gel analysis. These proteinases may hydrolyse the host mucosal plug within the buccal capsule for the provision of nutrients.

7.4.4. *Syngamus*

One member of this genus, *Syngamus trachea*, is of veterinary significance, as it parasitises the upper respiratory tract of game birds and domestic fowl, particularly turkey poults. The adult parasites secreted proteinase(s) which hydrolysed azocoll over a broad pH range (6 to 10, optimum pH 7.0 and 9.0) (Riga *et al.*, 1995). The authors proposed that the proteinases, along with secreted acetylcholinesterases, may have an immunomodulatory role.

7.5. Trichostrongyloidea

Some members of this superfamily are very important parasites of domestic animals. Infection occurs by ingestion of contaminated vegetation; third-stage larvae exsheath in the gut and, depending on the species, penetrate the mucosal surface to varying degrees. They develop to fourth-stage larvae, then emerge into the gut lumen where they establish themselves as adults. Disease symptoms result from mucosal damage caused by invading larvae, the feeding habits of the adults and the immune response to the parasite. *Haemonchus contortus*, for example, inhabits the true stomach (abomasum) in ruminants and, because it is a blood-feeder, is the most pathogenic species. The genera, *Ostertagia*, *Teldorsagia* as well as *Trichostrongylus axei*, live in or close to the abomasal glands, while other genera, including *Trichostrongylus* spp., *Cooperia* and *Nemtodirus*, inhabit the intestine.

Members of the family Metastrongylidae inhabit the respiratory system of mammals, and include the important bovine lungworm *Dictyocaulus viviparus* as well as genera that infect pigs, cats and deer. In the case of *Dictyocaulus*, larvae hatch in the bronchi and are either coughed out directly into the environment or are swallowed and passed in the faeces. They then moult twice to infective larvae (the shed cuticles being retained for some time) and, following ingestion by a host, migrate to the lungs via the lymphatics and the blood. L4 break out of the capillaries into the alveoli.

7.5.1. Dictyocaulus

Rege et al. (1989b) partially purified a 25 kDa cysteine proteinase from acidic extracts of adult *Dictyocaulus viviparus*, which could degrade type I haemoglobin and collagen, properties which may contribute to parasite nutrition and bronchial pathology in cattle. Besides cysteine proteinase activity, Britton et al. (1992) also detected serine and metallo proteinases in adult parasite extracts, and showed that the proteinases were expressed in a stage-specific manner. Antibodies harvested from both *D. viviparus* infected calves and calves vaccinated against infection with the irradiated larval vaccine, Dictol, inhibit L3 homogenate and adult ES proteinases and perhaps mediate protective immunity (Britton et al., 1992).

7.5.2. Haemonchus

The non-blood-feeding third-stage larvae of *Haemonchus contortus* release a different profile of proteinases from both L4 and adult parasites. *In vitro* cultured exsheathed L3 develop to fourth-stage larvae in 48 to 72 hours, and remain viable for 120 hours (Gamble et al., 1996). Leucine aminopeptidase (Rogers, 1970) and pseudocollagenase (Rogers, 1982) have been implicated in exsheathment, the role of the former activity being refuted in an earlier study (Ozerol and Silverman, 1969). Furthermore, although Gamble et al. (1989) did not detect either activity in exsheathing fluids, they did identify a 44 kDa zinc-metalloproteinase which mediated digestion of the second-moult sheath, a process inhibited by 1,10 o-phenathroline and enhanced by zinc ions. This proteinase hydrolysed several protein substrates, but not native collagen, and it induced refractile ring formation *in vitro*. Coincident with the third moult and the onset of feeding, larvae secreted a 46 kDa zinc metalloproteinase which could be distinguished from the 44 kDa proteinase on the basis of migration in SDS-PAGE gels and substrate specificity towards azocoll, azocasein and β-insulin (Gamble et al., 1996). The 46 kDa proteinase was released in increasing amounts after the third moult, and degraded a variety of high molecular weight substrates of host origin, including fibrinogen. The proteinase may have a dual function, in aiding ecdysis and in the inhibition of blood clotting to facilitate extracorporeal digestion (Gamble et al., 1996). Gamble and Mansfield (1996) also identified a cathepsin C-like proteinase in L4 ES using the substrate Gly-Phe-*p*-nitroanilide (see Section 8.4.2).

Histological evidence suggested that bleeding from damaged mucosal capillaries continued for a long period after detachment of the adult *H. contortus*, and indicated that these parasite release an anticoagulant. Knox and Jones (1990) demonstrated the presence of elastase-like activity in the

ES from adult parasites maintained *in vitro*, which may be analogous to the elastase-like anticoagulant in ES from *Ancylostoma caninum* described by Hotez and Cerami (1983).

The proteinases in a USA isolate of adult *H. contortus* ES were the subject of detailed analysis by Karanu *et al.* (1993). Their studies showed that ES contained at least four cysteine proteinases, which ranged in size from 32 to 40 kDa and had slightly differing pH sensitivities. In contrast, they observed that the predominant activity in ES from a Kenyan isolate migrated at 52 kDa, consistent with data from UK adult parasite extracts (Knox *et al.*, 1993) and ES (Knox, D.P. and Schallig, H.D.F, unpublished data). This potential geographical diversity may have implications for control strategies targeting secreted proteinases (Karanu *et al.*, 1993). Proteinase inhibitor studies revealed that *Haemonchus* ES also contained aspartic and metalloproteinase activities.

Cathepsin L-like proteinases were detected in ES from the L4 and adult stages, but not the L3 (Rhoads and Fetterer, 1995). This pattern of release correlates with the onset of blood feeding and, indeed, the proteinases hydrolysed fibrinogen, haemoglobin, collagen and IgG. Moreover, whole ES prevented blood clotting, and this property was inhibited only by the cysteine proteinase-specific inhibitor, E64. These ES cysteine proteinases were derived, at least in part, from the gut of *H. contortus*, and *in vitro* release was reduced in the presence of the cholinergic antagonist levamisole or the mitochondrial uncoupler rafonoxide, indicating that release was an active process, possibly associated with worm defaecation (Rhoads and Fetterer, 1995). Irrespective of the mechanism of release, fibrinogen-degrading and haemoglobinase activities are consistent with blood feeding, IgG cleavage with immune evasion and collagenase activity with penetration of the abomasal mucosa to gain access to blood capillaries.

A role for proteinases in feeding was supported by the observation that L4 and adult *Haemonchus*, maintained in the presence of a ^{3}H-labelled extracellular matrix which mimicked connective tissue, readily degraded the glycoprotein, elastin and collagen components of the matrix (Rhoads and Fetterer 1996, 1997a). Matrix degradation was blocked by Z-phe-ala-FMK, a specific cysteine proteinase inhibitor. In contrast, the *in vitro* uptake and incorporation of radiolabelled haemoglobin by adult parasites was not inhibited by the cysteine proteinase inhibitor Z-phe-ala-FMK at 0.1 mM, although haemoglobin breakdown in the culture medium was reduced by 50% (Fetterer and Rhoads, 1997b). Uptake was markedly reduced in the presence of inhibitor at 1.0 mM, but at this concentration the inhibitor also reduced parasite viability. Uptake was, however, reduced by 40% by the serine proteinase inhibitor AEBSF.

Substantial protection against *H. contortus* infection in both sheep and goats has been achieved by immunization with either crude extracts of intestines dissected from adult parasites (Jasmer and McGuire, 1991; Smith, 1993) or by using individual antigens (Munn, 1997) or antigen complexes (Smith et al., 1994a,b; Knox et al., 1995). In general terms, immunization reduced faecal egg output by at least 90% and final worm burdens by at least 53% and up to 90%. Of these preparations, a protein, designated H11, is the most effective antigen that has been described for a parasitic nematode. H11 is an integral membrane glycoprotein that is expressed in the intestinal microvilli of the parasite, and is a homologue of mammalian microsomal aminopeptidases (Smith et al., 1997). Activity was inhibited by the aminopeptidase inhibitors bestatin, amastatin and the metalloproteinase inhibitor 1,10 o-phenanthroline. The enzyme activity was also inhibited by immunoglobulins from sera of vaccinated lambs, and the level of inhibition correlated with the level of protection against challenge infection.

Recently, a cDNA encoding full-length H11 was isolated (Smith et al., 1997). The deduced amino acid sequence showed 52% similarity and 32% identity with those of mammalian microsomal aminopeptidases, and included an HEXXHXW sequence motif characteristic of microsomal aminopeptidases. These predictions also indicated that H11 possesses a short N-terminal cytoplasmic tail, a single transmembrane region and a long extracellular region which includes the active site as well as putative N-linked glycosylation sites. Recombinant H11 was expressed in the baculovirus-insect cell system as an enzymically active detergent-soluble 112 kDa glycoprotein. The effcicacy of this recombinant H11, and various bacterially expressed versions, as vaccines are currently under evaluation.

Smith et al. (1994a) described the isolation of a galactose-containing glycoprotein complex (H-gal-GP) from integral membrane protein extracts of the adult parasite. It contains strong haemoglobinase activity at acidic pH (pH 4.0) which is inhibited by pepstatin, and activity at neutral pH that is inhibited by chelating and reducing agents, characteristics shared by mammalian neutral endopeptidases. Expression of both these proteinases and H11 is developmentally regulated, being restricted to the blood-feeding L4 and adult stages (Smith et al., 1997). The aspartyl proteinase component of H-gal-GP is encoded by a single copy gene, the coding sequence (428 amino acids) of which showed significant homology (58% similarity, 34% identity) to human and porcine pepsinogen A precursors. Regions of homology included two conserved hydrophobic-hydrophobic-Asp-Thr-Gly active site domains, two potential peptidase cleavage sites and two cysteine residues involved in conformational disulphide bonding (Longbottom et al., 1997). Despite being purified with the integral membrane protein fraction, immunolocalization studies showed that the pepsinogen component was also released into the intestinal lumen, and this gene product may therefore be responsible for aspartyl proteinase activity detected in adult ES (Karanu et al., 1993).

The deduced amino acid sequence of the neutral proteinase showed maximal homology (50% similarity, 25% identity) to mammalian type II integral membrane protein neutral endopeptidase, and Southern blot analysis indicated it was a member of a multigene family (Redmond et al., 1997). The predicted amino acid sequence contained a characteristic VxxHExxH as VVGHELVH, which is involved in zinc binding and in catalysis. By analogy to bacterial thermolysin, the two histidine residues are zinc-co-ordinating ligands, while the glutamine plays a role in catalysis by polarizing a water molecule (Redmond et al., 1997).

Several studies suggest that cysteine proteinases may also be useful vaccine candidates for controlling *Haemonchus*. Cox et al. (1990) detected cysteine proteinase activity in water-soluble adult parasite extracts, which preferentially degraded fibrinogen and increased clotting time of sheep plasma *in vitro*. Lambs immunized with extracts from the adult parasite enriched for this proteinase activity were substantially protected against challenge infection (Boisvenue et al., 1992). Similarly, Knox et al. (1993) showed that cysteine proteinases in extracts from adult *Haemonchus* hydrolysed the blood proteins haemoglobin, albumin and fibrinogen, and that this activity was inhibited by sera from lambs which had been successfully vaccinated against haemonchosis with parasite gut extracts. In addition, these authors showed that the expression of different proteinase 'isoforms' in parasites capable of surviving in immunized lambs and those harvested from unvaccinated controls differed. Subsequently, Knox et al. (1995) demonstrated that vaccination of lambs with integral membrane protein extracts enriched for cysteine proteinase activity from adult *Haemonchus* conferred substantial protection against challenge infection. Analysis of the extracts by gelatin-substrate PAGE showed that they contained 35 kDa and 55 kDa cysteine proteinases and also weak serine and metalloproteinase activity in the 70 kDa to 100 kDa size range.

There has been substantial progress in isolating the genes encoding the cysteine proteinases present in the above protective extracts. Cox et al. (1990) isolated a cDNA encoding a 35 kDa cysteine proteinase associated with the anticoagulant properties of a host-protective extract from the adult parasite. The predicted coding sequence (AC1) comprised 342 aminoacids, including an N-terminal signal sequence of 15 hydrophobic residues, and showed 42% sequence identity to human cathepsin B. Genomic DNA analysis showed that the gene was a member of a small gene family, two of which (AC2 and AC3) are tandemly linked. Members of this family may have different substrate specificities and different physiological functions as predicted from differences in primary structure (Pratt et al., 1990, 1992a). At least four distinct gene family members are now known to be expressed by adult *H. contortus* (Pratt et al., 1992a).

Using a combination of immunoscreening and PCR, three distinct cathepsin B-like cysteine proteinase-encoding cDNAs have been isolated from adult *H. contortus* sequences (Skuce, Redmond, Stewart and Knox, unpublished) which show 60% to 85% homology to each other and to the previously described AC1–3. Two points are worth noting here: first, attempts to amplify the AC-1 coding sequence, which was isolated from a USA parasite strain (Cox *et al.*, 1990), from a UK isolate have been consistently unsuccessful (D.L. Redmond and D.P. Knox, unpublished observations) and may be a result of geographical strain divergence. This suggestion is supported by a recent report which identified inter- and intra-geographic isolate heterogeneity of proteinases detected in adult worm ES (Karanu *et al.*, 1997). Second, the cysteine proteinase genes isolated to date all show greatest homology to mammalian cathepsin Bs, although this is relatively low (< 50% aminoacid identity). Attempts to isolate cathepsin L-like proteinases from adult *Haemonchus*, using PCR as well as DNA hybridization screens with cathepsin L encoding sequences from schistosomes, have been singularly unsuccessful. The presence of genes encoding proteinases with cathepsin L-like substrate preferences is suggested by the substrate specificities shown by ES and gut-derived proteinases (Rhoads and Fetterer, 1995). Therefore, cathepsin B-like genes may encode proteinases with cathepsin L-like substrate specificities in the same way as described for the AcCP-1 cathepsin B from *A. caninum* (Brinkworth *et al.*, 1996) (see Section 8.4).

7.5.3. *Ostertagia and Teladorsagia*

These genera have only recently been considered as distinct, and will, for comparative purposes, be considered together here. *Ostertagia ostertagi* and *Teladorsagia circumcincta* (formerly *O. circumcincta*) reside in the abomasa of cattle and sheep respectively, and the damage caused to the abomasal glands modifies gastric physiology to the detriment of efficient digestion. As a result, weight gain and other animal production parameters are impaired. Natural host immunity to both species is characterized by stunting of parasite growth which may be in part attributable to immune inhibition of proteinases involved in feeding.

Knox and Jones (1990) demonstrated proteinase release by both infective third-stage larval and adult *T. circumcincta*, and stage-specificity of expression was indicated in a comparison of the molecular size and inhibitor sensitivity of proteinases released by L3, L4 and adult parasites (Young *et al.*, 1995). Metalloproteinases predominated in adult ES. L4 ES contained a serine proteinase and several other activities which could not be clearly defined, while cysteine proteinase activity was identified in L3 ES. Adult ES metalloproteinase(s) degraded the α- and β-fibrinogen chains at alkaline pH, while α-, β- and γ-fibrinogen chains were all degraded at acidic pH, apparently by an aspartyl proteinase. In contrast, IgG was not degraded by these enzymes. Similarly, studies by De Cock *et al.* (1993) showed that L3 and adult parasites of *O. ostertagi* express proteinases that were inhibited by PmsF and 4-hydroxymercuribenzoate but not by 1,10 phenanthroline and EDTA. L3 also expressed an activity enhanced by DTT, indicative of cysteine proteinase activity, while L4 ES contained serine proteinase activity.

Adult *T. circumcincta* and *O. ostertagi* contain proteins equivalent to H11, H-gal-GP and the cysteine proteinase extract as judged by protein profiles in SDS-PAGE and enzyme activity determinations (Smith *et al.*, 1994; Knox *et al.*, 1995). Integral membrane protein extracts from adult *Teladorsagia* prepared by Thiol-Sepharose affinity chromatography contained similar acid and alkaline proteinase activities to those found in adult *Haemonchus*. Faecal egg output and final worm burdens in lambs vaccinated with this extract were reduced by 85% and 65% respectively compared to challenge controls, although greater individual variation was observed compared to the equivalent experiment in *Haemonchus* (Knox *et al.*, 1995).

O. ostertagi contain at least three cysteine proteinase genes (CP1–3), of which two, as is the case in *H. contortus*, are tandemly linked (Pratt *et al.*, 1992b). Primary sequence comparisons indicated that CP1 and CP3 showed closer homology to the *Haemonchus* AC sequences than to human or *S. mansoni* cathepsin B (Sm31) cysteine proteinase.

7.5.4. *Trichostrongylus*

Trichostrongylus colubriformis and *Trichostrongylus vitrinus* both infect the proximal small intestine of small ruminants, and can cause significant disease problems in the host including diarrhoea, anorexia and, in the extreme, death. Both species released proteinase activity *in vitro* that can hydrolyse azocasein azocoll and elastin, and highest secretion was associated with infective third-stage larvae (Knox and Jones, 1990). Serine and metallo-proteinases were released *in vitro* by fourth-stage larvae and adult *T. vitrinus*, which could degrade a variety of macromolecules including fibrinogen, plasminogen and fibronectin, but not immunoglobulin (MacLennan *et al.*, in press). These specificities may indicate a role in the disruption of mucosal tissue barriers allied to a role in parasite feeding. Interestly, ES proteinases from the adult parasites were partially (43%) inhibited by lymph harvested from immune lambs.

7.5.5. *Trichostrongylid (laboratory models)*

The rodent nematode parasite infections, *Nippostrongylus brasiliensis* (rat) and *Heligmospiroides polygyrus* (mouse), are commonly used as laboratory model systems to study intestinal host–parasite interactions. *Nippostrongylus brasiliensis*, infective L3 penetrate the host's skin and migrate to the lungs, where they moult to the L4 stage. They subsequently pass to the digestive tract and mature to adulthood in the small intestine. Cathepsin D-like aspartic proteinase activity in adult worm extracts was related to feeding and morphogenesis (Bolla and Weinstein, 1980). The profile of L3 larval proteinases differed from those of adult parasites, and contained low levels of elastase-type proteinase activity (Knox and Jones, 1990). In a more extensive analysis of infective larval and adult proteinases, Healer *et al.* (1991) identified an abundant, stage-specific 51 kDa metalloproteinase, which they suggested may be a homologue of similar proteinases from larvae of *Strongyloides stercoralis* (McKerrow *et al.*, 1990), *S. ratti* (Lewert and Lee, 1956) and *S. ransomi* (Dresden *et al.*, 1985), all skin-penetrating nematodes. Adult ES also contained a 50 kDa proteinase which was not affected by any of the inhibitors tested, although high molecular weight (> 300 kDa) activity was reduced by 1,10 *o*-phenanthroline (Healer *et al.*, 1991). These authors also suggested that a 20 kDa activity they identified may be of host origin.

Kamata *et al.* (1995) purified a 16 kDa cysteine proteinase from adult parasite extracts prepared in citrate buffer, pH 5.5. The proteinase was also identified in adult ES, and elicited IgG1 and IgE, but not IgG2, isotype responses in infected rats. The proteinase may stimulate Th2-type helper T cell production with associated production of IL-4, IL-5 and IL-10 cytokines, resulting in an IgE mediated allergenic response.

Serine proteinases, optimally active at pH 8.0, were detected in ES of adult *Nematospiroides dubius* (renamed *Heligmosomoides polygyrus*) and gelatin SDS-PAGE identified three SDS-resistant proteinases of molecular weight 200, 105 and 48 kDa (Monroy *et al.*, 1989). These proteinases are immunogenic in infected mice, and inactivation by antibody may contribute to the retarded development and stunted growth associated with parasites harvested from immune mice.

7.6. Trichuroidea

This superfamily includes the genera *Trichuris*, species of which cause widespread but mainly chronic disease symptoms in children and domestic animals. The latter include pigs and dogs, in which heavy worm burdens can cause significant disease problems (Bundy and Cooper, 1989). The *Trichuroidea* also include *Trichinella spiralis*, which is the cause of an important world-wide zoonotic infection. The adult parasite in the intestine can cause enteritis, while subsequent larval invasion of the muscles results in a variety of symptoms, including myositis, myocarditis, oedema and ascites.

The *Trichuroidea* are characterized by the presence of a structure termed a stichosome, which is comprised of a fine capillary tube embedded in a column of single cell, and is thought to function as an oesophageal gland.

7.6.1. *Trichuris*

A zinc-metalloproteinase was isolated from the *in vitro* culture fluids of adult *T. suis* using cation exchange HPLC (Hill *et al.*, 1993). The proteinase had a molecular size of 45 kDa and a pH optimum of pH 7.0 against azocoll, and was inhibited by chelating agents such as 1,10 o-phenathroline. Rabbit antiserum, prepared against the proteinase excised from gels, bound to the stichocyte cells, suggesting that the enzyme is secreted. The proteinase may facilitate feeding and/or parasite migration through mucosal connective tissues in the caecum and colon (Hill *et al.*, 1993).

Proteolytic enzymes which hydrolysed casein and haemoglobin at acidic pH were identified in extracts of adult *T. muris* (Nimmo-Smith and Keeling, 1960). Drake *et al.* (1994) detected serine, metallo-aminopeptidase and cysteine proteinase activity in adult worm extracts, all of which were optimally active at or around neutral pH. Gelatin-gel analysis of adult *T. muris* ES revealed two major proteinases of 85 and 105 kDa and another activity at 55 kDa, with inhibitor sensitivities indicative of the serine proteinases (Drake *et al.*, 1994). Live adult worms degraded radio-labelled extracellular matrix proteins.

7.6.2. *Trichinella*

Infective *T. spiralis* larvae penetrate the small intestinal mucosa and pass, via the blood and lymphatics, to skeletal muscle cells, where they arrest and become encapsulated. The larvae do not possess a buccal stylet, so mucosal penetration may be facilitated by tissue-degrading proteinases (Criado-Fornelio *et al.*, 1992). Crude larval extracts hydrolysed azocoll over the pH range 2 to 8, with maximal hydrolysis at pH 5.0. Proteinases of 48, 54 and 62 kDa were detected in larval extracts, while proteinases at 33, 62 and 230 kDa were evident in larval *in vitro* ES. Inhibition studies indicated that larval extract activity was predominantly due to serine and metalloproteinases (Criado-Fornelio *et al.*, 1992).

Adult *T. spiralis* ES contained predominantly serine proteinase activities in the size range 14 to 100 kDa, although some inhibition was observed with cysteine and metalloproteinase inhibitors (Todorova *et al.*, 1995). ES proteinases degraded fibrinogen and plasminogen, possibly indicating an anticoagulant function which may facilitate parasite feeding. ES proteinase activity was markedly reduced in the presence of IgG isolated from serum of immune mice, suggesting that, in common with other nematode species (Knox and Kennedy, 1988; Britton *et al.*, 1992), inhibition of parasite proteinases by antibody may impair their survival.

7.7. Parasitic Nematodes of Plants and *Caenorhabditis elegans*

Like their animal parasite counterparts, nematodes that infect plants may require proteinases for egg hatching, larval moulting, tissue penetration and feeding.

7.7.1. Globodera, Meloidogyne and Heterodera

Perry *et al.* (1992) examined lipase and proteinase activities during the hatching of the root-knot nematode *Meloidogyne incognita* and the potato cyst nematode *Globodera pallida*. In the case of the former, azocasein- and azocoll-degrading activity at pH 5.0 increased as the percentage hatch increased, but leucine aminopeptidase activity, although detectable, did not appear to be related to this process. In contrast, proteinase activity did not appear to correlate with the hatching process in *G. pallida*. Second-stage juveniles of *M. incognita* were reported to contain a 23 kDa serine proteinase, which may play a role in hatching (Dasgupta and Ganguly, 1975). The growth rate and sexual fate of *G. pallida* grown on transgenic potato plants expressing cowpea trypsin inhibitor were adversely affected, implying an important role for serine proteinases in growth and development (Atkinson, 1993). Koritsas and Atkinson (1994) defined proteinase activity in young adult female *G. pallida* using a plant protein substrate, Rubisco. Substrate degradation was clearly evident at pH 5.5, and less so at pH 7.5. Activity at pH 5.5 was abolished by cysteine proteinase inhibitors and stimulated by DTT. It was suggested that a 64 kDa proteinase had an intra-intestinal role or was secreted into the syncytium for a role in extracorporeal digestion (Koritsas and Atkinson, 1994).

Two distinct proteinases were localized to the intestine of the soybean cyst-nematode *Heterodera glycines*. One had cathepsin L-like substrate specificity and was inhibited by an engineered variant of a rice cysteine proteinase inhibitor. Intestinal proteinase activity against the substrate Z-Phe-Arg-4MNA was completely inhibited by a combination of this inhibitor and cowpea trypsin inhibitor (Lilley *et al.*, 1996). A cDNA fragment encoding a cysteine proteinase putatively expressed in the gut of *H. glycines* was isolated using degenerate primers and PCR (Lilley *et al.*, 1996). The predicted amino acid sequence showed 82% and 78% identity to chick and human cathepsin Ls respectively. By contrast, a cDNA fragment isolated from *G. pallida* by the same approach showed 74% identity to a gut-expressed cathepsin B-like proteinase from *Caenorhabditis elegans*. Subsequently, Urwin *et al.* (1997) cloned two *H. glycines* full-length cDNAs encoding cysteine proteinases; HGCP-1 and HGCP-2 are 63% identical and 81% similar within their mature enzyme region. Although HGCP-1 was reported to be most similar to cathepsin L and HGCP-2 most similar to cathepsin S, the present study shows that they are both cathepsin L-like proteinases (see Section 8.5.2). Southern blot analyses indicated there may be additional *H. glycines* cysteine proteinase genes, although HGCP-1 was the most abundant transcript and may be responsible for intestinal activity described above (Lilley *et al.*, 1996).

7.7.2. *Caenorhabditis elegans*

Extracts of adult *C. elegans* exhibited strong proteinase activity from pH 3.0 to pH 6.0, with optimal activity between pH 4.5 and 5.5, depending on the protein substrate used (Sarkis *et al.*, 1988a). In this pH range, activity was almost totally inhibited by pepstatin, indicative of aspartyl proteinases. Indeed, a putative structural gene encoding a cathepsin D homologue in *C. elegans* has been reported (Jacobson *et al.*, 1988). In addition, casein-degrading activity sensitive to leupeptin could be resolved into two prominent and one small peak by DEAE-Sephadex chromatography (Sarkis *et al.*, 1988a). The two prominent peaks (Ce1 and Ce2) were designated as cathepsin B-like on the basis of substrate specificity, but their differing pIs and substrate affinities led the authors to suggest they were products of different genes (Sarkis *et al.*, 1988a). The third small peak (Ce3) contained a thiol-independent proteinase which was not inhibited by leupeptin and hydrolysed Z-Phe-Arg-MCA very slowly. In addition to Ce1, 2 and 3, Sarkis *et al.* (1988a) detected a proteinase which hydrolysed succinyl-Ala-Ala-Ala-3-MCA, an elastase substrate, and activity was inhibited by elastatinal. *C. elegans* proteinase activities tended to decline with age (Sarkis *et al.*, 1988b), and levels were reduced by acute or chronic starvation (Hawdon *et al.*, 1989).

Ray and McKerrow (1992) identified a gene (*gcp-1*) which encoded a gut-expressed cathepsin B-like proteinase, and showed that expression was temporally and spatially regulated. Expression was initiated at or near hatching, and was restricted to the intestinal cells, particularly the anterior gut. Therefore proteinases may be differentially expressed along the gut. Because *gcp-1* showed homology to *H. contortus* AC-1, Ray and McKerrow (1992) suggested that the two cysteine proteinases identified by Sarkis *et al.* (1988a) might be encoded by distinct cysteine proteinase genes. This suggestion was confirmed by recent work, where four developmentally regulated cathepsin B-like cysteine proteinase genes (*cpr-2* to *cpr-5*) were isolated and characterized from *C. elegans*. These genes were distinct from *gcp-1* and from each other, and had distinct temporal expression patterns. For example, mRNA transcript abundance for *cpr-3* tended to decline from L1 to adult stage, while that for *cpr-6* increased quite dramatically over the same period (Larminie and Johnstone, 1996). This apparent diversity of gut cysteine proteinases, presumably with a digestive function, is analogous to the situation in parasitic nematodes such as *Haemonchus contortus*. Therefore the diversity observed in the proteinases of *Haemonchus* may not be solely an adaptation to parasitism.

8. PHYLOGENETIC RELATIONSHIPS OF CYSTEINE PROTEINASES OF THE PAPAIN SUPERFAMILY

8.1. Introduction

The biological activity of the cysteine proteinases is dependent on the catalytic dyad of cysteine (Cys) and histidine (His) which, depending on the enzyme, could appear as Cys/His or His/Cys in the linear sequence of amino acids (Rawling and Barrett, 1994). These proteinases represent a large group of enzymes that have been separated into 27 *Clans* by Rawlings and Barrett, based on their sequence and structure relatedness (Rawlings and Barrett, 1997; http:/delphi.phys.univ-tours.fr/Prolysis/merops/history.htm). The Clans are further sub-divided into *families*. Members of the papain superfamily are placed into Clan CA, family C1. This family also contains the bacterial aminopeptidases C and the bleomycin hydrolases, whereas the calpains are members of a separate family (C2).

The similarity between the enzymes in Clan CA is largely restricted to regions in the immediate vicinity of the active site residues Cys25, His159 and Asn175 (papain numbering), and is suggestive of a common origin for these enzymes. However, the precise relationships and origin of the members of the Clan CA are not clear. The vertebrate bleomycin hydrolases show considerable sequence identity with the bacterial aminopeptidases C, which would indicate a common origin. Rawlings and Barrett (1994) estimated the divergence between the *Lactococcus* aminopeptidase and the yeast bleomycin hydrolase to be 2500 million years ago (MYA). On the other hand, Berti and Storer's (1995) phylogenetic analysis suggested that the bacterial aminopeptidase enzyme could have originated by lateral transfer of a eukaryotic gene for bleomycin hydrolase. None the less, both groups estimated that a cysteine proteinase from the Gram negative bacterium *Porphyromonas gingivalis* (family C10) and the calpains diverged some 2600 MYA, which is a similar value to that mentioned above.

Recently, a putative cysteine proteinase has been detected by sequencing of the complete genome of the archaeon *Archaeglobus fulgidus* (CYSP ARCHFU, GenBank 2648597). The predicted amino acid sequence of this proteinase showed marked similarity to members of the papain superfamily, although the region of homology is contained within a larger open reading frame, and the spacing between the conserved active site Cys and His is shorter than that observed in the eukaryotic genes. A tree constructed using conserved regions around the active site shows that the archeal sequence is most closely related to the eukaryotic proteinases of the papain superfamily, while the eubacterial cysteine proteinases of the eubacteria, such as *Lactobaccillus helveticus*, are more related to the bleomycin hydrolases (Figure 4). This analysis would place the divergence of these two types of proteinases before the divergence of archea and eubacteria. However, in another complete archeabacterial genome, that of *Methanobacterium thermoautotrophicum*, there is no evidence of similar cysteine proteinases, indicating the possibility of the lateral transfer of a cysteine proteinase into the genome of *A. fulgidus*.

Although the calpains may have an ancient origin, the various sequences of human, rat, chicken and schistosome calpains were not very well resolved in the phylogenetic trees of Berti and Storer (1995). Their analysis shows that the schistosome calpain diverged prior to the divergence of the m- and u-type calpains. Furthermore, they suggested that the m- and u-type calpains diverged at about the time of the avian/mammalian divergence, approximately 300 MYA. Sequence alignments by Karcz *et al.* (1991) and Andresen *et al.* (1991) showed that the cDNA encoding the large domain of the schistosome calpain showed highest similarity to the human u-calpain and chicken calpain. Therefore, it is possible that the schistosome enzyme is similar to the original calpain type.

8.2. The Papain Superfamily

As discussed earlier, parasite proteinases are suspected to be involved in host–parasite interactions such as tissue invasion, immuno-evasion and nutrition. For this reason a considerable amount of effort has been dedicated to the purification of these enzymes and the isolation of their corresponding genes. The majority of protist and invertebrate cysteine proteinase genes reported in the databases are derived from parasitic organisms. This simply reflects the fact that the proteinases of these species have been investigated, while those of their free-living relatives await characterization. Moreover, an overwhelming majority of these genes encode enzymes belonging to the papain superfamily (Clan CA, family C1), and therefore our phylogenetic analysis reported here concentrates on these. However, it must be kept in mind that the collection of sequences

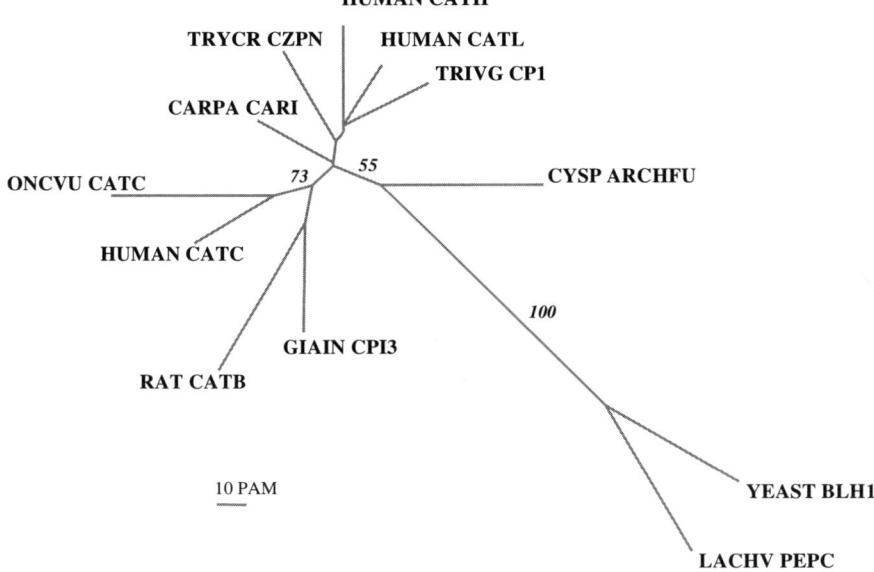

Figure 4 Relationship of a sequence of encoding a putative cysteine proteinase from the archaeon *Archaeglobus fulgidus* with the papain superfamily. A neighbour-joining tree of conserved regions of the archaeon *Archaeglobus fulgidus* proteinase (CYSP ARCHFU) and representative members of diverse cysteine proteinase groups in the papain superfamily is shown. Because of the large differences between yeast bleomycin hydrolase (YEAST-BLH1) and *Lactococcus helveticus* aminopeptidase C (LACHV-PEPC) with the other sequences, only conserved blocks around the active sites were used for alignment (as a consequence of taking such restricted regions, the position of certain sequences in the branches of the figure may differ from those in subsequent figures; see legend of Figure 5).

present in the databases may represent a sampling bias. Many cDNAs have been isolated by PCR using primers designed to hybridize to conserved regions of cysteine proteinase genes, or by the screening of genomic and cDNA libraries with homologous probes which would consequently increase the number of 'related' proteinases in the databases.

The evolution of the papain superfamily has been punctuated by numerous gene duplications. Consequently we are now confronted with an expanding suite of proteinase genes from diverse origins that have had their relationships obscured by millions of years of evolution at, most probably, unequal rates. Nonetheless, analysis of the evolutionary relationships of the known members of the papain superfamily can identify distinct groups of genes, and may give important clues to specific gene-function relationships. This information may be exploitable in the design of strategies for the control of parasitic diseases that are directed against these enzymes.

Table 4

Name	Species	Sequence ID	Database	Accession no.
CYSP ARCHFU	*Archaeglobus fulgidus*	2648597	gb	AE000969
ACTCH ACTN	*Actidinia chinensis*	113285	sp	P00785
AEDEG CATB	*Aedes aegypti*	1008858	gb	L41940
ANCCA*CPRO	*Ancylostoma caninum*	496968	gb	U02611
ANCCA CB1	*Ancylostoma caninum*	984960	gb	U18912
ANCCA CB2	*Ancylostoma caninum*	984958	gb	U18911
ASCSU CATB	*Ascaris suum*	1777779	embl	U51892
BOMMO CPRO	*Bombix morii*	1085124	pir	JX0366
CAEEL*CPR	*Caenorhabditis elegans*	508264	gb	U11245
CAEEL CPR3	*Caenorhabditis elegans*	1169083	sp	P43507
CAEEL CPR4	*Caenorhabditis elegans*	1169085	sp	P43508
CAEEL CPR5	*Caenorhabditis elegans*	1169086	sp	P43509
CAEEL CPR6	*Caenorhabditis elegans*	1169087	sp	P43510
CAEEL CYS1	*Caenorhabditis elegans*	118116	sp	P25807
CAEEL R07e	*Caenorhabditis elegans*	798824	gb	z49207
CEL C50F4	*Caenorhabditis elegans*	1262931	gb	z70750
CEL F15D4	*Caenorhabditis elegans*	1556385	gb	z80344
CEL F32B5	*Caenorhabditis elegans*	2088784	gb	af003148
CEL F41E6	*Caenorhabditis elegans*	2315454	gb	af016448
CEL M04G12	*Caenorhabditis elegans*	1903097	gb	z81103
CELF26E4	*Caenorhabditis elegans*	1813914	gb	z81070
CELF57F5	*Caenorhabditis elegans*	1429280	gb	z75953
CARPA CARI	*Carica papaya*	1709574	sp	P10056
DICDI CYS1	*Dictyostelium discoideum*	118117	sp	P04988
DICDI CYS2	*Dictyostelium discoideum*	118121	sp	P04989
ENTHI CP6	*Entamoeba hystolitica*	1246527	embl	X91645
ENTHI CPP1	*Entamoeba hystolitica*	544088	sp	Q01957
FASHE*FCP2	*Fasciola hepatica*	452254	embl	Z22763
FASHE*FCP4	*Fasciola hepatica*	452262	embl	Z22767
FASHE CL1	*Fasciola hepatica*	1809286	gb	U62288
FASHE CL2	*Fasciola hepatica*	1809288	gb	U62289
FASHE CLES	*Fasciola hepatica*	535600	gb	L33772
GIAIN CPI3	*Giardia muris*	1763663	gb	U83277
HAECO*PDM4	*Haemonchus contortus*	1181143	embl	Z69345
HAECO AC3	*Haemonchus contortus*	478099	pir	D48435
HAECO AC4	*Haemonchus contortus*	478007	pir	C48435
HAECO AC5	*Haemonchus contortus*	477808	pir	B48435
HAECO CP6	*Haemonchus contortus*	1644295	embl	Z81327
HAECO CYS1	*Haemonchus contortus*	118118	sp	P19092
HAECO CYS2	*Haemonchus contortus*	118122	sp	P25793
HAECO PDM5	*Haemonchus contortus*	1181145	embl	Z69346
HETGL CATL	*Heterodera glycines*	2239107	embl	Y09498

(*continued*)

Table 4 Continued

Name	Species	Sequence ID	Database	Accession no.
HETGL CATS	*Heterodera glycines*	2239109	embl	Y09499
HOMAM CYS1	*Homarus americanus*	118119	sp	P13277
HOMAM CYS2	*Homarus americanus*	118123	sp	P25782
HOMAM CYS3	*Homarus americanus*	118125	sp	P25784
HORVU ALEU	*Hordeum vulgare*	113603	sp	P05167
HORVU CYS1	*Hordeum vulgare*	118120	sp	P25249
HUM CATO	*Homo sapiens*	1168795	sp	P43234
HUMAN CATC	*Homo sapiens*	1006657	embl	X87212
HUMAN CATH	*Homo sapiens*	115728	sp	P09668
HUMAN CATK	*Homo sapiens*	1168793	sp	P43235
HUMAN CATL	*Homo sapiens*	115741	sp	P07711
HUMAN CATS	*Homo sapiens*	115748	sp	P25774
LACHV PEPC	*Lactobacillus helveticus*	629180	pir	S47225
LEIME CATB	*Leishmania mexicana*	728602	embl	Z48599
LEIME LCPA	*Leishmania mexicana*	126021	sp	P25775
LEIME LCPB	*Leishmania mexicana*	547835	sp	P36400
LYMP HUMAN	*Homo sapiens lymphopain*	2829471	sp	P56202
LYMP MOUSE	*Mus musculus lymphopain*	2829472	sp	P56203
MAIZE CYS1	*Zea mays*	1706260	sp	Q10716
MAIZE CYS2	*Zea mays*	1706261	sp	Q10717
NAEFO+CPRO	*Naegleria fowlerii*	1353726	gb	U42758
NEPNO CLE	*Nephrops norvegicus*	630815	pir	S47432
NEPNO CLS	*Nephrops norvegicus*	630816	pir	S47433
NITOB CATB	*Nicotiana tabacum*	1076610	pir	S52212
ONCVU CATC	*Onchocerca volvulus*	1680720	gb	U71150
OSTOS*CYS3	*Ostertagia ostertagii*	729283	sp	Q06544
OSTOS CYS1	*Ostertagia ostertagii*	1345924	sp	P25802
PARTE CTL1	*Paramecium tetraurelia*	1403087	embl	X91754
PARTE CTL2	*Paramecium tetraurelia*	1403089	embl	X91756
PARWM NTP	*Paragonimus westermani*	633096	ddbj	D21124
PENVA PCP1	*Penaeus vannamei*	1085687	pir	S53027
PENVA PCP2	*Penaeus vannamei*	1483570	embl	X99730
PLAFA CYSP	*Plasmodium falciparum*	118152	sp	P25805
RAT CATB	*Rattus norvegicus*	1524328	embl	X82396
SARPE CATB	*Sarcophaga perregrina*	481614	pir	s38939
SARPE CATL	*Sarcophaga perregrina*	1079183	pir	A53810
SCHJP CATC	*Schistosoma japonicum*	2599293	gb	U77932
SCHJP CB1	*Schistosoma japonicum*	1169189	sp	P43157
SCHJP CB2	*Schistosoma japonicum*	345308	pir	S31909
SCHJP CL2	*Schistosoma japonicum*	1185457	gb	U38476
SCHJP+CL1	*Schistosoma japonicum*	1185457	gb	U38475

(continued)

Table 4 Continued

Name	Species	Sequence ID	Database	Accession no.
SCHMA CATB	*Schistosoma mansoni*	118153	sp	P25792
SCHMA CATC	*Schistosoma mansoni*	1262412	embl	Z32531
SCHMA CL1	*Schistosoma mansoni*	1094710	prf	1094710
SCHMA+CL2	*Schistosoma mansoni*	630486	pir	S44151
SPIER CPRO	*Spirometra erinacei*	1834309	ddbj	D63671
SPIMA+CPRO	*Spirometra mansonoides*	1272388	gb	U51913
STRRA*CPRO	*Strongyloides ratti*	603044	gb	U09818
TETER SGC5	*Tetrahymena termophyla*	476938	pir	A47306
THEAN CYSP	*Theileria annulata*	118154	sp	P25781
THEPA CYSP	*Theileria parva*	118155	sp	P22497
TOXCA CATL	*Toxocara canis*	1279986	gb	U53172
TRIVG CP1	*Trichomonas vaginalis*	542419	pir	S41427
TRYCR CZPN	*Trypanosoma cruzi*	118157	sp	P25779
TTMFT CP1	*Tritichomonas foetus*	1141743	gb	U13153
URECA CATB	*Urechis caupo*	945054	gb	U30877
VICSA CPR1	*Vicia sativa*	457756	embl	Z30338
VICSA CPR2	*Vicia sativa*	600111	embl	Z34895
VICSA CPR3	*Vicia sativa*	535473	embl	X75749
YEAST $_{BLH1}$	*Saccharomyces cerevisiae*	267476	sp	Q01532

8.3. The ERFNIN Motif

Almost all proteinases are secreted as inactive proenzymes in order to protect cells from uncontrolled degradative activity. The cysteine proteinases of the papain superfamily have a N-terminal extension or propeptide which is cleaved during the formation of the active mature enzyme. This propeptide interacts closely with the mature portion of the enzyme, and is important in maintaining its stability and inactivity while it is trafficked through the Golgi apparatus. A close analysis of the amino acid sequence of cysteine proteinases by Karrer et al. (1993) allowed the subdivision of the papain superfamily into two groups, based on the presence or absence of a distinctive set of amino acids residues with the propeptide — the ERFNIN motif — that consisted of conserved residues separated by three amino acids and which was predicted to form an alpha helix. The ERFNIN motif was present in papain, cathepsin L-like and cathepsin H-like cysteine proteinases, but absent from the cathepsin B-like enzymes. The recent elucidation of the three-dimensional structure of the proenzymes of cathepsin L and caricain revealed that the ERFNIN motif was indeed contained within a long conserved alpha helix, constituting the core of a globular portion of the propeptide (Groves *et al.*, 1996; Coulombe *et al.*, 1996).

Although the differentiation of the papain superfamily into two main groups (cathepsin B-like and ERFNIN containing proteinases) by Karrer *et al.* (1993) was based on a restricted sequence in the pro-peptide, it was subsequently supported by the phylogenetic analysis of Hughes (1994) and Berti and Storer (1995), which examined complete proenzymes and mature enzymes respectively. The latter authors referred to the two main groups as cathepsin Bs and non-cathepsin Bs.

Our recent analysis of more than 150 sequences (which considered conserved blocks within the propeptide and the mature enzymes) also positioned the cathepsin Bs into a separate branch of the gene tree from all other sequences (Figure 5, Tables 4 and 5). However, in agreement with Berti and Storer (1995), our study reliably (bootstrap support 91%) places the cathepsin C proteinases in the branch with the cathepsin Bs (Figures 5 and 6). An alignment of the schistosome cathepsin C with the mammalian cathepsin C genes revealed that these contain an ERFNIN-like sequence within their unusually long pro-peptides (Hola-Jamriska, *et al.*, 1998). Moreover, our phylogenetic analysis has identified a new class — here termed cathepsin X because it includes bovine cathepsin X (Gay and Walker, 1985) — which is related to, but different from, the cathepsin Cs (see Section 8.4.3). As very few complete sequences of members of this group are available, it is not yet clear if they contain ERFNIN motifs in their propeptides.

In this report we refer to the two main branches of the papain family simply as Branch A and Branch B (Table 5). Branch A is clearly resolved, and includes the vertebrate cathepsin Bs, cathepsin Cs, cathepsin Xs and related enzymes from invertebrates, plants and protists (Figures 5 and 6). Branch B, which comprises most of the proteinases in the papain superfamily, is poorly resolved in the deep branches of the phylogenetic tree. The low resolution was explained by Berti and Storer (1995) by a period of rapid gene duplication and divergence, giving rise to many different enzymes in a short period of evolutionary time. Nevertheless, at least three classes can be identified reliably within this branch. The first of these consisted of sequences from different taxonomic groups related to cruzipain (bootstrap 71%); a second is constituted exclusively of sequences from plants (papain class) (bootstrap 77%); and a third includes sequences exclusively from animals (cathepsin L class) (bootstrap 58%). The cathepsin H/aleurain genes form a reliable group (bootstrap 100%) and can be considered as a fourth class, although their relationship with the other classes within this branch is obscure (Figure 7). Other genes form consistent clusters, such as a sequence from *Toxocara canis* with genes from *C. elegans* or the grouping of several baculovirus sequences, which may indicate that

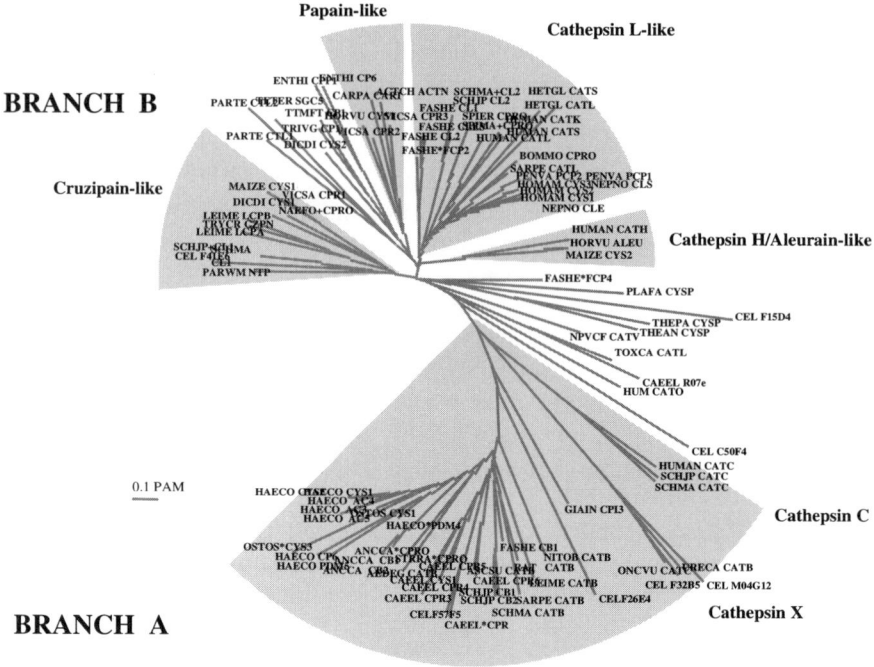

Figure 5 Unrooted neighbour-joining phylogenetic tree of representatives of the papain superfamily. Sequences were analysed by progressive alignment using the program ClustalW (Thompson *et al.*, 1994). As the general fold of the members of the papain superfamily is well conserved, the available data on secondary structure of several members of the family was considered in order to produce a profile for the alignment. This methodology permits the assignment of higher gap penalties to discontinuities in the alignment that interrupt the secondary structure features. The final profile was based on the crystal structure data available for three members of the superfamily, namely rat procathepsin B (Cygler *et al.*, 1996, PDB: 1mir), human procathepsin L (Coulombe *et al.*, 1996, PDB: 1cjl) and procaricain (Groves *et al.*, 1996 PDB: 1pci). The alignments were visualized and manually corrected using Genedoc (Nicholas *et al.*, 1997). Phylogenetic trees were calculated using the neighbour-joining method of Saitou and Nei (1987) based on a matrix of distances between all sequences. The distances were corrected using the Kimura formula for multiple substitutions (Kimura, 1983). Subsets of optimized alignments were also analysed by parsimony methods using the program PROTPARS from the Phylip package, and gave similar results (not shown) (Felsenstein, 1985). Trees were visualized with TreeView (Page, 1996). The major classes are indicated. The sequences used are listed in Table 4.

Table 5 Classification of papain superfamily (Clan CA, Family C1[1]) based on gene sequence relationships.

BRANCH	Sub-family	Group	Example
BRANCH A	Cathepsin B	Giardia	G. muris and G. lamblia cathepsin B
		Trematode	Fasciola hepatica cathepsin B S. mansomi and S. japonicum cathepsin B
		Nematode	H. contortus, O. ostertagia, Ancylostoma caninum, Strongyloides ratti, and C. elegans cathepsins Bs.
		Plant	Tobacco and wheat cysteine proteinases
		Vertebrate	Human, bovine, rat and mouse cathepsin B
	Cathepsin C	Trematode	S. mansoni and S. japonicum cathepsin C, human and rat cathepsin C.
		Vertebrate	
	Cathepsin X	Nematode	O. volvulus cathepsin X
		Vertebrate	Bovine cathepsin X.
BRANCH B		Trichomonas-like	T. vaginalis cysteine proteinases, Tritrichomonas foetus cysteine proteinases
		Entamoeba-like	E. histolytica, E. dispar and E. invadens proteinases
		Plasmodium-like	P. falciparum, P. vivax, P. malariae, P. fragile, P. ovale, P. berghei and P. gallinaceum cysteine proteinases.
		Theileria-like	T. parva and T. annulata cysteine proteinases.
	Cruzipain	Cruzipains	T. cruzi cruzipain, L. mexicana cysteine proteinases,
		Trematode	S. mansoni and S. japonicum cathepsin L1, P. westermani cysteine proteinase
		Nematode	C. elegans cysteine proteinase
		Plant	Maize and soya bean cysteine proteinase
		Vertebrate	Mammalian lymphopains.

(continued)

Table 5 Continued

BRANCH	Sub-family	Group	Example
	Cathepsin L	Trematode	*Fasciola hepatica* gene family *S. mansoni* and *S. japonicum* cathepsin L2,
		Cestode	*Spirometra mansoni* and *S. erinacei* cathepsin Ls
		Nematode	*Heterodera glycines* cathepsin Ls
		Crustacean	*Homarus americanus* (lobster) digestive proteases
		Vertebrate	Vertebrate cathepsin L, cathepsin S and cathepsin K
	Papain	Papain	Papain, chymopapain, stem bromelain
		Vignain	Vignain
		Orizain	Orizain a.
	Cathepsin H/ Aleurain	Vertebrate	Human, rat and mouse cathepsin H
		Plant	Aleurain, orizain gamma, maize cysteine proteinase 2
	Incertae saedis	*Toxocara* family	*T. canis* and *C. elegans* cysteine proteinase
		Baculovirus cathepsins	*Autophaga california* viral cathepsin

[1] See Rawlings and Barret, 1997; http://delphi.phys.univ-tours.fr/Prolysis/merops.history.htm

these also represent defined families. Again, their relationships with other proteinases in Branch B are not clear, and more sequences are needed to clarify their status.

The presence of a single cysteine proteinase similar to members of the papain superfamily in archaea suggests that this superfamily has an ancient origin. The gene duplication(s) that gave rise to the enzymes in the two branches of the papain superfamily might have occurred very early in eukaryotic evolution, and possibly before the prokaryotic/eukaryotic divergence. Evidence for this suggestion is provided by the presence of members of the papain superfamily in diplomonads and triplomonads, amitochondriate taxa considered to be among the earliest diverging eukaryotes (Sogin, 1991; Knoll, 1992). Three cysteine proteinase genes were identified in *Giardia muris*, and are closely related to the vertebrate cathepsin B (only one shown in Figures 4, 5 and 6, Table 5; Ward *et al.*, 1997). In contrast, the cysteine proteinase genes of *Trichomonas vaginalis* and *Tritichomonas foetus* are more closely related to the vertebrate cathepsin

BRANCH A

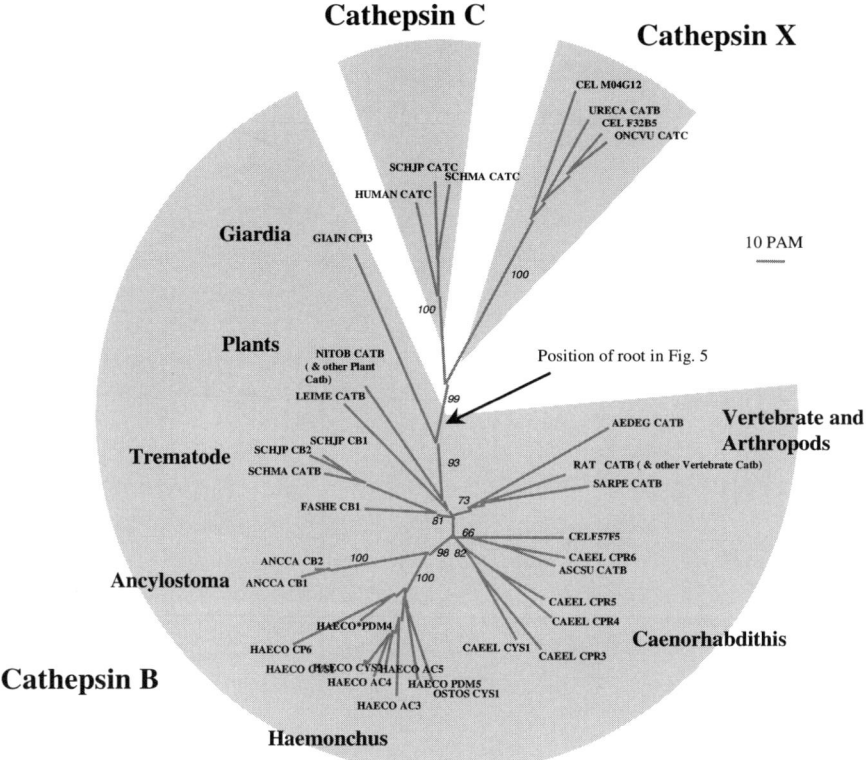

Figure 6 Unrooted neighbour-joining phylogenetic tree of papain superfamily members in Branch A. Conserved regions of the propeptide and the mature enzymes were used to generate the tree as for Figure 5. The major classes are indicated.

Ls (Mallinson *et al.*, 1994, 1995). The cysteine proteinase genes of the protist *Entamoeba* are most closely related to the triplomonad genes (Figures 5 and 7). (Although Entamoebae are amitochondriate, it was proposed by Clark and Roger (1995) that they are phylogenetically related to mitochondrial eukaryotes.)

The relationships of the cysteine proteinases of protists with metazoan or plant homologues are not clear. After the divergence of the two main branches there is evidence of unequal rates of evolution in different lines which obviously would affect their position in the phylogenetic tree (Berti and Storer, 1995). The several cysteine proteinase genes of *Entamoeba* consitute a well-defined node, and tend to cluster with the virulence factor

BRANCH B

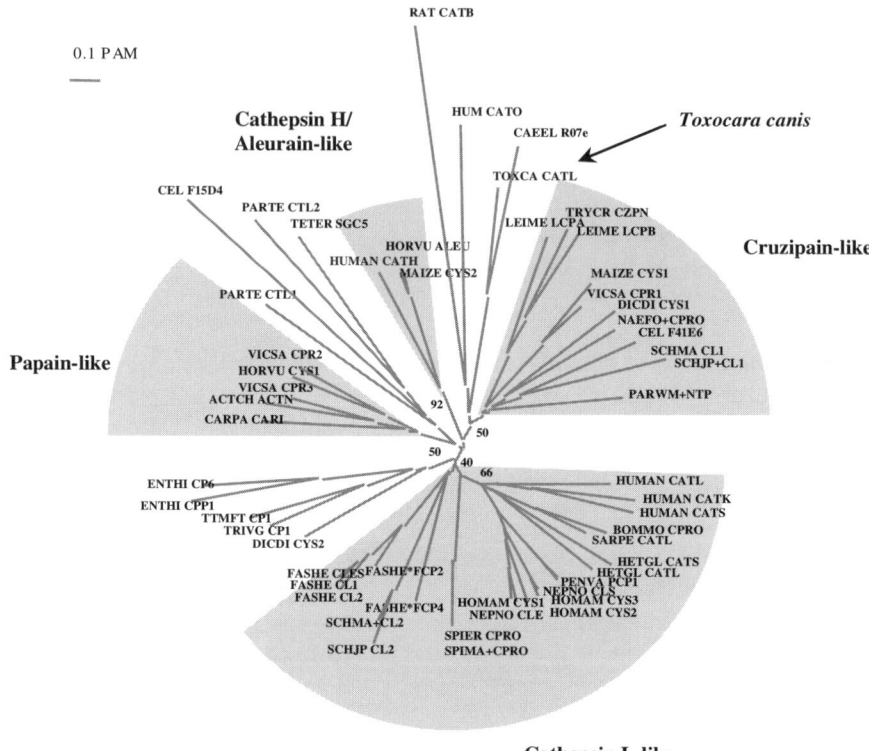

Figure 7 Unrooted neighbour-joining phylogenetic tree of papain superfamily members in Branch B. Conserved regions of the propeptide and the mature enzymes were used to generate the tree as for Figure 5. The major classes are indicated.

cysteine proteinases of *Trichomonas vaginalis* and *Tritrichomonas foetus* in Branch B; although the bootstrap values for the clustering of these groups were low, no alternative topologies were observed. The cysteine proteinase of the apicoplexan *Plasmodium* spp. cluster together as expected (one representative, PLAFA CYSP, is shown in Figure 5), but separately from the cluster of *Theileria* cysteine proteinase genes (see also Ward *et al.*, 1997), while the cysteine proteinase genes of the ciliates *Paramecium* and *Tetrahymena* are not well resolved. However, although all the genes of the Alveolata (apicoplexans and ciliates) were poorly resolved, in all cases they were located in Branch B (see Table 5).

The presence of multicopy gene families is common in protists (for example, the cysteine proteinases of *E. histolytica*, *L. mexicana* and *T. cruzi*) and in helminths (for example, the *F. hepatica* cathepsin L genes and the *H. contortus* and *C. elegans* cathepsin B genes). However, the genes from the different lineages generally find their closest relatives within each lineage, indicating that gene duplication and diversification took place recently and independently in the different groups. This observation implies that the proteinases of diverse taxonomic groups are subjected to different selective pressures. Consequently, the papain superfamily is a mixture of paralogous (which arise by gene duplication and exist in the same organism) and orthologous (similar genes in different organisms) sequences.

8.4. Branch A of the Papain Superfamily

Within Branch A, many genes from helminths are to be found, together with sequences from other metazoans, plants and protists. This branch has been identified previously by Berti and Storer (1995) as cathepsin B class. However, three main families constitute the branch: cathepsin Bs, cathepsin Cs and cathepsin Xs (Figures 5 and 6; Table 5) (Tort, Wolfe and Dalton, unpublished data).

8.4.1. *The Cathepsin B-like Proteinases*

Most of the cysteine proteinase genes isolated from nematodes are related to vertebrate cathepsin B enzymes, and in several cases are members of multigene families. For example, a developmentally regulated family of cathepsin B-like enzymes has been identified in *H. contortus*. The enzymes are more highly expressed in the adult compared to the larval stages. At least two of these genes are arranged in tandem on the *H. contortus* genome (Cox *et al.*, 1990, Pratt *et al.*, 1990, 1992a). Similarly, three cathepsin B-like genes have been reported from *Ostertagia ostertagi* (Pratt *et al.*, 1992b), and a family of developmentally regulated cathepsin B-like genes has been detected in *C. elegans* (Ray and McKerrow, 1992; Larminie and Johnstone, 1996). Partial cathepsin B proteinase genes from the nematodes *Strongyloides ratti*, *S. stercoralis*, *Ancylostoma caninum* (two genes) and *Caenorhabditis elegans* were isolated by Harrop *et al.* (1995a) using PCR which employed primers designed to anneal to conserved regions that encoded the active site regions.

Our phylogenetic analysis shows that the genes from *H. contortus*, *O. ostertagi*, *A. caninum* and partial sequences from *S. ratti* and *S. stercoralis* separate into a well-defined clade within the cathepsin B class (Figure 6).

The *C. elegans* cathepsin B genes constitute a second clade; however, these show greater variation between each other compared to the genes in the *H. contortus* clade. Two of the *C. elegans* cysteine proteinases, CPR 6 and CELF57F5 (Figure 6), appear to be quite different and separate early from the *C. elegans* node; these two genes do not group together, and CPR 6 is more closely related to a cysteine proteinase gene from *Ascaris suum*. The proteinase genes found in the two different nematode clusters may have arisen as a result of a gene duplication which occurred at least as early as the earliest metazoans, and subsequently underwent further diversification to give rise to a family of paralogous genes in each species (Hughes, 1994; Tort, 1997).

Cathepsin B-like sequences have also been isolated from the trematodes *S. mansoni*, *S. japonicum* and *F. hepatica*. It can be seen that these sequences constitute a separate clade (Figure 6). This position of the trematode genes in our analysis does not allow us to establish precisely when the duplication that led to the nematode's genes took place, but it suggests that this event occurred after the diversification of the flatworm genes.

Hughes (1994) analysed the rates for synonymous and non-synonymous nucleotide substitutions per site in the *H. contortus* genes, and found that non-synonymous substitutions that cause a change in amino acid residue charge in helical regions occurred at higher rates than expected for random substitutions. Therefore, these genes appear to have been subjected to positive diversifying natural selection. Some of these changes are in helices that make up the substrate binding cleft and may have led to proteinases that could be used by the parasite to cleave new proteins. Assuming a similar evolutionary rate for the *H. contortus* and mammalian cathepsin Bs, Hughes (1994) estimated that the diversification of cysteine proteinases in this parasite took place relatively recently, at the period when the eutherian mammals arose and diversified. Therefore, the author suggested, this diversification might have occurred as an adaptation to parasitism of the mammalian host. However, although no data on the rate of non-synonymous that occurred in the *C. elegans* representatives is available, the diversification of cathepsin B-like cysteine proteinases in this free-living nematode might indicate that this process could be more general and not restricted to parasitic nematodes.

The cathepsin B enzymes usually possess an insertion in their primary sequence that is not present in other cysteine proteinases of the papain superfamily. This insertion encodes for a structural element, the occluding loop, which accepts the negative charge of the substrate P2′ carboxylate and is responsible for the carboxy-terminal dipeptidyl-peptidase activity that is characteristic of these enzymes (Cyler *et al.*, 1996). Sequence alignments show that this loop is conserved in the cysteine proteinases from vertebrates, the parasitic nematodes *A. suum*, *H. contortus*, *O. ostertagi*, *A. caninum*, three of the *C. elegans* cysteine proteinases (CPR4, CPR5 and CPR6), each of the trematode sequences, and the parasitic trypanosomatid *Leishmania mexicana*. In contrast, the loop is missing in the cathepsin B-like genes of *Giardia*, and a shorter loop is present in the plant cathepsin Bs, the *C. elegans* genes CYS1, CYS2 and CPR3 and an enzyme from the yellow fever mosquito vector *Aedes aegypti*. The two histidine residues involved in the exopeptidase activity (His^{110}, His^{111} — mammalian cathepsin B numbering) are conserved in the loop-bearing enzymes, the only exception being two members of the *C. elegans* cathepsin B gene family (CPR4 and CPR5) and the *F. hepatica* cathepsin B. In the two *C. elegans* proteinases, these residues have been substituted by a Glu and a Thr respectively, and in the *F. hepatica* enzyme the His^{111} is replaced by a Val. The plant homologues, which have a shorter loop, maintain only one of the histidine residues. It is reasonable to suspect that in those enzymes where the occluding loop is missing, shortened, or substituted in the His residues, there is a lack of carboxy-exopeptidase activity. However, it is difficult to assess if this region and the enzymatic activity related to it are ancestral, or if it originated later in evolution. The absence of the loop in diplomonads and plants, the first diverging groups within the cathepsin B family, support the idea of the absence of this region in the ancestral gene. On the other hand, the presence of the loop in *L. mexicana* suggests the occluding loop originated at least as early as the first mitochondrial eukaryotes and was subsequently lost in plants.

8.4.2. *The Cathepsin C-like Proteinases*

Cathepsin C (dipeptidyl-peptidase I, DPP I) activity and genes have only been reported from vertebrates and from the trematodes *S. mansoni* and *S. japonicum*. A cathepsin C-like activity was also detected in the excretory/secretory products of the L4 larva of *H. contortus*, but a corresponding gene has not yet been isolated (Gamble and Mansfield, 1996). Cathepsin Cs are oligomeric enzymes with unusually long propeptides. It is suspected that portions of the long propeptide constitute one of the subunits of the mature enzyme (Dolenc *et al.*, 1995).

Additional information is required to establish the origin of the cathepsin C proteinases. However, Ward *et al.* (1997) postulated that the *Giardia* cysteine proteinase genes were the first to diverge in the cathepsin B-like branch, and that therefore they precede the divergence of the cathepsin Bs and Cs. In our analysis (Figure 5), both distance and parsimony methods indicate that the *Giardia* genes branched after the cathepsin B and cathepsin C genes diverged. The phylogenetic study of Ward *et al.* (1997) compared only the regions corresponding to the catalytic domains of the enzymes, whereas in the present study conserved blocks within the propeptide were also taken into account. Our observation opens the possibility that cathepsin

C enzymes may be present in primitive organisms such as *Giardia*. The presence of cathepsin Cs in trematodes and mammals indicates that they exist in all metazoans. It is therefore certain that more helminth cathepsin C genes will be identified in the future.

8.4.3. *The Cathepsin X-like Proteinases*

The cathepsin X cysteine proteinases that constitute a third family within the cathepsin B-like branch (Figures 5 and 6; Table 5). This group is related to the cathepsin C family, and is constituted by a gene encoding a cysteine proteinase suggested to be involved in the moulting process of *Onchocerca volvulus* (Lustigman *et al.*, 1996), a maternal gene of the echiurid worm *Urechis caupo* (Rosenthal, 1993), two *C. elegans* genomic clones (CELF32B5 cds5, CELM04G12 cds2), a cDNA from a bovine heart (bovine cathepsin X, SwissProt P05689), an EST from a mouse embryo (GenBank AA030446), an EST from a human breast cell line (GenBank H42663) and three other vertebrate partial sequences. A pair of cysteine residues present in cathepsin Bs (positions Cys^{100} and Cys^{132} in human cathepsin B) are shared by cathepsin Cs and the members of this new family. A short insertion in the same position of the occluding loop of cathepsin Bs is present in this new group of enzymes. Two cysteine residues within this region, which correspond with Cys108 and Cys119 in cathepsin B, are conserved in the new family. The presence of members of this family in nematodes and vertebrates suggest that it originated early in the evolution of metazoans.

8.5. Branch B of the Papain Superfamily

Enzymes belonging to Branch B of the papain superfamily are abundant, and widespread in all the eukaryotic lineages. Determining the early evolutionary history of this group of enzymes is difficult because, as discussed before, extensive diversification of these enzymes seems to have occurred very early in this eukaryotic lineage. Nevertheless, within Branch B different classes can be distinguished; cruzipain-like, cathepsin L-like, papain-like and cathepsin H/aleurain-like (Figure 7; Table 5).

8.5.1. *The Cruzipain-like Proteinases*

The cruzipain-like class is the first apparently coherent group in Branch B (Figure 7). This class includes all the kinetoplastid sequences (with the exception of the cathepsin B-like genes of *L. mexicana*), a sequence from the slime mould *Dictyostelium discoideum* (Dicdi-cys1), a sequence from the heterolobosean amoeba *Naegleria fowleri*, and several sequences from plants and metazoans. Berti and Storer (1995) originally termed this the Ddis 1 class because of the presence of the *D. discoideum* sequence. However, since kinetoplastids have diverged earlier in the mitochondrial eukaryote lineage than the Dictyostellida, and because they are the most characterized members of this class, we have renamed it the cruzipain-like class, after the best-known member of *T. cruzi*.

Although the *D. discoideum* Dicdi-cys1 gene is a member of this class, other slime mould sequences (Dicdi cys2–cys7) are more closely related to the vertebrate cathepsin Ls. Based on this observation, Hughes (1994) suggested that the duplication that gave rise to these separate classes preceded the divergence of slime moulds. Consistent with this hypothesis, our study shows that cruzipain-like genes are found in plants and animals. Genes from trematodes fall into this class, i.e. the *S. mansoni* and *S. japonicum* 'cathepsin L1' (Smith *et al.*, 1994; Day *et al.*, 1995; Dalton *et al.*, 1996a) and the neutral thiol proteinase from the lung fluke *Paragonimus westermani* (Yamamoto *et al.*, 1994). A sequence from the free-living nematode *C.elegans* (CELF41E6) is also related to members of this class (Figure 7). Very recently, cysteine proteinases from natural killer and T-lymphocytes of humans and rodents have been isolated (GenBank 2829472). These genes, termed lymphopains, find their closest relatives with the schistosome and *P. westermani* sequences, and thus constitute the first examples of vertebrate genes belonging to this cruzipain class.

8.5.2. *The Cathepsin L-like Proteinases*

The cathepsin L-like class comprises the vertebrate cathepsins L, S and K, and similar sequences from insects, crustaceans, nematodes, cestodes and trematodes. The majority of the cysteine proteinase sequences obtained from adult *F. hepatica* correspond to a family of cathepsin L-like enzymes (Yamasaki and Aoki, 1993; Heussler and Dobbelaere, 1994; Wijffels *et al.*, 1994; Panaccio *et al.*, 1994; Roche *et al.*, 1997; Dowd *et al.*, 1997). At least five different cathepsin L-like genes constitute this gene family present in *F. hepatica* (Heussler and Dobbelaere, 1994; Tort and Dalton, unpublished). The cathepsins L2 of *S. mansoni* and *S. japonicum* (Day *et al.*, 1995; Michel *et al.*, 1995; Dalton *et al.*, 1996a) are closely related to the *F. hepatica* cluster. The only sequences available from cestodes — *Spirometra erinacei* (Liu *et al.*, 1996) and *Spirometra mansoni* (Phares, 1996) — branch after the trematode cathepsin L-like enzymes and before similar sequences from arthropods and vertebrates (Figure 7).

For a long time, genes encoding nematode cysteine proteinases other than cathepsin B-like enzymes were absent from the databases (although biochemical studies employing fluorogenic peptide substrates had suggested the presence of cathepsin L-like activities in parasitic nematodes). However, more recently, non-cathepsin B-like cysteine proteinase genes have been isolated. Two sequences from the plant cyst nematode *Heterodera glycines* were reported recently as cathepsin L and cathepsin S (EMBL Y09498 and Y09499). However, our analysis of these sequences in the present study showed that both were related to other invertebrate cathepsin L-like cysteine proteinases, and that they branch after the platyhelminth genes but before the arthropod homologues (Figure 7). A search into the expressed sequence tag (EST) databases identified several partial sequences from nematodes with similarities to members of the cathepsin L and cruzipain classes.

Our phylogenetic analysis shows that the helminth cathepsin L-like genes branched before the vertebrate cathepsin Ls, cathepsin Ss and cathepsin Ks (Figures 7 and 8). The distances of two representative sequences of the *F. hepatica* family to the three vertebrate cathepsins are similar (Tort, 1997). Likewise, the 'cathepsin L2' from *S. mansoni* and *S. japonicum* are equally related to vertebrate cathepsins L, S and K. These flatworm genes are the first in a line that included the nematode cathepsin L-like genes and the digestive enzymes of decapod and other arthropod sequences, and that gave rise to at least three different genes as a consequence of recent duplications in the vertebrate genome.

A comparative analysis of the substitution rates of human and rodent cathepsins Bs, Hs and Ls indicated that the cathepsin L enzymes have a higher rate of mutation (Hughes, 1994). Based on comparisons between rat and human cathepsin S and L to an outgroup (Dicdi cys2), Berti and Storer (1995) proposed that these enzymes diverged relatively recently by gene duplication of an original cathepsin L, and that the cathepsin Ss have since then undergone a faster evolution. The cathepsin Ks constitute a well-defined node with cathepsin Ss, and are equally distant from the cathepsin Ls (Tort, 1997). Based on these data, it appears that two rapid duplications gave rise to the cathepsin S family and the cathepsin K family (Figure 8). The presence of a cathepsin L, cathepsin S and cathepsin K in bony fishes indicates that these duplications took place early in the vertebrate lineage, at least before bony fishes appeared. If the duplication that gave rise to the vertebrate cathepsin Ss and Ks occurred in the chordate lineage, then cathepsin S- or K-like genes would not be expected to be found in the invertebrates.

8.5.3. *The Papain-like Proteinases*

Discussion of the papain-like class is beyond the scope of this review (for further information see Berti and Storer, 1995). However, it is worth noting

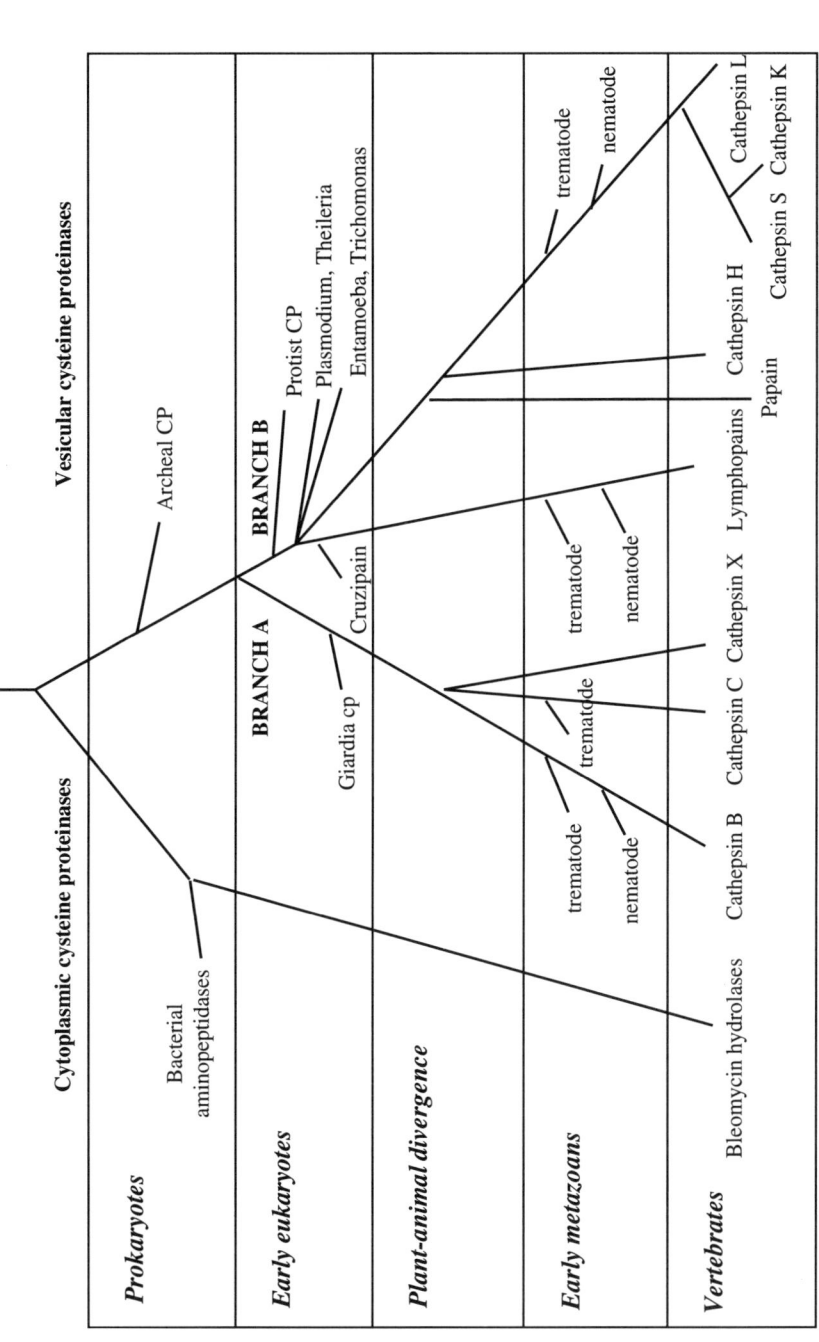

Figure 8 Hypothetical diversification of the cysteine proteinase members of the papain superfamily based on the relationships observed between gene sequences.

that there is a complete absence of animal sequences in this class. Conversely, there is an absence of plant sequences in the cathepsin L-like class. This observation suggests that independent gene duplications could have occurred within the plant and metazoan lineages to give rise to the papain-like and cathepsin L-like classes, i.e. that the papain-like and cathepsin L-like genes are orthologous.

8.5.4. *The Cathepsin H/Aleurain-like Proteinases*

Several studies have shown that the cathepsin Hs from vertebrates, and several sequences from plants (including aleurain, orizain gamma, maize cysteine proteinase 2 and tomato cysteine proteinase 3), constitute a defined group with strong statistical support (Hughes, 1994; Berti and Storer, 1995; Kirschke *et al.*, 1995; Tort, 1997) (Figures 5 and 7). The class appears to have diverged before the divergence of the cathepsin L-like and papain-like genes. This class, the cruzipain-like class and the cathepsin B-like class are the only groups with both animal and plant members, indicating that before animals and plants diverged at least three different classes of cysteine proteinases of the papain superfamily had emerged (Tort, 1997). Accordingly, it is surprising that no cathepsin H-like genes have been isolated from invertebrates, or, more relevantly to the present study, from helminths.

8.5.5. *Incertae saedis*

A gene encoding a cysteine proteinase secreted by infective larvae of *Toxocara canis* (Loukas *et al.*, 1998) shows limited similarity to the other members of the Branch B of the papain superfamily. A related EST has been obtained from a *C. elegans* library (R07E3.1, EMBL Z49207). Because these sequences show a low level of homology to the cathepsin Ls and other characterized families, they may constitute a new group of cysteine proteinases (Figures 5 and 7). However, more sequences of the *T. canis* type are needed to determine if it will be a well-supported family. Similarly, several genes of baculovirus constitute a well-defined node, but the relationships with other members of the branch are obscure.

9. CONCLUDING REMARKS

Parasites live in a nutritionally rich environment where food can be gathered with ease. Indeed, it is considered that this easy procurement of food has been the main evolutionary driving force that initiated host–parasite association. Consequently, some of the most significant alterations that have evolved from the basic turbellarian plan relate to the structures involved in the acquisition of food (Halton, 1997). In the monogenen ectoparasites, such as *Calicotyle kroyeri* and *Diclidophora merlangi*, digestion of host tissues (skin or blood) takes place in lysosomal-like vesicles within the gastrodermal cells lining the digestive tract. In contrast, in the digenean trematodes such as *Fasciola hepatica* and *Schistosoma* spp., digestion of the blood meal takes place extracellularly in the lumen of the gut. In the latter cases, the contents of the lysosomal-like vesicles are secreted into the gut lumen (Halton, 1997). Our phylogenetic analysis has shown that by the time the digestive tract had appeared in flatworms, four main types of cysteine proteinases of the papain superfamily, namely cathepsin B, cathepsin C, cathepsin L and cruzipain, had evolved (Figure 8). Many histochemical and immunolocalization studies have demonstrated that these proteinases are packaged into the lysosomal-like vesicles of the gastrodermal cells of helminths. To take the schistosomes as an example, cysteine proteinases of all four types described above have been identified and localized to these cells. Cysteine proteinases, therefore, play an important role in the acquisition of nutrients by helminths. Likewise, cysteine proteinases play a digestive role in other organisms such as the American lobster (Laycock *et al.*, 1991), *Drosophila melanogaster* (Matsumoto *et al.*, 1995) and the corn pest *Sitophilius zeamais* (Matsumoto *et al.*, 1997).

But the parasitic way of life presents other challenges. The parasite must overcome the immune defence mechanisms of their host, and must make their way to its final destination where they can mature and produce offspring. Moreover, most helminths have complex life cycles which often include two or more hosts. In the preceding sections it was shown that parasites may employ proteinases for functions other than digestive, such as host penetration, tissue migration, immune evasion and moulting. In keeping with the notion of cysteine proteinases carrying out a variety of functions, phylogenetic analysis shows that a positive diversification of cysteine proteinase genes occured independently in many helminths. The net result of this diversification is that these parasites possess a complex panel of proteinases with unique substrate specificities and functions that are under independant temporal and spatial control.

In mammalian cells, lysosomal enzymes precursors are synthesized on ribosomes attached to the ER. As they are translocated into the lumen of the ER, the signal peptide is removed. Subsequently, during their trafficking in the Golgi apparatus on their way to the lysosomes, the enzymes are post-translationally modified by glycosylation and phosphorylation. A mannose-6-phosphate group added to the *N*-linked carbohydrates helps target the enzymes to the lysosomes. Finally, activation of the precursor enzyme involves the removal of the pro-peptide at acid pH. Although the helminth cysteine proteinases are also synthesized as inactive precursors, it is not yet known whether they progress through similar trafficking and processing events as the orthologous mammalian enzymes. Moreover, the importance of glycolysation in directing the enzymes to the intracellular lysosomal vesicles is not understood. However, recent studies by Roche et al. (1997) and Dowd et al. (1997) demonstrated that Saccharomyces cerevisiae transformed with vectors carrying the F. hepatica cathepsin L1 or cathepsin L2 genes synthesized and trafficked the enzymes through the normal secretory pathways, and secreted fully processed enzymes into culture medium. Although these F. hepatica enzymes lack potential N-glycosylation sites, more recent studies with schistosome cathepsin Ls, which do contain these sites, have yielded similar results (Brady, Dowd, Brindley and Dalton, unpublished data).

The propeptide plays an important role in maintaining the stability and correct folding of the precursor enzyme while it is trafficked through the Golgi apparatus. Although we and others (Hughes, 1994) have found that the level of conservation in the propeptide is lower than that observed in the mature protein, the secondary structure features that are essential in the interaction of the propeptide and mature enzyme were conserved. Variation observed in the propeptides of the various members of the papain superfamily correlated with the branches and classes described here (Table 5). This observation is of interest, since free pro-peptides are potent and selective inhibitors of their cognate enzymes and can therefore form the basis for the design of parasite-specific enzyme inhibitors (see Cyler et al., 1996; Tort et al., 1997).

Our analysis, and those of others, of parasite genes encoding cysteine proteinases of the papain superfamily has shown that these can be classified according to their sequence and secondary structure relatedness (Table 5; Berti and Storer, 1995; Rawlings and Barrett, 1997). The challenge now will be to define these various groups by means of their substrate specificity and biological functions. Although many helminth cysteine proteinases have been characterized in parasite extracts or purified fractions using the peptide substrates classically used to distinquish classes of mammalian enzymes (and thus have been described in the same framework as the mammalian enzymes, e.g. cathepsin B, cathepsin L, etc.), few genes have been assigned to precise activities. Assigning genes to precise activities is complicated by the fact that a large number of helminths express families of related proteinases with similar activities and reactivity to antibody reagents. However, with the advent of many systems for expressing these enzymes in a functionally active form, this information should be forthcoming, hopefully in the near future.

More information is available on cysteine proteinases of the papain superfamily compared to the other major groups of proteinases which occur in the cells of helminth proteinases and their hosts. As more sequences become available for helminth members of these other groups of proteinases, such as the metallo-proteinases, it will be of interest to examine their phylogenetic relationships in similar ways to those reported herein and elsewhere for the papain-like cysteine proteinases.

REFERENCES

Acosta, D., Goni, F. and Carmona, C. (1998). Characterization and partial purification of a leucine aminopeptidase from *Fasciola hepatica*. *Journal of Parasitology* **84**, 1–7.

Amiri, P., Sakanari, J., Basch, P., Newport, G. and McKerrow, J. H. (1988). The *Schistosomatium douthitti* cercarial elastase is biochemically and structurally distinct from that of *Schistosoma mansoni*. *Mololecular and Biochemical Parasitology* **28**, 113–120.

Andresen, K., Tom, T. and Strand, M. (1991). Characterization of cDNA clones encoding a novel calcium-activated neutral proteinase from *Schistosoma mansoni*. *Journal of Biological Chemistry* **266**, 15085–15090.

Asch, H.L. and Dresden, M.L. (1979). Acidic thiol proteinase activity of *Schistosoma mansoni* egg extracts. *Journal of Parasitology* **65**, 543–549.

Atkinson, H.J. (1993). Opportunities for improved control of plant parasitic nematodes via plant biotechnology. In: *Opportunities for Molecular Biology in Crop Production*. (D.J. Beadle, D.H. Bishop, L.G. Copping, G.K. Dixon and D.W. Holloman, eds.) pp. 257–266. British Crop Protection Council.

Auriault, C., Ouaissi, M. A., Torpier, G., Eisen, H. and Capron, A. (1981). Proteolytic cleavage of IgG bound to the Fc receptor of *Schistosoma mansoni* schistosomula. *Parasite Immunology* **3**, 33–44.

Auriault, C., Pierce, R., Cesari, I.M. and Capron, A. (1982). Neutral protease activities at different developmental stages of *Schistosoma mansoni* in mammalian hosts. *Comparative and Biochemical Physiology* **72B**, 337–384.

Barrett, A.J. and Rawlings, N.D. (1996). Families and clans of cysteine peptidases. *Perspectives in Drug Discovery and Design* **6**, 1–32.

Barrett, A.J., Khembhavi, A.A., Brown, M.A., Kirschke, H., Knight, C. G., Tamai, M. and Hanada, K. (1982). L-*trans*-expoxysuccinyl-leucylamido(4-guanidino)-butane (E-64) and its analogues as inhibitors of cysteine proteinases including cathepsins B, H and L. *Biochemical Journal* **201**, 189–198.

Barrett, J. (1981). Nutrition and biosynthesis. In: *Biochemistry of Parasitic Helminths*. pp. 154–155, London and Basingstoke: Macmillan Publishers Ltd.

Becker, M.M., Harrop, S.A., Dalton, J.P., Kalinna, B.H., McManus, D.P. and Brindley, P.J. (1995). Cloning and characterization of the *Schistosoma japonicum* aspartic proteinase involved in haemoglobin degradation. *Journal of Biological Chemistry* **270**, 24496–24501.

Berasain, P., Goni, F., McGonigle, S., Dowd, A., Dalton, J. P., Frangione, B. and Carmona, C. (1997). Proteinases secreted by *Fasciola hepatica* degrade extracellular matrix and basement membrane components. *Journal of Parasitology* **83**, 1–5.

Berti, P. and Storer, A.C. (1995). Alignment /Phylogeny of the papain superfamily of cysteine proteinases. *Journal of Molecular Biology* **246**, 273–283.

Bogitsh, B.J. (1983). Peptidase activity in the egg-shell-enclosed embryo of

Schistosoma mansoni using fluorescent histochemisrty. *Transactions of the American Microscopical Society* **102**, 169–172.

Bogitsh, B.J. and Dresden, M.D. (1983). *Schistosoma japonicum*: biochemistry and cytochemistry of dipeptidyl aminopeptidase-II-like activity in adults. *Experimental Parasitology* **60**, 163–170.

Bogitsh, B.J. and Kirschner, K.F. (1986). *Schistosoma japonicum*: ultrastructural localization of a haemoglobinase using mercury labeled pepstatin. *Experimental Parasitology* **62**, 211–215.

Bogitsh, B.J. and Kirschner, K.F. (1987). *Schistosoma japonicum*: immunocytochemistry of adults using heterologous antiserum to bovine cathepsin D. *Experimental Parasitology* **64**, 213–218.

Bogitsh, B.J., Kirschner, K.F. and Rotmans, J.P. (1992). *Schistosoma japonicum*: immunoinhibitory studies on haemoglobin using heterologous antiserum to bovine cathepsin D. *Journal of Parasitology* **78**, 454–459.

Boisvenue, R.J., Tonkinson, L.V., Cox, G.N. and Hageman, R. (1992). Fibrinogen-degrading proteins from *Haemonchus contortus* used to vaccinate sheep. *American Journal for Veterinary Research* **53**, 1263–1265.

Bolla, R. and Weinstein, P.P. (1980). Acid protease activity during development and aging of *Nippostrongylus brasiliensis*. *Comparative Biochemistry and Physiology*, **66B**, 475–481.

Brindley, P.J., Gam, A.A., McKerrow, J.H. and Neva, F.A. (1995a) Ss40: the zinc endopeptidase secreted by infective larvae of *Strongyloides stercoralis*. *Experimental Parasitology* **80**, 1–7.

Brindley, P.J., Ramirez, B., Tiu, W.U., Wu, G., Wu, H.W. and Yi, X. (1995b). Networking schistosomiasis japonica. *Parasitology Today* **11**, 163–165.

Brindley, P.J., Kalinna, B.M., Dalton, J.P., Day, S.R., Wong, J.Y.M., Smythe, M.L. and McManus, D.P. (1997). Proteolytic degradation of host hemoglobin by schistosomes. *Molecular and Biochemical Parasitology* **89**, 1–9.

Brinkworth, R.I., Brindley, P.J and Harrop S.A. (1996). Structural analysis of the catalytic site of AcCP-1, a cysteine proteinase secreted by the hookworm *Ancylostoma caninum*. *Biochemica and Biophysica Acta* **1298**, 4–8.

Britton, C., Knox, D.P., Urquhart, G. and Kennedy, M.W. (1992). Characterisation of proteinases of *Dictyocaulus viviparus*, and their inhibition by antibody from infected and vaccinated calves. *Parasitology* **105**, 325–333.

Bundy, D.A.P. and Cooper, E.S. (1989). *Trichuris* and Trichuriasis in humans. *Advances in Parasitology* **28**, 107–173.

Butler, R., Michel, A., Kunz, W. and Klinkert, M-Q. (1995). Sequence of *Schistosoma mansoni* cathepsin C and its structural comparison with papain and cathepsins B and L of the parasite. *Protein and Peptide Letters* **2**, 313–320.

Caffrey, C.R. and Ryan, M.F. (1994). Characterisation of proteolytic activity of excretory-secretory products from adult *Strongylus vulgaris*. *Veterinary Parasitology* **52**, 285–296.

Carmona, C., Dowd, A.J., Smith, A.M. and Dalton, J.P. (1993). Cathepsin L proteinase secreted by *Fasciola hepatica in vitro* prevents antibody-mediated eosinophil attachment to newly excysted juveniles. *Molecular and Biochemical Parasitology* **62**, 9–18.

Carmona, C., McGonigle, S., Dowd, A.J., Smith, A. M., Coughlan, S., MacGowran, E. and Dalton, J.P. (1994). A dipeptidylpeptidase secreted by *Fasciola hepatica*. *Parasitology* **109**, 113–118.

Chapman, C.B. and Mitchell, G.F. (1982). Proteolytic cleavage of immunoglobulin by enzymes released by *Fasciola hepactica*. *Veterinary Parasitology* **11**, 165–178.

Chappell, C.L. and Dresden, M.H. (1986). *Schistosoma mansoni*: protease activity of 'haemoglobinase' from the digestive tract of adult worms. *Experimental Parasitology* **61**, 160–167.

Chavez-Olortegui, C., Resende, M. and Tavares, C.A.P. (1992). Purification and characterization of a 47 kDa protease from *Schistosoma mansoni* cercarial secretion. *Parasitology* **105**, 211–218.

Chen, J.-M., Dando, P.M., Rawlings, N.D., Brown, M.A., Young, N.E., Stevens, R.A., Hewitt, E., Watts, C. and Barrett, A.J. (1997). Cloning, isolation, and characterization of mammalian legumain, an asparaginyl endopeptidase. *Journal of Biological Chemistry* **272**, 8090–8096.

Chung, Y.B., Yong, Y., Joo, I.J., Cho, S.Y. and Kang, S.Y. (1995). Excystment of *Paragonimus westermani* metacercariae by endogenous cysteine proteases. *Journal of Parasitology* **81**, 137–142.

Clark, C.G. and Rogers, A.J. (1995). Direct evidence for a secondary loss of mitochondria in *Entamoeba histolytica*. *Proceedings of the National Academy of Sciences of the USA* **92**(14), 6518–6521.

Clegg, J.A. and Smith M.A. (1978). Dead vaccines against helminths. *Advances in Parasitology* **16**, 165–206.

Cocude, C., Pierrot, C., Cetre, C., Godin, C., Capron, A.A. and Khalife, J. (1997). Molecular characterization of a partial sequence encoding a novel *Schistosoma mansoni* serine protease. *Parasitology* **115**, 395–402.

Cohen, F.E., Gregoret, L. M., Amiri, P., Aldape, K., Railey, J., and McKerrow, J.H. (1991). Arresting the tissue invasion of a parasite by protease inhibitors chosen with the aid of computer modelling. *Biochemstry* **30**, 11221–11229.

Coles, G.C. (1970). A comparison of some isoenzymes of *Schistosoma mansoni* and *Schistosoma haematobium*. *Comparative and Biochemical Physiology* **33**, 549–558.

Coulombe, R., Grochulski, P., Sivaraman, J., Menard, R., Mort J.S. and Cygler, M. (1996). Structure of human procathepsin L reveals the molecular basis of inhibition by the prosegment. *EMBO Journal* **15**(20), 5492–5503.

Cox, G.N., Pratt, D., Hageman, R. and Boisvenue, R.J. (1990). Molecular cloning and primary sequences of a cysteine protease expressed by *Haemonchus contortus* adult worms. *Molecular and Biochemical Parasitology* **41**, 25–34.

Creaney, J., Wilson, L., Dosen, M., Sandeman, M., Spithill, T. W. and Parson, J. C. (1996). *Fasciola hepatica*: irradiation-induced alterations in carbohydrate and cathepsin-B protease expression in newly excysted juvenile liver fluke. *Experimental Parasitology* **83**, 202–215.

Creighton, T.E. and Darby, N.J. (1989). Functional evolutionary divergence of proteolytic enzymes and their inhibitors. *Trends in Biochemical Science* **14**, 319–324.

Criado-Fornelio, A., de Armas-Serra, C., Gimenez-Pardo, C., Casado-Escribano, C., Jimenez-Gonzalez, A. and Rodriquez-Caabeiro, F. (1992). Proteolytic enzymes from *Trichinella spiralis* larvae. *Veterinary Parasitology* **45**, 133–140.

Croall, D.E. and Demartino, G.N. (1991). Calcium-activated neutral protease (calpain) system: structure, function, and regulation. *Physiological Reviews* **71**, 813–847.

Croall, D.E. and McGrody, K.S. (1994). Domain structure of calpain: mapping the site for calpastatin. *Biochemistry* **33**, 13223–13230.

Curtis, R.H., Fallon, P.G. and Doenhoff, M.J. (1996). Sm480: a high molecular weight *Schistosoma mansoni* antigen associated with protective immunity. *Parasite Immunology* **18**, 149–157.

Cygler, M., Sivaraman, J., Grochulski, P., Coulombe, R., Storer, A.C. and Mort, J.S. (1996). Structure of rat procathepsin B: model for inhibition of cysteine protease activity by the proregion. *Structure* **15**(4), 405–416.
Dalton, J.P. and Brindley, P.J. (1996). Schistosome asparaginyl endopeptidase Sm 32 in hemoglobin digestion. *Parasitology Today* **12**, 125.
Dalton, J.P. and Brindley, P.J. (1997). Proteases of trematodes. In: *Advances in Trematode Biology* (Gracyzyk, T.K and Fried, B., eds.) pp. 265–304. Boca Raton, Florida: CRC Press.
Dalton, J.P. and Brindley, P.J. (in press). *Schistosome legumain*. In: *Handbook of Proteolytic Enzymes* (A.J. Barrett, N.D. Rawlings and J.F. Woessner, jr., eds.). London: Academic Press.
Dalton, J. P. and Heffernan, M. (1989). Thiol proteases released *in vitro* by *Fasciola hepatica*. *Molecular and Biochemical Parasitology* **35**, 161–166.
Dalton, J. P., Dowd, A. J. and Carmona, C. (1994). Cathepsin L proteinases secreted by the parasite *Fasciola hepatica*. In: *The Biology of Parasitism* (R. Ehrlich and A. Nieto, eds.), Montevideo: Ediciones TRILCE.
Dalton, J.P., Smith, A.M., Clough, K.A. and Brindley, P.J. (1995a). Digestion of haemoglobin by schistosomes: 35 years on. *Parasitology Today* **12**, 299–303.
Dalton, J.P., Hola-Jamriska, L. and Brindley, P.J. (1995b). Asparaginyl endopeptidase activity in adult *Schistosoma mansoni*. *Parasitology* **111**, 575–580.
Dalton, J.D., Clough, K.A., Jones, M.K. and Brindley, P.J. (1996a).Characterization of the cathepsin-like cysteine proteinases of *Schistosoma mansoni*. *Infection and Immunity* **64**, 1328–1334.
Dalton, J.P., McGonigle, S., Rolph, T.P. and Andrews, S.J. (1996b). Induction of protective immunity in cattle against infection with *Fasciola hepatica* by vaccination with cathepsin L proteinases and with haemoglobin. *Infection and Immunity* **64**, 5066–5074.
Dalton, J.P., Clough, K.A., Jones, M.K. and Brindley, P.J. (1997). Characterization of the cysteine proteinases of *Schistosoma mansoni* cercariae. *Parasitology* **114**, 105–112.
Darani, H.Y., Curtis, R.H.C., McNeice, C., Price, H.P., Sayers, J.R. and Doenhoff, M.J. (1997). *Schistosoma mansoni*: anomalous immunogenic properties of a 27 kDa larval serine protease associated with protective immunity. *Parasitology* **115**, 237–247.
Dasgupta, D.R. and Ganguly, A.K. (1975). Isolation, purification and characterisation of a trypsin-like protease from the root-knot nematode, *Meloidogyne incognita*. *Nemtologica* **21**, 370–384.
Davis, A.H., Nanduri, J. and Watson, D.C. (1987). Cloning and gene expression of *Schistosoma mansoni* protease. *Journal of Biological Chemistry* **262**, 12851–12855.
Day, S.R., Dalton, J.P., Clough, K.A., Leonardo, L., Tiu, W.U., and Brindley, P.J. (1995). Cloning and characterization of cathepsin L proteinases of *Schistosoma japonicum*. *Biochemical and Biophysical Research Communications* **217**, 1–9.
DeCock, H., Knox, D.P., Claerebout, E. and De Graaf, D.C. (1993). Partial characterisation of proteolytic enzymes in different developmental stages of *Ostertagia ostertagi*. *Journal of Helminthology* **67**, 271–278.
Doenhoff, M.J. (1998). A vaccine for schistosomiasis: alternative approaches. *Parasitology Today* **14**, 105–109.
Doenhoff, M.J., Modha, J., Curtis, R.H.C. and Adeoye, G.O. (1988). Immunological identification of *Schistosoma mansoni* peptidases. *Molecular and Biochemical Parasitology* **31**, 233–240.
Dolenc, I., Turk, B., Pungercic, G., Ritonja, A. and Turk, V. (1995). Oligomeric

structure and substrate induced inhibition of human cathepsin C. *Journal of Biological Chemistry* **270**, 21626–21631.
Dowd, A.J., Dalton, J.P., Loukas, A.C., Prociv, P. and Brindley, P.J. (1994a). Secretion of cysteine proteinase activity by the zoonotic hookworm *Ancylostoma caninum*. *American Journal of Tropical Medicine and Hygeine* **51**, 341–347.
Dowd, A.J., Smith, A.M., McGongle, S., Dalton, J.P. (1994b). Purification and characterisation of a second cathepsin L proteinase secreted by the parasitic trematode *Fasciola hepatica*. *European Journal of Biochemistry* **223**, 91–98.
Dowd, A.J., McGongle, S. and Dalton, J.P. (1995). *Fasciola hepatica* cathepsin L proteinase cleaves fibrinogen and produces a novel type of fibrin clot. *European Journal of Biochemistry* **232**, 241–246.
Dowd, A.J., Tort, J., Roche, L. and Dalton, J.P. (1997). Isolation of a cDNA encoding *Fasciola hepatica* cathepsin L2 and functional expression in *Saccharomyces cerevisiae*. *Molecular and Biochemical Parasitology* **88**, 163–174.
Drake, L.J., Bianco, A.E., Bundy, D.A.P. and Ashall, F. (1994). Characterisation of peptidases of adult *Trichuris muris*. *Parasitology* **109**, 623–630.
Dresden, M.H. and Deelder, A.M. (1979). *Schistosoma mansoni*: Thiol protease properties of adult worm 'hemoglobinase'. *Experimental Parasitology* **48**, 190–197.
Dresden, M.L., Sung, C.K. and Deelder, A.M. (1983). A monoclonal antibody from infected mice to a *Schistosoma mansoni* egg proteinase. *Journal of Immunology* **130**, 1–3.
Dresden, M.H., Rege, A.A. and Murrell, K.D. (1985). *Strongyloides ransomi*: Proteolytic enzymes from larvae. *Experimental Parasitology* **59**, 257–263.
Eakin, A.E., Bouvier, J., Sakanari, J.A., Craik, C.S. and McKerrow, J.H. (1990). Amplification and sequencing of genomic DNA fragments encoding cysteine proteases from protozoan parasites. *Molecular and Biochemical Parasitology* **39**, 1–8.
El Meanawy, M.A., Aji, T., Phillips, N.F.B., Davis, R.E., Salata, R.A., Malhorta, L., McClain, D., Aikawa, M. and Davis, A.H. (1990). Definition of the complete *Schistosoma mansoni* hemoglobinase mRNA sequence and gene expression in developing parasites. *American Journal of Tropical Medicine and Hygeine* **43**, 67–78.
Elliot, E. and Sloane, B.F. (1996). The cysteine protease cathepsin B in cancer. *Perspectives in Drug Discovery and Design* **6**, 1–32.
Fagbemi, B.O. and Hillyer, G.V. (1991). Partial purification and characterisation of the proteolytic enzymes of *Fasciola gigantica* adult worms. *Veterinary Parasitology* **40**, 217–226.
Fagbemi, B.O. and Hillyer, G.V. (1992). The purification and characterisation of a cysteine protease of *Fasciola gigantica* adult worms. *Veterinary Parasitology* **43**, 223–232.
Felsenstein, J. (1989). PHYLIP Phylogeny Inference Package (version 3.2). *Cladistics* **5**, 164–166.
Fallon, P.G., Teixeira, M.M., Niece, C.M., Williams, T.J., Hellewell, P.G. and Doenhoff, M.J. (1996). Enhancement of *Schistosoma mansoni* infectivity by intradermal injections of larval extracts: a putative role for larval proteases. *Journal of Infectious Diseases* **173**, 1460–1466.
Felleisen, R. and Klinkert, M.-Q. (1990). *In vitro* translation and processing of cathepsin B of *Schistosoma mansoni*. *EMBO Journal* **9**, 371–377.
Fetterer, R.H. and Rhoads, M.L. (1997a). The *in vitro* uptake and incorporation of haemoglobin by adult *Haemonchus contortus*, *Veterinary Parasitology* **69**, 77–87.

Fetterer, R.H. and Rhoads, M.L. (1997b). The *in vitro* uptake of albumin by adult *Haemonchus contortus* is altered by extracorporeal digestion. *Veterinary Parasitology* **73**, 249–256.

Finken, M., Sobek, A., Symmons, P. and Kunz, W. (1994). Characterization of the complete protein disulfide isomerase gene of *Schistosoma mansoni* and identification of the tissues of its expression. *Molecular and Biochemical Parasitology* **64**, 135–144.

Finken-Eigen, M. and Kunz, W. (1997). *Schistosoma mansoni*: gene structure and localization of a homologue to cysteine protease ER 60. *Experimental Parasitology* **86**, 1–7.

Fishelson, Z. (1989). Complement and parasitic trematodes. *Parasitology Today* **5**, 19–25.

Fishelson, Z., Amiri, P., Friend, D.S., Marikovsky, M., Petitt, M., Newport, G. and McKerrow, J.H. (1992). *Schistosoma mansoni*: cell-specific expression and secretion of a serine protease during development of cercariae. *Experimental Parasitology* **75**, 87–98.

Freedman, R.B., Hirst, T.R. and Tutte, M.F. (1994). Protein disuphide isomerase: building bridges in protein folding. *Trends in Biochemical Sciences* **19**, 331–336.

Fripp, P.J. (1967). The histochemical localization of leucine aminopeptidase in *Schistosoma rodhaini*. *Comparative and Biochemical Physiology* **20**, 307–309.

Fukase, T., Matsuda, Y., Akihama, S. and Itagaki, H. (1985). Purification and some properties of cysteine protease of *Spirometra erinacei* plerocercoid. *Japanese Journal of Parasitology* **34**, 351–360.

Gamble, H.R., Purcell, J.P. and Fetterer, R.H. (1989). Purification of a 44 kilodalton protease which mediates the ecdysis of infective *Haemonchus contortus*. *Molecular and Biochemical Parasitology* **33**, 49–58.

Gamble, H.R. and Mansfield, L.S. (1996). Characterisation of excretory-secretory products from larval stages of *Haemonchus contortus*. *Veterinary Parasitology* **62**, 291–305.

Gamble, H.R., Fetterer, R.H. and Mansfield, L.S. (1996). Developmentally regulated zinc metalloproteinases from third- and fourth-stage larvae of the ovine nematode *Haemonchus contortus*. *Journal for Parasitology* **82**, 197–202.

Gay, N.J and Walker, J.E. (1985). Molecular cloning of a bovine cathepsin. *Biochemical Journal* **225**, 707–712.

Gelb, B.D., Shi, G.P., Heller, M., Weremowicz, S., Morton, C., Desnick, R.J. and Chapman, H.A. (1997). Structure and chromosomal assignment of the human cathepsin K gene. *Genomics* **41**(2), 258–262.

Ghendler, Y., Arnon, R. and Fishelson, Z. (1994). Isolation and characterization of Smpi56, a novel serine proteinase inhibitor. *Experimental Parasitology* **78**, 121–131.

Ghendler, Y., Parizade, M., Arnon, R., McKerrow, J.H. and Fishelson, Z. (1996). *Schistosoma mansoni*: evidence for a 28-kDa membrane-anchored protease on schistosomula. *Experimental Parasitology* **83**, 73–82.

Ghoneim, H. and Klinkert, M-Q. (1995). Biochemical properties of purified cathepsin B from *Schistosoma mansoni*. *International Journal for Parasitology* **25**, 1515–1519.

Glickman, L.T. and Schantz, P.M. (1981). Epidemiology and pathogenesis of zoonotic toxocariasis. *Epidemiology Reviews* **3**, 230–250.

Goll, D.E., Thompson, V.F., Taylor, R.G. and Zalewska, T. (1992). Is calpain activity regulated by membranes and autolysis or by calcium and calpastatin? *Bioessays* **14**, 549–556.

Gotz, B. and Klinkert, M-Q. (1993). Expression and partial characterization of a cathepsin B-like enzyme (Sm31) and a proposed 'haemoglobinase'(Sm32) from *Schistosoma mansoni*. *Biochemical Journal* **290**, 801–806.

Groves, M.R., Taylor, M.A.J., Scott, M., Cummings, N.J., Pickersgill, R.W. and Jenkins, J.A. (1996). The prosequence of procaricain forms an a-helical domain that prevents access to the substrate binding cleft. *Structure* **15** (4), 1193–1203.

Halton, D.W. (1967). Observations on the nutrition of digenetic trematodes. *Parasitology* **57**, 639–660.

Halton, D.W. (1997). Nutritional adaptations to parasitism within the platyhelminths. *International Journal for Parasitology* **27**, 693–704.

Hamajima, F., Yamakami, K. and Tsuru, S. (1986). Experimental eosinophil accumulation in mice by a thiol protease from metacercariae of *Paragonimus westermani*. *Japanese Journal of Parasitology* **35**, 63–64.

Hamajima, F., Yamakami, K., Fukuda, K., Tsuru, S., Fujino, T., Hamajima, H. (1991). The cell content of peritoneal exudates following the injection of neutral thiol protease into guinea pigs. *Journal of Parasitology* **77**, 635–637.

Hamajima, F., Yamamoto, M., Tsuru, S., Yamakami, K., Fujino, T., Hamajima, H. and Katsura, Y. (1994). Immunosuppression by a neutral thiol protease from parasitic helminth larvae in mice. *Parasite Immunology* **16**, 261–273.

Hara-Nishimura, I., Takeuchi, Y. and Nishimura, M. (1993). Molecular characterization of a vacuolar processing enzyme related to a putative cysteine proteinase of *Schistosoma mansoni*. *Plant Cell* **5**, 1651–1659.

Harrop, S.A., Prociv, P. and Brindley, P.J. (1995a). Amplification and characterization of cysteine proteinase genes from nematodes. *Tropical Medicine and Parasitology* **46**, 119–122.

Harrop, S.A., Sawangjaroen, N., Prociv, P. and Brindley, P.J. (1995b). Characterisation and localization of cathepsin B proteinases expressed by adult *Ancylostoma caninum* hookworms. *Molecular and Biochemical Parasitology* **71**, 163–171.

Harrop, S.A., Prociv, P. and Brindley, P.J. (1996). Acasp, a gene encoding a cathepsin D-like aspartic protease expressed by adult *Ancylostoma caninum* hookworms. *Biochemical and Biophysical Research Communications* **227**, 294–302.

Hawdon, J.M., Emmons, S.W. and Jacobson, L.A. (1989). Regulation of proteinase levels in the nematode *Caenorhabditis elegans* — preferential depression by acute or chronic starvation. *Biochemical Journal* **264**, 161–165.

Hawdon, J.M., Jones, B.F., Perregaux, M.A. and Hotez, P.J. (1995). *Ancylostoma caninum*: metalloprotease release coincides with activiation of infective larvae *in vitro*. *Experimental Parasitology* **80**, 205–211.

Hawthorne, S.J., Halton, D.W. and Walker, B. (1994). Identification and characterization of the cysteine and serine proteinases of the trematode, *Haplometra cylindracea* and determination of their hemoglobinase activity. *Parasitology* **108**, 595–601.

Healer, J., Ashall, F. and Maizels, R.M. (1991). Characterisation of proteolytic enzymes from larval and adult *Nippostrongylus brasiliensis*. *Parasitology* **103**, 305–314,

Heussler, V.T. and Dobbelaere, D.A.E. (1994). Cloning of a protease gene family of *Fasciola hepatica* by the polymerase chain reaction. *Molecular and Biochemical Parasitology* **64**, 11–23.

Hill, D.E., Gamble, H.R., Rhoads, M.L., Fetterer, R.H. and Urban, J.F. (1993). *Trichuris suis*: a zinc metalloprotease from culture fluids of adult parasites. *Experimental Parasitology* **77**, 170–178.

Hoeppli, R. (1933). On histolytic changes and extra-intestinal digestion in parasitic infection. *Lingnam Science Journal* **12**, Suppl. 1–11.

Hola-Jamrisksa, L., Tort, J.F., Dalton, J.P., Day, S.R, Fan, J., Aaskov, J. and Brindley, P.J. (1998). Cathepsin C of *Schistosoma japonicum*: cDNA encoding the preproenzyme and its phylogenetic relationships. *European Journal of Biochemistry* (in press).

Homma, K., Kurata, S. and Natori, S. (1994). Purification, characterization and cDNA cloning of procathepsin L from the culture medium of NIH-Spae-4, an embrionic cell line of *Sarcophaga peregrina* (flesh fly), and its involvement in the differentiation of imaginal discs. *Journal of Biological Chemistry* **269**, 15258–15264.

Hong, X.Q., Bouvier, J., Wong, M.M., Yamagata, G.Y.L. and McKerrow, J.H. (1993). *Brugia pahangi*: identification and characterisation of an aminopeptidase associated with larval molting. *Experimental Parasitology* **76**, 127–133.

Hota-Mitchell, S., Siddiqui, A.A., Dekabin, G.A., Smith, J., Tognon, C. and Podesta, R. (1997). Protection against *Schistosoma mansoni* infection with a recombinant baculovirus-expressed subunit of calpain. *Vaccine* **15**, 1631–1640.

Hotez, P.J. and Cerami, A. (1983). Secretion of a proteolytic anticoagulant by *Ancylostoma* hookworms. *Journal of Experimental Medicine* **157**, 1594–1603.

Hotez, P.J., Le Trang, N., McKerrow, J.H. and Cerami, A. (1985). Isolation and characterisation of a proteolytic enzyme from the adult hookworm *Ancylostoma caninum*. *Journal Biological Chemistry* **260**, 7343–7348.

Hotez, P., Haggerty, J., Hawdon, J., Milstone, L., Gamble, H.R., Schad, G. and Richards, F. (1990). Metalloproteases of infective *Ancylostoma* hookworm larvae and their possible functions in tissue invasion and ecdysis. *Infection and Immunity* **58**, 3883–3892.

Howell, M.J. (1973). Localisation of proteolytic activity in *Fasciola hepatica*. *Journal of Parasitology* **59**, 454–456.

Howell, R.M. (1966). Collagenase activity of immature *Fasciola hepatica*. *Nature* **209**, 713–714.

Hughes, A.L. (1994). The evolution of functionally novel proteins after gene duplication. *Proceedings of the Royal Society of London B* **256**, 119–124.

Hsu, H.F. (1933). Study on the the oesophageal glands of the parasitic nematoda superfamily Ascaroidea. *Chinese Medical Journal* **47**, 1247–1288.

Hui, K-S. (1988). A novel dipeptidyl aminopeptidase in rat brain membranes. Its isolation, purification, and characterization. *Journal of Biological Chemistry* **263**, 6613–6618.

Ischii, S.-I. (1994). Legumain: asparaginyl endopeptidase. *Methods in Enzymology* **244**, 604–615

Iwata, M., Munoz, J.J. and Ishizaka, K. (1983). Modulation of the biologic activities of IgE-binding factor. IV Identification of glycosylation-enhancing factor as a kallikrein-like. *Journal of Immunology* **131**, 1954–1960.

Jacobson, L.A., Jacobson, L, Hawdon, J.M. and Owens, G.P. (1988). Identification of a putative structural gene for cathepsin D in *Caenorhabditis elegans*. *Genetics*, **119**, 355–363.

Jankovic, D., Åslund, L., Oswald, I.P., Caspar, P., Champion, C., Pearce, E., Colligan, J. E., Strand, M., Sher, A. and James, S.L. (1996). Calpain is the target antigen of a Th1 clone that transfers protective immunity against *Schistosoma mansoni*. *Journal of Immunology* **157**, 806–814.

Jasmer, D.P. and McGuire, T.C. (1991). Protective immunity to a blood-feeding nematode (*Haemonchus contortus*) induced by parasite gut antigens. *Infection and Immunity* **59**, 4412–4417.

Jensen, R.A. (1976). Enzyme recruitment in evolution of new function. *Annual Review in Microbiology* **30**, 409–425.

Johnston, D.A.J. (1997). The WHO/UNDP/Word Bank *Schistosoma* Genome initiative: current status. *Parasitology Today* **13**, 45–46.

Jolodar, A. and Miller, D.J. (1997). Preliminary characterisation of an *Onchocerca volvulus* aspartyl protease. *International Journal for Parasitology* **27**, 1087–1090.

Kamata, I., Yamada, M., Uchikawa, R., Matsuda, R. and Arizono, N. (1995). Cysteine protease of the nematode *Nippostrongylus brasiliensis* evokes an IgE/IgG1 antibody response in rats. *Clinical and Experimental Immunology* **102**, 71–77.

Kane, S.E. (1993). Mouse procathepsin L lacking a functional glycosylation site is properly folded, stable and secreted by NIH 3T3 cells. *Journal of Biological Chemistry* **268**, 11456–11462.

Kang, S.Y., Cho, M.S., Chung, Y.B., Kong, Y. and Cho, S.Y. (1995). A cysteine protease of *Paragonimus westermani* eggs. *Korean Journal of Parasitology* **33**, 323–330.

Karanu, F.N., Rurangirwa, F.R., McGuire, T.C. and Jasmer, D.P. (1993). *Haemonchus contortus*: identification of proteases with diverse characteristics in adult worm excretory-secretory products. *Experimental Parasitology* **77**, 362–371.

Karanu, F.N., Rurangirwa, F.R., McGuire, T.C. and Jasmer, D.P. (1997). *Haemonchus contortus*: inter- and intrageographic isolate heterogeneity of proteases in adult worm excretory-secretory products. *Experimental Parasitology* **86**, 88–91.

Karcz, S.R., Podesta, R.B., Siddiqui, A.A., Dekaban, G.A, Strejan, G.H. and Clarke, M.W. (1991). Molecular cloning and sequence analysis of a calcium-activated neutral protease (calpain) from *Schistosoma mansoni*. *Molecular and Biochemical Parasitology* **49**, 333–336

Karrer, K.M., Peiffer, S.L. and DiTomas, M.E. (1993). Two distinct gene subfamilies within the family of cysteine protease genes. *Proceedings of the National Academy of Sciences of the USA* **90**, 3063–3067.

Kennedy, M.W. and Qureshi, F. (1986). Stage-specific antigens secreted by the parasitic larval stages of the nematode *Ascaris*. *Immunology* **58**, 515–522.

Kembhavi, A.A., Buttle, D.J., Knight, C.G. and Barrett, A.J. (1993). The two cysteine endopeptidases of legume seeds: purification and characterization by use of specific fluorometric assays. *Archives in Biochemistry and Biophysics* **303**, 208–213.

Kimura, M. (1983). *The Neutral Theory of Molecular Evolution*. Cambridge, Cambridge University Press.

Kirschke, H., Barrett, A.J. and Rawlings, N.D. (1995). Proteinases 1: lysosomal cysteine proteinases. In: *Protein Profile* (P. Shetterline, ed.) Vol. 2, issue 14. London: Academic Press.

Klinkert, M.-Q., Cioli, D., Shaw, E., Turk, V., Bode, W. and Butler, R. (1994). Sequence and structure similarities of cathepsin B from the parasite *Schistosoma mansoni* and human liver. *FEBS Letters* **351**, 397–400.

Klinkert, M.-Q., Ruppel, A. and Beck, E. (1987). Cloning of diagnostic 31/32 kilodalton antigens of *Schistosoma mansoni*. *Molecular and Biochemical Parasitology* **25**, 247–255.

Klinkert, M.-Q., Felleisen, R., Link, G., Ruppel, A. and Beck, E. (1989). Primary structures of Sm31/32 diagnostic proteins of *Schistosoma mansoni* and their identification as proteases. *Molecular and Biochemical Parasitology* **33**, 113–122.

Knoll, A.H. (1992). The early evolution of eukaryotes: a geological perspective. *Science* **256**, 622–627.
Knox, D.P. and Kennedy, M.W. (1988). Proteinases released by the parasitic larval stages of *Ascaris suum* and their inhibition by antibody. *Molecular and Biochemical Parasitology* **28**, 207–216.
Knox, D.P. and Jones, D.G. (1990). Studies on the presence and release of proteolytic enzymes (proteinases) in gastro-intestinal nematodes of ruminants. *International Journal for Parasitology* **20**, 243–249.
Knox, D.P., Redmond, D.L. and Jones, D.G. (1993). Characterisation of proteinases in extracts of adult *Haemonchus contortus*, the ovine abomasal nematode. *Parasitology* **106**, 395–404.
Knox, D.P., Smith, S.K., Smith, W.D., Redmond, D.L. and Murray, J.M. (1995). Thiol Binding Proteins. International patent application number PCT/GB95/00665.
Kong, Y., Chung, Y.B., Cho, S.Y., Choi, S.H. and Kang, S.Y. (1994). Characterisation of three neutral proteases of *Spirometra mansoni* plerocercoid. *Parasitology* **108**, 359–368.
Kong, Y., Kang, S.Y., Kim, S.H., Chung, Y.B. and Cho, S.Y. (1997). A neutral cysteine protease of *Spirometra mansoni* plerocercoid invoking and IgE response. *Parasitology* **114**, 263–271.
Korsitas, V.M. and Atkinson, H.J. (1994). Proteinases of females of the phytoparasite *Globodera pallida* (potato cyst nematode). *Parasitology* **109**, 357–365.
Kramer, J.D. and Bogitsh, B.J. (1985). *Schistosoma japonicum*: biochemistry and cytochemistry of dipeptidyl aminopeptidase-II-like activity in adults. *Experimental Parasitology* **60**, 163–170.
Kumar, S. and Pritchard, D.I. (1992). The partial characterisation of proteases present in the excretory/secretory products and exsheathing fluid of the infective (L3) larva of *Necator americanus*. *International Journal for Parasitology* **22**, 563–572.
Kwa, B.H. (1972). Studies on the sparganum of *Spirometra erinacei*. II Proteolytic enzymes in the scolex. *International Journal for Parasitology* **2**, 29–33.
Lackey, A., James, E.R., Sakanari, J.A., Resnick, S.D., Brown, M., Bianco, A.E. and McKerrow, J.H. (1989). Extracellular proteases of *Onchocerca*. *Experimental Parasitology* **68**, 176–185.
Landsperger, W.J., Stirewalt, M.A. and Dresden, M.H. (1982). Purification and properties of a proteolytic enzyme from the cercariae of the human trematode parasite *Schistosoma mansoni*. *Biochemical Journal* **201**, 137–144.
Larminie, C.G.C. and Johnstone, I.L. (1996). Isolation and characterization of four developmentally regulated cathepsin B-like cysteine protease genes from the nematode *Caenorhabditis elegans*. *DNA and Cell Biology* **15**, 75–82.
Lawrence, J.D. (1973). The ingestion of red blood cells by *Schistosoma mansoni*. *Journal of Parasitology* **59**, 60–63.
Lee, H.F., Chen, I-Li. and Lin, R.P. (1973). Ultrastructure of the excretory system of *Anisakis* larvae (Nematoda: *Anisakidae*) *Journal of Parasitology* **59**, 289–298.
Lewert, R.M. and Lee, C.L. (1954). Studies on the passage of helminth larvae through host tissues. Histochemical studies on the extracellular changes caused by penetrating larvae II Enzymatic activity of larvae *in vivo* and *in vitro*. *Journal of Infectious Diseases* **95**, 13–51.
Lewert, R.M. and Lee. C.L. (1956). Quantitative studies on the collagenase-like enzymes of cercariae of *Schistosoma mansoni* and the larvae of *Strongyloides ratti*. *Journal of Infectious Diseases* **99**, 1–14.

Li, R., Chen, X., Gong, B., Selzer, P.M., Li, Z., Davidson, E., Kurzan, G., Miller, R.E., Nuzum, E.O., McKerrow, J.H., Fletterick, R.J., Gillmor, S.A., Craik, C.S., Kuntz, I.D., Cohen, F.E. and Kenyon, G.L. (1996). Structure-based design of parasitic protease inhibitors. *Bioorganic and Medicinal Chemistry* **4**, 1421–1427.

Lilley, C.J., Urwin, P.E., McPherson, M.J. and Atkinson, H.J. (1996). Characterisation of intestinally active proteinases of cyst-nematodes. *Parasitology* **113**, 415–424.

Lindquist, R.N., Senft, A.W., Petitt, M. and McKerrow, J.H. (1986). *Schistosoma mansoni*: purification and characterization of the major acidic proteinase from adult worms. *Experimental Parasitology* **61**, 160–167.

Lipps, G., Fullkrug, R. and Beck, E. (1996). Cathepsin B of *Schistosoma mansoni*. Purification and activation of the recombinant proenzyme secreted by *Saccharomyces cerevisiae*. *Journal of Biological Chemistry* **271**, 1717–1725.

Liu, D.W., Kato, H., Nakamura, T. and Sugane, K. (1996). Molecular cloning and expression of the gene encoding a cysteine proteinase of *Spirometra erinacei*. *Molecular and Biochemical Parasitology* **76**, 11–21.

Loeb, L. and Smith, A.J. (1904). The presence of a substance inhibiting the coagulation of the blood in Ancylostomiasis. *Proceedings of the Pathological Society of Philadelphia*, New Series **7**, 173–178.

Longbottom, D., Redmond, D.L., Russell, M., Liddell, S., Smith, W.D. and Knox, D.P. (1997). Molecular cloning and characterisation of a putative aspartate proteinase associated with a gut membrane protein complex from adult *Haemonchus contortus*. *Molecular and Biochemical Parasitology* **88**, 63–72.

Loukas, A., Selzer, P.M. and Maizels, R.M. (1998). Characterisation of Tc-cpl-1, a cathepsin L-like cysteine proteases from *Toxocara canis* infective larvae. *Molecular and Biochemical Parasitology* **92**, 275–289

Lustigman, S., McKerrow, J.H., Shah, K., Lui, J., Huima, T., Hough, M. and Brotman, B. (1996). Cloning of a cysteine protease required for the molting of *Onchocerca volvulus* third stage larvae. *Journal of Biological Chemistry* **271**, 30181–30189.

MacLennan, K., Gallagher, M.P. and Knox, D.P. (in press). Stage-specific serine and metallo-proteinase release by adult and larval *Trichostrongylus vitrinus*. *International Journal for Parasitology*.

Maki, J., Furuhashi, A. and Yanagisawa, T. (1982). The activity of acid proteases hydrolysing haemoglobin in parasitic helminths with special reference to interspecific and intraspecific distribution. *Parasitology* **84**, 137–147.

Maki, J. and Yanagisawa, T. (1986). Demonstration of carboxyl and thiol protease activities in adult *Schistosoma mansoni, Dirofilaria immitis, Angiostrongylus cantonensis* and *Ascaris suum. Journal of Helminthology* **60**, 31–37.

Mallison, D.J., Lockwood, B.C., Coombs, G.H. and North, M.J. (1994). Identification and molecular cloning of four cysteine proteinase genes from the pathogenic protozoon *Trichomonas vaginalis*. *Microbiology* **140**, 2725–2735.

Mallison, D.J., Livingstone, J., Appleton, K.M, Lees, S.J., Coombs, G.H. and North M.J. (1995). Multiple cysteine proteinases of the pathogenic protozoon *Tritrichomonas foetus*: identification of seven diverse and differentially expressed genes. *Microbiology* **141**, 3077–3085.

Marco, M. and Nieto, A. (1991). Metalloproteinases in the larvae of *Echinococcus granulosus*. *International Journal for Parasitology* **21**, 743–746.

Marikovsky, M., Arnon, R. and Fishelson, Z. (1988a). Proteases secreted by transforming schistosomula of *Schistosoma mansoni* promote resistance to killing by complement. *Journal of Immunology* **141**, 273–278.

Marikovsky, M., Fishelson, Z. and Arnon, R. (1988b). Purification and characterization of proteases secreted by transforming schistosomula of *Schistosoma mansoni*. *Molecular and Biochemical Parasitology* **30**, 45–54.

Marikovsky, M., Parizade, M., Arnon, R. and Fishelson, Z. (1990a). Complement regulation on the surface of cultured schistosomula and adult worms of *Schistosoma mansoni*. *European Journal of Immunology* **20**, 221–227.

Marikovsky, M., Arnon, R. and Fishelson, Z. (1990b). *Schistosoma mansoni*: localization of the 28 kDa secreted protease in cercaria. *Parasite Immunology* **12**, 389–401.

Matthews, B.E. (1977). The passage of larval helminths through tissue barriers. *Symposium of the British Society for Parasitology* **15**, 93–119.

Matthews, B.E (1982a). Behaviour and enzyme release by *Anisakis* spp. Larvae. *Journal of Helminthology* **56**, 177–183

Matthews, B.E. (1982b). Skin penetration by *Necator americanus* larvae. *Zietschrift fur Parsitenkunde* **68**, 81–86.

Matthews, B.E. (1984). The source, release and specificity of proteolytic enzyme activity produced by *Anisakis simplex* larvae. *Journal of Helminthology* **58**, 175–185.

Matsumoto, I., Watanabe, H., Abe, K., Arai, S. and Emori, Y. (1995). A putative digestive cysteine proteinase from *Drosophila melanogaster* is predominantly expressed in the embryonic and larval midgut. *European Journal of Biochemistry* **227**, 582–587.

Matsumoto, I., Emori, Y., Abe, K. and Arai, S. (1997). Characterization of a family encoding cysteine proteinases of *Sitophilus zeamais* (maize weevil), and analysis of the protein distribution in various tissues including alimentary tract and germ cells. *Journal of Biochemistry* **121**, 464–476.

McGinty, A., Moore, M., Halton, D.W. and Walker, B. (1993). Characterization of the cysteine proteinases of the common liver fluke *Fasciola hepatica* using novel, active-site directed affinity labels. *Parasitology* **106**, 487–493.

McGuire, M.J., Lipsky, P.E. and Thiele, D.L. (1997). Cloning and characterization of the cDNA encoding mouse dipeptidyl peptidase I (Cathepsin C). *Biochimica et Biophysica Acta* **1351**, 267–273.

McManus, D.P. and Barrett, N.J. (1985). Isolation, fractionation and partial characterisation of the tegumental surface from the protoscoloces of the hydatid organism, *Echinococcus granulosus*. *Parasitology* **90**, 111–129.

McKerrow, J.H. and Doenhoff, M.J. (1988). Schistosome proteases. *Parasitology Today* **4**, 334–340.

McKerrow, J.H., Keene, W.E., Jeong, K.H. and Werb, Z. (1983). Degradation of extracellular matrix by larvae of *Schistosoma mansoni*. I. Degradation by cercariae as a model of initial parasite invasion of the host. *Laboratory Investigation* **49**, 195–200.

McKerrow, J.H., Pino-Heiss, S., Lindquist, R. and Werb, Z. (1985). Purification and characterization of an elastinolytic proteinase secreted by cercariae of *Schistosoma mansoni*. *Journal of Biological Chemistry* **260**, 3703–3707.

McKerrow, J.H., Brindley, P., Brown, M., Gam, A.A., Staunton, C. and Neva, F.A. (1990). *Strongyloides stercoralis*: identification of a protease that facilitates penetration of the skin by infective larvae. *Experimental Parasitology* **70**, 134–143.

McKerrow, J. H., Newport, G. and Fishelson, Z. (1991). Recent insights into the structure and function of the larval proteinase involved in host infection by a multicellular parasite. *Proceedings of the Society of Experimental Biology and Medicine* **192**, 119–124.

Merckelbach, A., Hasse, S., Dell, R. and Ruppel, A. (1994). cDNA sequences of *Schistosoma japonicum* coding for two cathepsin B-like proteins and Sj32. *Tropical Medicine and Parasitology* **45**, 193–198.

Michel, A., Ghoneim, H., Resto, M., Klinkert, M.-Q. and Kunz, W. (1995). Sequence, characterization and localization of a cysteine proteinase cathepsin L in *Schistosoma mansoni*. *Molecular and Biochemical Parasitology* **73**, 7–18.

Minard, P., Murrell, K.D., and Stirewalt, M.A. (1977). Proteolytic, antigenic and immunogenic properties of *Schistsoma mansoni* cercarial secretion material. *American Journal of Tropical Medicine and Hygiene* **26**, 491–499.

Moczon, T. (1994a). Histochemistry of proteinases in the cercariae of *Diplostomium pseudospathaceum* (Trematoda, Diplostomatidae). *Parasitology Research* **80**, 680–683.

Moczon, T. (1994b). A cysteine proteinase in the cercariae of *Diplostomium pseudospathaceum* (Trematoda, Diplostomatidae). *Parasitology Research* **80**, 684–686.

Modha, J., Parikh, V., Gauldie, J. and Doenhoff, M.J. (1988). An association between schistosomes and contrapsin, a mouse serine protease inhibitor (serpin). *Parasitology* **96**, 99–109.

Modha, J., Roberts, M.C. and Kusel, J.R. (1996). Schistosomes and serpins: a complex business. *Parasitology Today* **12**, 119–122.

Moncrief, N.D., Kretsinger, R.H. and Goodman, M. (1990). Evolution of EF-Hand calcium-modulated proteins. 1. Relationships based on amino acid sequences. *Journal of Molecular Evolution* **30**, 522–562.

Morris, S.R. and Sakanari, J.A. (1994). Characterisation of the serine protease and serine protease inhibitor from the tissue-penetrating nematode *Anisakis simplex*. *Journal of Biological Chemistry* **269**, 27650–27656.

Mulcahy, G., Clery, D., O'Connor, F., Dowd, A.J., McGonigle, S., Andrews, S. and Dalton, J.P. (in press). Correlation of specific antibody titre and avidity with protection in cattle immunized against *Fasciola hepatica*. *Vaccine*.

Munroy, F.G., Cayzer, C.J.R., Adams, J.H. and Dobson, C. (1988). Proteolytic enzyymes in excretory-secretory products from adult *Nematospiroides dubius*. *International Journal for Parasitology* **19**, 129–131.

Neurath, H. (1984). Evolution of proteolytic enzymes. *Science* **224**, 350–357.

Newport, G.R., McKerrow, J.H., Hedstrom, R., Petitt, M., McGarrigle, L., Barr, P.J. and Agabian, N. (1988). Cloning of the protease that facilitates infection by schistosome parasites. *Journal of Biological Chemistry* **263**, 13179–13184.

Nicholas, K.B, Nicholas, H.B. and Deerfield, D.W. (1997). GeneDoc: Analysis and Visualization of Genetic Variation. Embnet.news 4, 2 (http://www2.ebi.ac.uk/embnet.news/).

Nimmo-Smith, R.H. and Keeling, J.E.D. (1960). Some hydrolytic enzymes of the parasitic nematode *Trichuris muris*. *Experimental Parasitology* **10**, 337–355.

Ohja, M. (1996). Ca2+-dependent protease I from *Allomyces arbuscula*. *Biochemical and Biophysical Research Communications* **218**, 22–29.

Ohno, S., Emori, Y., Imajoh, S., Kawasaki, H., Kisaragu, M. and Suzuki, K. (1984). Evolutionary origin of a calcium-dependent protease by fusion of genes for a thiol protease and a calcium-binding protein? *Nature* **312**, 566–570.

Ozerol, N.H. and Silverman, P.H. (1970). Partial characterisation of *Haemonchus contortus* exsheathing fluid. *Journal of Parasitology* **55**, 79–87.

Page, R.D.M. (1996). TREEVIEW: An application to display phylogenetic trees on personal computers. *Computer Applications in the Biosciences* **12**, 357–358.

Park, H., Ko, M.Y., Paik, M.K., Soh, C.T., Seo, J.H. and Im, K.I. (1995).

Cytotoxicity of cysteine proteinases of adult *Clonorchis sinensis*. *Korean Journal of Parasitology* **33**, 211–218.

Perry, R.N., Knox, D.P. and Beane, J. (1992). Enzymes released during hatching of *Globodera rostochiensis* and *Meloidogyne incognita*. *Fundamental and Applied Nematology* **15**, 283–288.

Petralanda, I., Yarzabal, L. and Piessens, W.F. (1986). Studies on a filarial antigen with collagenase activity. *Molecular and Biochemical Parasitology* **19**, 51–59.

Petralanda, I. and Piessens, W.F. (1994). Pathogenesis of onchocercal dermatitids — possible role of parasite proteases and autoantibodies to extracellular matrix proteins. *Experimental Parasitology* **79**, 177–186.

Phares, K (1996). An unusual host-parasite relationship: the growth hormone-like factor from plerocercoids of spirometrid tapeworms. *International Journal of Parasitology* **26**, 575–588.

Pierrot, C., Capron, A. and Khalife, J. (1995). Cloning and characterization of two genes encoding *Schistosoma mansoni* elastase. *Molecular and Biochemical Parasitology* **75**, 113–117.

Pierrot, C., Godin, C., Liu, J.L., Capron, A. and Khalife, J. (1996). *Schistosoma mansoni* elastase: an immune target regulated during the parasite life-cycle. *Parasitology* **113**, 519–526.

Pino-Heiss, S., Brown, M. and McKerrow, J.H. (1985). *Schistosoma mansoni*: degradation of host extracellular matrix by eggs and miracidia. *Experimental Parasitology* **59**, 217–221.

Polzer, M., Overstreet, R.M. and Taraschewski, H. (1994). Proteinase activity in the plerocercoid of *Proteocephalus ambloplitis (Cestoda)*. *Parasitology* **109**, 209–213.

Polzer, M. and Conradt, U. (1994). Identification and characterisation of the proteases from different developmental stages of *Schistocephalus solidus*. *International Journal for Parasitology* **24**, 967–973.

Pratt, D., Cox, G.N., Milhausen, M.J. and Boisvenue, R.J. (1990). A developmentally regulated cysteine protease gene family in *Haemonchus contortus*. *Molecular and Biochemical Parasitology* **43**, 181–192.

Pratt, D., Armes, L.G., Hageman, R., Reynolds, V., Boisvenue, R. and Cox, G.N. (1992a). Cloning and sequence comparison of four distinct cysteine proteases expressed by *Haemonchus contortus* adult worms. *Molecular and Biochemical Parasitology* **51**, 209–218.

Pratt, D., Boisvenue, R.J. and Cox, G.N. (1992b). Isolation of putative cysteine protease genes of *Ostertagia ostertagi*. *Molecular and Biochemical Parasitology* **56**, 39–48.

Price, H.P., Doenhoff, M.J. and Sayers, J.R. (1997). Cloning, heterologous expression and antigenicity of a schistosome cercarial protease. *Parasitology* **114**, 447–453.

Pritchard, D.I., McKean, P.G. and Schad, G.A. (1990). An immunological and biochemical comparison of hookworm species. *Parasitology Today* **6**, 154–156.

Pritchard, D.I. (1990). *Necator americanus*: antigens and immunological targets. In: *Hookworm Disease: Current Status and New Directions*. (Warren, K. and Schad, G.A.,eds) pp 340–350. London: Taylor and Francis.

Rawlings, N.D. and Barrett, A.J. (1994). Families of cysteine peptidases. *Methods in Enzymology* **244**, 461–486.

Rawlings, N.D. and Barrett, A.J. (1997). Classification of peptidases by comparison of primary and secondary structures. In: *Proteolysis in Cell Functions* (V.K. Hopsu-Havu, M. Jarvinen and H. Kirschke, eds.), pp. 13–21. Amsterdam: IOS Press.

Ray, C. and McKerrow, J.H. (1992). Gut-specific and developmental expression of a *Caenorhabditis elegans* cysteine protease gene. *Molecular and Biochemical Parasitology* **51**, 239–250.

Redmond, D.L., Knox, D.P., Newlands, G. and Smith, W.D. (1997). Molecular cloning and characterisation of a developmentally regulated putative metallopeptidase present in a host protective extract of *Haemonchus contortus*. *Molecular and Biochemical Parasitology* **85**, 77–87.

Rege, A.A. and Dresden, M.H. (1987). *Strongyloides* spp.: Demonstration and partial characterisation of acidic collagenolytic activity from infective larvae. *Experimental Parasitology* **64**, 275–280.

Rege, A. A., Herrera, P. R., Lopez, M. and Dresden, M. H. (1989a). Isolation and characterization of a cysteine proteinase from *Fasciola hepatica* adult worms. *Molecular and Biochemical Parasitology* **35**, 89–96.

Rege, A.A., Song, C., Bos, H.J. and Dresden, M.H. (1989b). Isolation and partial characterisation of a potentially pathogenic cysteine proteinase from adult *Dictyocaulus viviparus*. *Veterinary Parasitology* **34**, 95–102.

Rhoads, M.L. and Fetterer, R.H. (1995). Developmentally regulated secretion of cathepsin L-like cysteine proteases by *Haemonchus contortus*. *Journal for Parasitology* **81**, 505–512.

Rhoads, M.L. and Fetterer, R.H. (1997). Extracellular matrix: a tool for defining the extracorporeal function of parasite proteases. *Parasitology Today* **13**, 119–122.

Richer, J.K., Sakanari, J.A., Frank, G.R. and Grieve, R.B. (1992). *Dirofilaria immitis*: proteases produced by third- and fourth-stage larvae. *Experimental Parasitology* **75**, 213–222.

Richer, J.K., Hunt, W.G., Sakanari, J.A. and Grieve, R.B. (1993). *Dirofilaria immitis*—the effect of fluoromethylketone (FMK) cysteine protease inhibitors on the third stage to fourth stage moult. *Experimental Parasitology* **76**, 221–231.

Riga, E., Perry, R.N., Barrett, J. and Johnstone, M.R.L. (1995). Biochemical analyses on single amphidial cells, excretory-secretory gland cells, pharyngeal glands and their secretions from the avian nematode *Syngamus trachea*. *International Journal for Parasitology* **25**, 1151–1158.

Ring, C.S., Sun, E., McKerrow, J.H., Lee, G.K., Rosenthal, P.J., Kuntz, I.D. and Cohen, F.E. (1993). Structure-based inhibitor design by using protein models for the development of antiparasitic agents. *Proceedings of the National Academy of Sciences of the USA* **90**, 3583–3587.

Robertson, C.D., Bianco, A.E., McKerrow, J.H. and Maizels, R.M. (1989). *Toxocara canis*: proteolytic enzymes secreted by the infective larvae *in vitro*. *Experimental Parasitology* **69**, 30–36.

Robertson, C.D., Coombs, G.H., North, M.J., and Mottram, J.C. (1996). Parasite cysteine proteinases. *Perspectives in Drug Discovery and Design* **6**, 33–46.

Roche, L., Dowd, A. J., Tort, J., McGonigle, S., McSweeney, A., Curley, P., Ryan, T., and Dalton, J.P. (1997). Functional expression of *Fasciola hepatica* cathepsin L1 in *Saccharomyces cerevisiae*. *European Journal of Biochemistry* **245**, 373–380.

Rogers, W.P. (1941). Digestion in parasitic nematodes III: The digestion of proteins. *Journal of Helminthology*, **19**, 47–58.

Rogers, W.P. (1970). The function of leucine aminopeptidase in exsheathing fluid. *Journal of Parasitology* **56**, 138–143.

Rogers, W.P. (1982). Enzymes in the exsheathing fluid of nematodes and their biological significance. *International Journal for Parasitology* **12**, 495–502.

Rood, J.A., van Horn, S., Drake, F.H., Gowen M. and Debouck, C. (1997).

Genomic organization and chromosome localization of the human cathepsin K gene (CTSK). *Genomics* **41**(2), 169–176.
Rosenthal, E. (1993). Sequence analysis of translationally controlled maternal mRNA from *Urechis caupo*. *Developmental Genetics* **14**, 485–491.
Ruitenberg, E.J. and Loendersloot, H.J. (1971). Histochemical properties of the excretory organ of *Anisakis* sp. larvae. *Journal of Parasitology* **57**, 1149–1150.
Rupova, L., Dragneva, N., Bankov, I. and Ossikovski, E. (1984). Investigations on a protease complex from adult *Ascaris suum*. I. Isolation of an acid protease and characterisation of its properties. *Helminthologia* **21**, 257–265.
Ruppel, A., Breternitz, U. and Burger, R. (1987). Diagnostic Mr 31 000 *Schistosoma mansoni* proteins: requirement of infection, but not immunization, and use of the 'miniblot' technique for the production of monoclonal antibodies. *Journal of Helminthology* **61**, 95–101.
Ruppel, A.L. Diesfeld, H.J. and Rother, U. (1985a). Immunoblot analysis of *Schistosoma mansoni* antigens with sera of schistosomiasis patients: diagnostic potential of an adult schistosome polypeptide. *Clinical and Experimental Immunology* **62**, 499–506.
Ruppel, A., Rother, U., Vongrichten, H., Lucius, R. and Diesfeld, H.J. (1985b). *Schistosoma mansoni*: immunoblot analysis of adult worm proteins. *Experimental Parasitology* **60**, 195–206
Saitou, N. and Nei, M. (1987). The neighbour-joining method: a new method for reconstructing phylogenetic trees. *Molecular and Biological Evolution* **4**, 406–425.
Sakanari, J.A. and McKerrow, J.H. (1990). Identification of secreted neutral proteases from *Anisakis simplex*. *Journal for Parasitology* **76**, 625–630.
Sakanari, J.A., Staunton, C.E., Eakin, A.E., Craik, C.S. and McKerrow, J.H. (1989). Serine proteases from nematode and protozoan parasites: isolation of sequence homologs using generic molecular probes. *Proceedings of the National Academy of Sciences of the USA* **86**, 4863–4867.
Sarkis, G.J., Kurpiewski, M.R., Ashcom, J.D., Jen-Jacobson, L. and Jacobson, L.A. (1988a). Proteases of the nematode *Caenorhabditis elegans*. *Archives of Biochemistry and Biophysics* **261**, 80–90.
Sarkis, G.J., Ashcom, J.D., Hawdon, J.M. and Jacobson, L.A. (1988b). Decline in protease activities with age in the nematode *Caenorhabditis elegans*. *Mechanisms of Aging and Development* **45**, 191–201.
Sato, K., Nagai, Y. and Suzuki, M. (1993). Purification and partial characterisation of an acid protease from *Dirofilaria immitis*. *Molecular and Biochemical Parasitology* **58**, 293–299.
Sato, K., Yamaguchi, H., Waki, S., Suzuki, M and Nagai, Y. (1995). *Dirofilaria immitis*: immunohistochemical localization of acid proteinase in the adult worm. *Experimental Parasitology* **81**, 63–71.
Siddiqui, A.A., Zhou, Y., Podesta, R.B., Karcz, S.R., Tognon, C.E., Strejan, G.H., Dekaban, G.A. and Clarke, M.W. (1993). Characterization of Ca^{++}-dependent neutral protease (calpain) from human blood flukes, *Schistosoma mansoni*. *Biochimica et Biophysica Acta* **1181**, 37–44.
Simpkin, K.G., Chapman, C.R. and Coles, G.C. (1980). *Fasciola hepatica*: a proteolytic digestive enzyme. *Experimental Parasitology* **49**, 281–287.
Smith, A.M, Dowd, A.J., McGonigle, S., Keegan, P.S., Brennan, G., Trudgett, A. and Dalton, J.P. (1993a). Purification of a cathepsin L-like proteinase secreted by adult *Fasciola hepatica*. *Molecular and Biochemical Parasitology* **62**, 1–8.
Smith, A.M., Dowd, A.J., Heffernan, M., Robertson, C.D. and Dalton, J. P. (1993b).

Fasciola hepatica: a secreted cathepsin L-like proteinase cleaves host immunoglobulin. *International Journal of Parasitology* **23**, 977–983.
Smith, A.M., Dalton, J.P., Clough, K.A., Kilbane, C.L., Harrop, S.A., Hole, N. and Brindley, P.J. (1994). Adult *Schistosoma mansoni* express cathepsin L proteinase. *Molecular and Biochemical Parasitology* **67**, 11–19.
Smith, W.D. (1993). Protection in lambs immunised with *Haemonchus contortus* gut membrane proteins. *Research in Veterinary Science* **54**, 94–101.
Smith, W.D., Smith, S., Murray, J. and Knox, D.P. (1994a). Aspartyl proteinases as vaccines. European patent application No. 93916110.5–2116.
Smith, W.D, Smith, S.K. and Murray, J.M. (1994b). Protection studies with integral gut membrane protein fractions of *Haemonchus contortus*. *Parasite Immunology* **16**, 231–241.
Smith, T.S., Munn, E.A., Graham, M., Tavernor, A.S. and Greenwood, C.A. (1993). Purification and evaluation of the integral membrane protein H11 as a protective antigen against *Haemonchus contortus*. *International Journal for Parasitology* **23**, 271–280.
Smith, T.S., Graham, M., Munn, E.A., Newton, S.E., Knox, D.P., Coadwell, W.J., McMichael-Phillips, D., Smith, H., Smith, W.D. and Oliver, J.J. (1997). Cloning and characterisation of H11, a microsomal aminopeptidase that induces protective immunity against the parasitic nematode *Haemonchus contortus*. *Biochimica et Biophysica Acta* **1338**, 295–306
Sogin, M.L. (1991). Early evolution and the origin of eukaryotes. *Current Opinion in Genetics and Development* **1**, 457–463.
Song, C.Y. and Chappell, C.L (1993). Purification and partial characterisation of cysteine proteinase from *Spirometra mansoni* plerocercoids. *Journal of Parasitology* **79**, 517–524.
Song, C.Y. and Dresden, M.H. (1990). Partial purification and characterization of cysteine proteinases from various developmental stages of *Paragonimus westermani*. *Comparative Biochemistry and Physiology* **95**, 473–476.
Song, C.Y. and Kim, T.S. (1994). Characterization of a cysteine from adult worms of *Paragonimus westermani*. *Korean Journal of Parasitology* **32**, 231–241.
Song, C.Y. and Rege, A.A. (1991). Cysteine proteinase activity in various developmental stages of *Clonorchis sinensis*: a comparative analysis. *Comparative Biochemistry and Physiology* **99**, 137–140.
Song, C.Y., Dresden, M.H. and Rege, A.A. (1990). *Clonorchis sinensis*: purification and characterization of a cysteine proteinase from adult worms. *Comparative Biochemistry and Physiology* **97**, 825–829.
Song, C.Y., Choi, D.H., Kim, T.S. and Lee, S.H. (1992). Isolation and partial characterisation of cysteine proteinase from saprganum. *Korean Journal of Parasitology* **30**, 191–200.
Sorimachi, H., Tsukahara, T., Okada-Ban, M., Sugita, H., Ishiura, S. and Suzuki, K. (1995). Identification of a third ubiquitous calpain species — chicken muscle expresses four distinct calpains. *Biochimica et Biophysica Acta* **1261**, 381–393.
Spithill, T. W. and Dalton, J.P. (in press). Progress in the development of vaccines against Fasciolosis. *Parasitology Today*.
Spithill, T.W. and Morrison, C.A. (1997). Molecular vaccines for control of *Fasciola hepatica* infection in ruminants. In: *Immunology, pathophysiology and control of fasciolosis*. (J. Boray, ed.) Merck AGVET, Rahway, New Jersey.
Stirewalt, M.A. (1974). *Schistosoma mansoni*: cercaria to schistosomule. *Advances in Parasitology* **12**, 115–182.

Sung, C.K. and Dresden, M.L. (1986). Cysteinyl proteinases of *Schistosoma mansoni* eggs: purification and partial characterization. *Journal of Parasitology* **72**, 891–900.

Swofford, D.L. and Olsen, G.J. (1990). Phylogeny reconstruction. In: *Molecular Systematics* (Hillis, D.M and Moritz, C., eds). Massachusetts, USA: Sinauer Associates.

Takeda, O., Miura, Y., Mitta, M., Matsushita, H., Kato, I., Abe, Y., Yokosawa, H. and Ischii, S.-I. (1994). Isolation and analysis of cDNA encoding a precursor of *Canavalia ensiformis* asparginyl endopeptidase (legumain). *Journal of Biochemistry* **116**, 541–546.

Tao, K., Stearns, N.A., Dong, J., Wu, Q. and Sahagian, G. (1994). The proregion of cathepsin L is required for proper folding, stability and ER exit. *Archives in Biochemistry and Biophysics* **311**, 19–27.

Tamashiro, W.K., Rao, M. and Scott, A.L. (1987). Proteolytic cleavage of IgG and other protein substrates by *Dirofilaria immitis* microfilarial enzymes. *Journal for Parasitology* **73**, 149–154.

Tang, J., James, M.N.G., Hsu, I.N., Jenkins, J.A. and Blundell, T.L. (1978). Structural evidence for gene duplication in the evolution of acid proteases. *Nature* **271**, 618–621.

Taylor, M.A.K., Baker, K.C., Briggs, G.S., Connerton, I.F., Cummings, N.J., Pratt, K.A., Revell, D.F., Freedman, R.B. and Goodenough, P.W. (1995). Recombinant pro-regions of papain and papaya proteinase IV are selective high affinity inhibitors of the mature papaya enzymes. *Protein Engineering* **8**, 59–62.

Teixeira, M.M., Doenhoff, M.J., McNeice, C., Williams, T.J. and Hellewell, P.G. (1993). Mechanisms of the inflammatory response induced by extracts of *Schistosoma mansoni* larvae in guinea pig skin. *Journal of Immunology* **151**, 5525–5534.

Thompson, J.D., Higgins, D.G. and Gibson, T.J. (1994). CLUSTAL W: improving the sensitivity of progressive multiple sequence alignment through sequence weighting, positions-specific gap penalties and weight matrix choice. *Nucleic Acids Research* **22**, 4673–4680.

Thorson, R.E. (1956a). The effect of extracts of amphidial glands, excretory glands and oesophagus on coagulation of dogs blood. *Journal of Parasitology* **42**, 26–30

Thorson, R.E. (1956b). Proteolytic activity in extracts of the oesophagus of adult *Ancylostoma caninum* and the effect of immune serum on this activity. *Journal of Parasitology* **42**, 21–25.

Timms, A. R. and Bueding, E. (1959). Studies of a proteolytic enzyme from *Schistosoma mansoni*. *British Journal of Pharmacology* **14**, 68–73.

Tkalcevic, J., Ashman, K. and Meeusen, E. (1995). *Fasciola hepatica*: rapid identification of newly excysted juvenile proteins. *Biochemical and Biophysical Research Communications* **213**, 169–174.

Todorova, V.K., Knox, D.P. and Kennedy, M.W. (1995). Proteinases in the excretory/secretory products (ES) of adult *Trichinella spiralis*. *Parasitology* **111**, 201–208.

Tort, J. (1997). Cloning, yeast expression, mutagenesis and phylogenetic analysis of a novel member of the *Fasciola hepatica* cathepsin L-like family. PhD Thesis. School of Biological Sciences, Dublin City University, Dublin 9, Republic of Ireland.

Toy, L., Pettit, M., Wang, Y.F., Hedstrom, R. and McKerrow, J.H. (1987). The immune response to stage specific proteolytic enzymes of *Schistosoma mansoni*. In: *UCLA Symposia on Molecular and Cellular Biology, Molecular Paradigms for*

Eradicating Helminth Parasites, Vol. 60, pp. 85–103, New York: Alan R. Liss, Inc.
Turk, D., Podobnik, M., Kuhelj, R., Dolinar, M. and Turk, V. (1996). Crystal structures of human procathepsin B at 3.2 and 3.3 Å resolution reveal an interaction motif between a papain-like cysteine protease and its propeptide. *FEBS Letters* **384**, 211–214.
Urwin, P.E., Lilley, C.J., McPherson, M.J. and Atkinson, C.J. (1997). Characterisation of two cDNAs encoding cysteine proteinases from the soybean cyst nematode *Heterodera glycines*. *Parasitology* **114**, 605–613.
Verwaerde, C., Auriault, C., Damonneville, M., Neyrinck, J.L., Vendeville, C. and Capron, A. (1986). Role of serine proteases of *Schistosoma mansoni* in the regulation of IgE synthesis. *Scandinavian Journal of Immunology* **24**, 509–516.
Verwaerde, C., Auriault, C., Neyrinck, J.L. and Capron, A. (1988). Properties of serine proteases of *Schistosoma mansoni* schistosomula involved in the regulation of IgE synthesis. *Scandinavian Journal of Immunology* **27**, 17–24.
Ward, W., Alvarado, L., Rawlings, N.D., Engel, J.C., Franklin, C. and McKerrow, J.H. (1997). A primitive enzyme for a primitive cell: the protease required for excystation of *Giardia*. *Cell* **89**, 437–444.
Wasilewski, M.M., Lim, K.C., Phillips, J. and McKerrow, J.H. (1996). Cysteine protease inhibitors block schistosome hemoglobin degradation *in vitro* and decrease worm burden and egg production *in vivo*. *Molecular and Biochemical Parasitology* **81**, 179–189.
Wertheim, G., Lustigman, S., Silberman, H. and Shoshan, S. (1983). Demonstration of collagenase activity in adult *Strongyloides ratti* and its absence in the infective larvae. *Journal of Helminthology* **57**, 241–246.
White, A.C., Molinari, J.L., Pillai, A.V. and Rege, A.A. (1992). Detection and preliminary characterisation of *Taenia solium* metacestode proteases. *Journal for Parasitology* **78**, 281–287.
White, A.C., Baig, S. and Robinson, P. (1996). *Taenia saginata* onchosphere excretory/secretory peptidases. *Journal of Parasitology* **82**, 7–10.
White, A.C., Baig, S. and Chappell, C.L. (1997). Characterisation of a cysteine proteinase from *Taenia crassiceps* cysts. *Molecular and Biochemical Parasitology* **85**, 243–253.
Wijffels, G.L., Salvatore, L., Dosen, M., Waddington, J., Wilson, L., Thompson, C., Campbell, N., Sexton, J., Wicker, J., Bowan, F., Friedel, T., and Spithill, T.W. (1994a). Vaccination of sheep with purified cysteine proteinases of *Fasciola hepatica* decreases worm fecundity. *Experimental Parasitology* **78**, 132–148.
Wijffels, G.L., Panaccio, M., Salvatore, L., Wilson, L., Walker, I.D., and Spithill, T.W. (1994b). The secreted cathepsin L-like proteinases of the trematode, *Fasciola hepatica*, contain 3-hydroxyproline residues. *Biochemical Journal* **299**, 781–790.
Willis, A.C., Diaz, A., Nieto, A. and Sim, R.B. (1997). *Echinococcus granulosus* antigen 5 may be a serine protease. *Parasite Immunology* **19**, 385.
Wilson, L.R., Good, R.T., Panaccio, M., Wijffels, G.L., Creaney, J., Bozas, S.E., Parsons, J.C., Sandeman, M. and Spithill, T.W. (in press). Characterisation and cloning of the major cathepsin B protease secreted by newly excysted juvenile *Fasciola hepatica*. *Experimental Parasitology*.
Wong, J.Y.M, Harrop, S.A., Day, S.R. and Brindley, P.J. (1997). Schistosomes express two forms of cathepsin D. *Biochimica et Biophysica Acta* **1338**, 156–160.
World Health Organization (1996). WHO Fact Sheet N112. World Health Organization, Geneva, Switzerland.

Xu, Y.-Z. and Dresden, M.H. (1986). Leucine aminopeptidase and hatching of *Schistosoma mansoni* eggs. *Journal of Parasitology* **72**, 507–511.
Yamakami, K. and Hamajima, F. (1988). Proteolytic activity of secretions from newly excysted metacercariae of *Paragonimus westermani*. *Japanese Journal of Parasitology* **37**, 380–384.
Yamakami, K., and Hamajima, F. (1990). A neutral thiol protease secreted from newly excysted metacercariae of trematode parasite *Paragonimus westermani*: purification and characterization. *Comparative Biochemistry and Physiology* **95**, 755–758.
Yamakami, K., Hamajima, F., Akao, S. and Tadakuma, T. (1995). Purification and characterisation of an acid protease from metacercariae of the mammalian trematode parasite *Paragonimus westermani*. *European Journal of Biochemistry*, **233**, 490–497.
Yamamoto, M., Yamakami, K. and Hamajima, F. (1994). Cloning of a cDNA encoding a neutral thiol protease *Paragonimus westermani* metacercariae. *Molecular and Biochemical Parasitology* **64**, 345–348.
Yamasaki, H., Aoki, T. and Oya, H. (1989). A cysteine proteinase from the liver fluke *Fasciola* sp.: purification, characterization, localization and application to immunodiagnosis. *Japanese Journal of Parasitology* **38**, 373–384.
Yamasaki, H., Kominami, E. and Aoki, T. (1992). Immunocytochemical localization of a cysteine protease in adult worms of the liver fluke *Fasciola* sp., *Parasitology Research* **78**, 574–580.
Yamasaki, H. and Aoki, T. (1993). Cloning and sequence analysis of the major cysteine protease expressed in the trematode parasite *Fasciola* sp.. *Biochemistry and Molecular Biology International* **31**, 537–542.
Yoshino, T.P., Lodes, M.J., Rege, A.A., and Chappell, C.L. (1993). Proteinase activity in miracidia, transformation excretory-secretory products, and primary sporocysts of *Schistosoma mansoni*. *Journal of Parasitology* **79**, 23–31.
Young, C.J., McKeand, J.B. and Knox, D.P. (1995). Proteinases released *in vitro* by the parasitic stages of *Teladorsagia circumcincta*, an ovine abomasal parasite. *Parasitology* **110**, 465–471.
Zerda, K.S., Dresden, M.H. and Chappell, C.L. (1988). *Schistosoma mansoni*: expression and role of cysteine proteinases in developing schistosomula. *Experimental Parasitology* **67**, 238–246.
Zhong, C., Skelly, P.J., Leaffer, D., Cohn, R.G., Caulfield, J.P. and Shoemaker, C.B. (1995). Immunolocalization of a *Schistosoma mansoni* facilitated glucose transporter to the basal, but not the apical, membranes of the surface syncytium. *Parasitology* **110**, 383–394.

Parasitic Fungi and their Interactions with the Insect Immune System

Andreas Vilcinskas and Peter Götz

Institute of Zoology, Free University of Berlin, Königin-Luise-Str. 1–3, 14195 Berlin, Germany

Abstract	268
1. Insect Immunity	268
2. Entomopathogenic Fungi	270
2.1. General aspects	270
2.2. Pathogenesis	271
3. Fungal Factors Determining Virulence	280
3.1. Fungal toxins and their effects on insects	280
3.2. Interactions of fungal proteases with the insect immune system	288
4. Humoral Immune Reactions of Insects Against Fungi	290
4.1. Inducible antifungal proteins	291
4.2. Inducible protease inhibitors	295
4.3. Detoxification proteins	299
5. Comparison with Other Pathogens and Parasites of Insects	301
References	303

ABSTRACT

Recent advances in research on parasitic fungi and their molecular and cellular interactions with the insect immune system have led to new insights into the complex relationships and mechanisms involved in fungal pathogenesis. This review focuses on molecules which mediate virulence of the producing fungi (fungal proteases and toxins) and on molecules contributing to the antifungal humoral immune responses of insects (antifungal proteins, protease inhibitors and detoxification proteins). Among the molecules produced by parasitic fungi during infection, at least proteolytic enzymes and toxins interfere with the host immune system in a sophisticated manner. Some of these compounds support fungal development within the infected host by suppression of its potent immune system. In particular, they impair circulating haemocytes, e.g. by damaging cytoskeletal structures and inducing apoptosis. Resistance of insects to microorganisms relies on cellular and humoral defence reactions. Although research on insect immunity attracted much attention in past decades, studies on the identification, characterization and mode of action of antifungal compounds, inducible protease inhibitors and proteins which detoxify fungal toxins in insects are more recent. Present data suggest a synergistic contribution of these proteins to antifungal defence. Successful development of parasitic fungi in infected host insects obviously requires combined activity of enzymes and immunosuppressive toxins. Current frontiers concerning the research on insect–fungal interactions at the cellular and molecular levels are outlined, with emphasis on applied aspects. Finally, a comparison of parasitic fungi with other groups of entomopathogens reveals parallels in strategies employed for evasion of host immune responses.

1. INSECT IMMUNITY

Immunity is the innate or acquired resistance of an organism to particular poisons, pathogens or parasites. Insects possess a potent immune system that facilitates their survival even in habitats that are highly contaminated with microorganisms, and this contributes to their conspicuous evolutionary success. Research on insect immunity has attracted much attention because insects are considered to be the most important competitors with humans for nutrition and, as vectors of many diseases, they pose a considerable threat to human health. Thus basic research on the interactions between insects and their pathogens was mostly inspired and supported by the expectation that new insights into the internal defence system of insects would provide new strategies for biological control of insect pests or vectors. Attempts to exploit insects as a source for novel antibiotics have so far had limited success.

The immune system comprises molecules, cells and mechanisms which provide non-self recognition and contribute to rejection or inactivation of poisons, pathogens and parasites. It is usually divided into the cellular and humoral immune response. The latter relies on a battery of constitutive or inducible proteins which exhibit antimicrobial activity and inactivate or detoxify poisons such as microbial toxins (Götz, 1988).

In comparison to that of vertebrates, the insect immune system lacks both antibody-mediated specificity and memory. Immune competent cells in insects do not exhibit specific recognition sites for particular foreign structures, but share recognition sites for a group of specific structures which have been designated as pattern recognition receptors (Janeway, 1994). A homologue of vertebrate cell surface marker CD36 has recently been cloned in vector insects and represents a candidate for pattern recognition (Richman and Kafatos, 1995).

The humoral immune response in insects is characterized by rapid and transient synthesis of immune proteins. Some of the induced proteins exhibit antimicrobial activity, and have therefore been assumed to mediate resistance against pathogens (Boman and Hultmark, 1987). The fat body has been recognized as the major site of immune protein synthesis (Trenczek and Faye, 1988). Numerous immune proteins from insect sources which have been identified, characterized and cloned directly or indirectly inhibit microorganisms *in vitro*. Most of them exhibit antibacterial activity, whereas others were demonstrated also, or exclusively, to affect fungi. Insect genes encoding for antimicrobial proteins bear upstream deoxyribonucleic acid (DNA) sequence elements resembling binding sites for transcription factors such as the *rel* family member NF-κB, which in mammals is associated with the acute phase response (Cociancich *et al.*, 1994; Hoffmann, 1995; Hoffmann *et al.*, 1996; Lemaitre *et al.*, 1996; Gillespie *et al.*, 1997).

The insect cellular defence system is based on haemocytes circulating in the haemocoel, which are able to distinguish between self and non-self. A great number of different haemocyte types has been identified, but their classification is still not consistent and their distinction under experimental conditions is often difficult because of the occurrence of intermediate stages with characteristics of more than one type. The biological function and origin of the different haemocyte types is not always clear. However, two basic types have been recognized, representing the typical immune competent haemocytes of insects. These are the so-called plasmatocytes and granular cells, which exhibit the capacity to recognize and attach to foreign surfaces. Granular cells are frequently also designated as granulocytes, but this term suggests homology with vertebrate granulocytes and should therefore be avoided (Götz and Boman, 1985; Ratcliffe, 1993; Gillespie et al., 1997).

The cellular immune response of insects against pathogens or parasites comprises two distinct mechanisms, phagocytosis of invading organisms by single haemocytes and multicellular encapsulation and nodule formation. Plasmatocytes and granular cells phagocytose or encapsulate recognized invaders, although their relative contribution differs among the insect species investigated (Trenczek, 1992; Ratcliffe, 1993). For example, blastospores or conidia of *Beauveria bassiana* produced *in vitro* are phagocytosed by granular cells of the beet army worm *Spodoptera exigua* (see Hung et al., 1993). In the greater wax moth *Galleria mellonella*, however, blastospores and conidia of the entomopathogenic fungi *Metarhizium anisopliae* and *B. bassiana* are predominantly ingested by plasmatocytes (Vilcinskas et al., 1997a).

Multicellular encapsulation and nodule formation are complex processes. They take place when invading pathogens or parasites are too large in number or size to be ingested by phagocytic cells. Nodules represent aggregates of haemocytes with particles such as bacteria entrapped in a sticky extracellular material. Granular cells are known to discharge such material from their intracellular granules upon contact with foreign structures. Aggregates consisting of degranulated granular cells, extracellular material and entrapped particles are often melanized because of the activation of phenoloxidase (Gillespie et al., 1997). Cellular encapsulation is related to nodule formation, but occurs as a defence reaction against larger invaders such as nematodes or eggs and larvae of insect parasitoids. Immune competent haemocytes attach to and spread over pathogens or parasites, which are too large to be phagocytosed. Such attachment and spreading is considered an unsuccessful effort by the phagocytic cells to ingest an impossibly large foreign body (Schliwa, 1986). Encapsulation also results in a multiple layer of attached, extremely flattened haemocytes. This tight and often melanized cellular envelope restricts the metabolism of the encapsulated invader, leading to inhibition of its development and final death (Götz and Boman, 1985). Melanization of foreign structures may also occur without direct participation of haemocytes. This so-called humoral or melanotic encapsulation is typical of certain species of Diptera (Götz, 1986).

2. ENTOMOPATHOGENIC FUNGI

2.1. General Aspects

Fungi play a major role as saprophytes in the natural food chain. Many soil-living fungi are capable of digesting insect cadavers. Specialization for this life style requires many adaptations, such as the capacity to release suitable digestive enzymes such as chitinases and proteases. This cluster of adaptations has resulted in an evolutionary 'plateau' favouring the development of fungal species that are pathogenic to living insects. The degree of specialization varies among recent fungi from opportunistic to facultative or obligate pathogens. Up to 1000 entomopathogenic fungal species have been identified. Most belong to the Deuteromycetes and Zygomycetes. They are distributed among different systematic groups, suggesting multiple and independent development of parasitic life cycles (St Leger and Bidochka, 1996).

This review focuses on entomopathogenic fungi that have been found to cause the early death of the insect by penetrating and proliferating inside the host, which is killed by being deprived of soluble nutrients in its haemolymph, by the invasion or digestion of vital tissues, and/or by the release of fungal toxins (Roberts and Humber, 1981; Roberts and Hajek, 1992). A minority of fungal pathogens has a restricted host range, such as *Aschersonia aleyrodis*, which infects only scale insects and whiteflies. Fungal species with a wide host range, such as *B. bassiana* and *M. anisopliae*, differ in their virulence to particular insect hosts. There are at least 200 insect species susceptible to *M. anisopliae*, and nearly 500 species susceptible to *B. bassiana* (see Khachatourians, 1991).

We will not attempt to provide an overview of host ranges and life cycles in different systematic groups of entomopathogenic fungi, but rather to outline recent advances in understanding the molecular interactions between fungal pathogens and their insect hosts (Figure 1).

2.2. Pathogenesis

2.2.1. *Adhesion and Germination of Fungal Spores*

Entomopathogenic fungi usually infect insect hosts via spores of sexual or asexual origin. The so-called *fungi imperfecti* include all fungal species without identified sexual spores, e.g. *B. bassiana* and *M. anisopliae*. They produce asexual spores (conidia) with a well-developed cell wall which provides long-term persistence in an infective form and is partly responsible for host recognition (Boucias and Pendland, 1991a, b). These conidia are usually formed on the tip or side of specialized hyphae (conidiophores). Conidia usually have a hydrophobic surface that facilitates attachment to the cuticle of potential host insects and maintains contact long enough for germination and subsequent invasion (Charnley, 1984). Fungal secondary metabolites which exhibit insecticidal activity and a broad spectrum of digestive enzymes such as proteases, esterases and lipases have been located

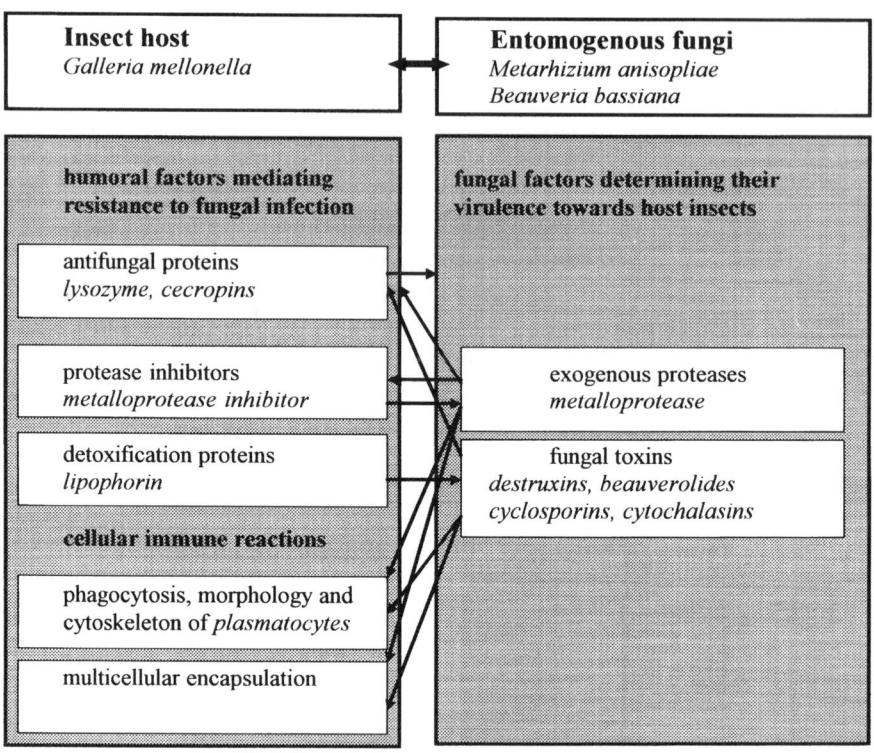

Figure 1 Model of the molecular interactions between entomogenous fungi and their host insects. The insect immune system comprises humoral factors such as antifungal proteins, inducible protease inhibitors and detoxification proteins, and cellular reactions such as phagocytosis and multicellular encapsulation. Among the factors produced by entomopathogenic fungi, at least exogenous proteases and secondary metabolites have been recognized to interfere with the insect immune system. Interactions between objects in italic letters are discussed in this review.

at the surface of pre-germinated conidia (David, 1967; Zacharuk, 1970; Ratault and Vey, 1977; Goetel *et al.*, 1989; Boucias and Pendland, 1991a; St Leger *et al.*, 1991; Matha *et al.*, 1992). Therefore we can expect that persistent conidia can utilize the insect integument surface as a source of nutrients. The composition of the conidial surface is influenced by the environmental conditions in which the spores developed. For example, *M. anisopliae* conidia collected from insect cuticle exhibit higher proteolytic activity on their surface than those harvested from Sabouraud agar, suggesting that the nutritional environment during development of conidia can pre-adapt them for the pathogenic life. This is crucial regarding the possible loss of virulence in the course of mass production of virulent fungal spores *in vitro*. Large-scale production of infectious material in fermentors is one of the prerequisites for the use of particular entomopathogenic fungi in the biological control of insect pests.

The cuticle-degrading activity of digestive enzymes is limited to the vicinity of the adhered fungal spores, but provides nutrients for their germination. Attachment of conidia to hydrophobic surfaces under appropriate conditions of nutrition and humidity is sufficient to induce germination, which results in formation of germ tubes and appressoria. The latter represent a specialized infectious structure (St Leger *et al.*, 1991). However, some cuticular compounds, particularly fatty acids, exhibit an inhibitory effect on the germination of conidia (Koidsumi, 1957; Smith and Grula, 1982), suggesting more complex molecular interactions between the insect integument and germination of adherent fungal spores. It has been established that the chemical composition of the integument surface differs among different insect species, and that epicuticular compounds which are soluble in lipid solvents play a role in cuticular resistance (Koidsumi, 1957; Ferron, 1981).

Hence, the exoskeleton of insects forms a first defence line against invading microorganisms; its protective role could be enhanced by exogenous antimicrobial fluids coating the cuticle surface. Indeed, an antifungal compound has recently been identified in exocrine glandular secretions of the mustard leaf beetle *Phaedon cochleariae*. Larvae and adults of this species possess exocrine glands producing secretions which are well known to be defensive devices against predators. A major component of the secretion, the iridoid monoterpene (epi)chrysomelidial, was found to inhibit germination and growth of *B. bassiana*, a recognized fungal pathogen of this insect species. The participation of (epi)chrysomelidial in antifungal defence is probably restricted to the prevention of fungal cell germination on the insect cuticle. The cytotoxic effect of this compound on insect cells precludes its employment in the internal defence mechanism against invading pathogens (Gross *et al.*, 1998).

2.2.2. *Penetration of the Host Cuticle*

The insect cuticle forms an efficient primary barrier against microorganisms. It is actively penetrated only by well adapted pathogens or parasites. Entomopathogenic fungi utilize both physical forces and chemical degradation to gain access to the haemocoel. Attached and germinating fungal conidia can penetrate the cuticle with their outgrowing hyphae immediately or following errant growth on the surface (Figure 2). They produce an array of enzymes capable of digesting protein, chitin and lipid components of the cuticle (Leopold and Samsinakova, 1970; Persson *et al.*,

1984). The release of such enzymes is controlled by catabolic regulation of messenger ribonucleic acid (mRNA) synthesis (Bidochka and Khachatourians, 1988; St Leger et al., 1992). The predominant role of fungal exogenous proteases in the penetration process and their interactions with protease inhibitors of the host will be discussed in later sections.

Since it has been established that cuticle lesions alone are sufficient to induce enhanced levels of antimicrobial activity in the neighbouring cuticle matrix and fat body, the insect exoskeleton appears to participate actively in humoral immune responses (Brey et al., 1993). This assumption is further strengthened by the observation that cells of the epidermis, even in the tracheal system, contain high levels of antimicrobial proteins such as lysozyme (Vilcinskas, 1994). However, neither increased lysozyme titres nor detectable cecropin-like activity were found in the haemolymph of G. mellonella larvae infected with B. bassiana when mycosis was initiated by topical application of conidia. Cuticle lesions caused by invading fungal cells are probably insufficient to elicit detectable humoral immune responses (Vilcinskas and Matha, 1997b). But other defence reactions are frequently observed. Localized melanization around penetrating fungal hyphae causes black spots on the insect cuticle, suggesting activation of prophenoloxidase (Charnley, 1984; Vey and Götz, 1986). It has been generally assumed that phenoloxidase-mediated reactions have antifungal effects, although in relatively few instances only are these reactions sufficient to prevent penetration (St Leger et al., 1988). Melanization reactions are regulated by proteolytic cascades that involve activation of proenzymes (Sugumaran and Kanost, 1993; Söderhäll et al., 1996). Injection of metalloproteases such as thermolysin or extracellular proteases released by B. bassiana or M. anisopliae into G. mellonella larvae resulted in strong melanization of the haemolymph (Griesch and Vilcinskas, 1998; Wedde et al. 1998). Thus, the locally restricted melanization around penetrating fungal hyphae is probably caused by exogenous fungal proteases which, as a side effect, activate proenzymes of the prophenoloxidase activating cascade because of the lack of corresponding inhibitors. Since evidence is available that exogenous M. anisopliae proteases remain, in part, associated with the fungal cell surface (St Leger et al., 1991), their limited release may avoid excessive melanization around invading fungal cells.

Upon penetration of the chitinous part of the insect integument, invading fungal cells have to pass the ectodermal epidermis, which is formed by cells with a high content of antimicrobial proteins such as lysozyme (Vilcinskas, 1994). The capacity of virulent fungi to penetrate the epidermal cells is probably also mediated by exogenous proteases capable of digesting antifungal proteins. Treatment of virulent B. bassiana or M. anisopliae strains in vitro with high concentrations of lysozyme induced the release of proteases which hydrolysed this antimicrobial protein (Vilcinskas, 1994).

2.2.3. Interactions of Parasitic Fungi with the Insect Immune System

Once entomopathogenic fungi have penetrated the insect integument and gained access to the haemocoel, they are there exposed to further humoral and cellular defence reactions of the infected host. Most, if not all, entomogenous fungi differentiate and propagate within the insect haemocoel by formation of cells that lack a fully developed cell wall. The nomenclature in the literature of fungal propagation bodies produced *in vivo* is confusing. Several terms, such as protoplasts, hyphal bodies and blastospores, are commonly used in an inconsistent manner. The term protoplast is attributed mainly to fungal cells without cell walls, typically produced in the haemocoel of infected insects by several members of the order Entomophthorales (Roberts and Humber, 1981). Hyphal bodies of other entomopathogenic fungi such as *B. bassiana* are generally considered to be produced *in vivo*; *M. anisopliae* hyphal bodies also do not occur outside hosts and are functional analogues of protoplasts. In contrast to hyphal bodies, blastospores produced *in vitro* have a well-developed cell wall and differ in size and shape. The lack of a complete cell wall is considered to be an adaptation to avoid non-self recognition by the infected hosts. Protoplasts and hyphal bodies have been demonstrated to lack sugar residues on their surfaces which are required for their recognition by host phagocytic cells (Boucias and Pendland, 1991a,b; Götz, 1991; Pendland *et al.*, 1993; Butt *et al.*, 1996).

Our knowledge concerning non-self-recognition of fungal cells in insects remains fragmentary. Present literature provides information about both non-specific and specific opsonin-mediated recognition of fungal cells. The adhesion and ingestion of fungal conidia is probably triggered by their hydrophobic surface, as with bacteria (Absolom, 1988). Fungal cell walls usually bear particular sugars such as β-1,3-glucans which represent targets for opsonins and activate immune responses in insects (Gunnarson and Lackie, 1985). Until now, β-1,3-glucan-binding proteins have been isolated from the haemolymph of the cockroach *Blaberus craniifer* (see Söderhäll *et al.*, 1988) and the silkworm *Bombyx mori* (see Ochiai and Ashida, 1988). Matha *et al.* (1990a,b) identified a β-1,3-glucan-specific lectin on the surface of plasmatocytes, immune-competent cells of the greater wax moth *G. mellonella*. The detection of such a cell membrane-associated lectin explains the ability of plasmatocytes to recognize and ingest fungal cells such as injected blastospores or conidia of *B. bassiana* or *M. anisopliae* (see Vilcinskas *et al.*, 1997a). A galactose-specific lectin purified from haemolymph of the beet army worm *Spodoptera exigua* has been shown to opsonize fungal cells, which are then endocytosed by haemocytes (Pendland and Boucias, 1996). The lack of chitin, galactomannan and mannoprotein residues on the surface of hyphal bodies is probably responsible for the failure of host haemocytes to recognize such fungal cells produced *in vivo* (Pendland *et al.*, 1993, 1994).

However, phagocytosis of hyphal bodies has been documented in an early stage of mycosis. Following topical application of *M. anisopliae* conidia, a limited number of ingested hyphal bodies was detected in infected *G. mellonella* larvae before free circulating hyphal bodies were seen in the haemolymph (Vilcinskas *et al.*, 1997a), suggesting that phagocytosis of hyphal bodies took place immediately after penetration of the integument (Figures 2 and 3(C and D)). This hypothesis is supported by the observation that haemocytes aggregate at sites on the inner integument surface where *M. anisopliae* hyphae penetrate the insect cuticle (Gunnarsson, 1988). Positive chemotactic behaviour of haemoyctes is probably due to released compounds of fungal or host origin.

The ability of haemocytes to recognize and ingest hyphal bodies changes dramatically at a later stage of mycosis when free circulating fungal cells occur. Plasmatocytes isolated from *G. mellonella* larvae at this stage of pathogenesis do not ingest neighbouring hyphal bodies, but they do phagocytose silica beads or blastospores of entomogenous fungi offered as particles for ingestion. A continuous decrease of phagocytic activity of plasmatocytes against fungal cells during mycosis has been observed.

Hyphal bodies ingested by *G. mellonella* plasmatocytes are not killed, but propagate and grow within the endocytic vacuole (Figure 3D). Such phagocytic host cells are probably used as vehicles for dispersion within the haemocoel (Vilcinskas *et al.*, 1997a). The survival of ingested hyphal bodies may be mediated by secondary fungal metabolites, as discussed in Section 3.1. Injected *B. bassiana* blastospores grown *in vitro* also survived phagocytosis by *Spodoptera exigua* granular cells. After ingestion of blastospores, granular cells adhere to the basement membranes of different tissues, and progeny hyphal bodies are produced (Boucias and Pendland, 1991b).

Invading cells of entomopathogenic fungi not only survive phagocytosis by host haemocytes, they also overcome multicellular encapsulation. Injection of a large number of blastospores of different fungal species into *G. mellonella* larvae resulted in the formation of melanized nodules within the haemocoel. Spores of virulent fungi not only survived encapsulation by haemocytes, they continued to grow and escaped from transplanted nodules when they were transferred into culture medium (Vey and Götz, 1986; Götz, 1991).

In a later stage of *B. bassiana* or *M. anisopliae* pathogenesis, only free circulating hyphal bodies occur and propagate extensively within the haemocoel; they are neither attached nor ingested by neighbouring haemocytes. In spite of the documented lack of typical fungal surface structures required for non-self-recognition by host haemocytes, the

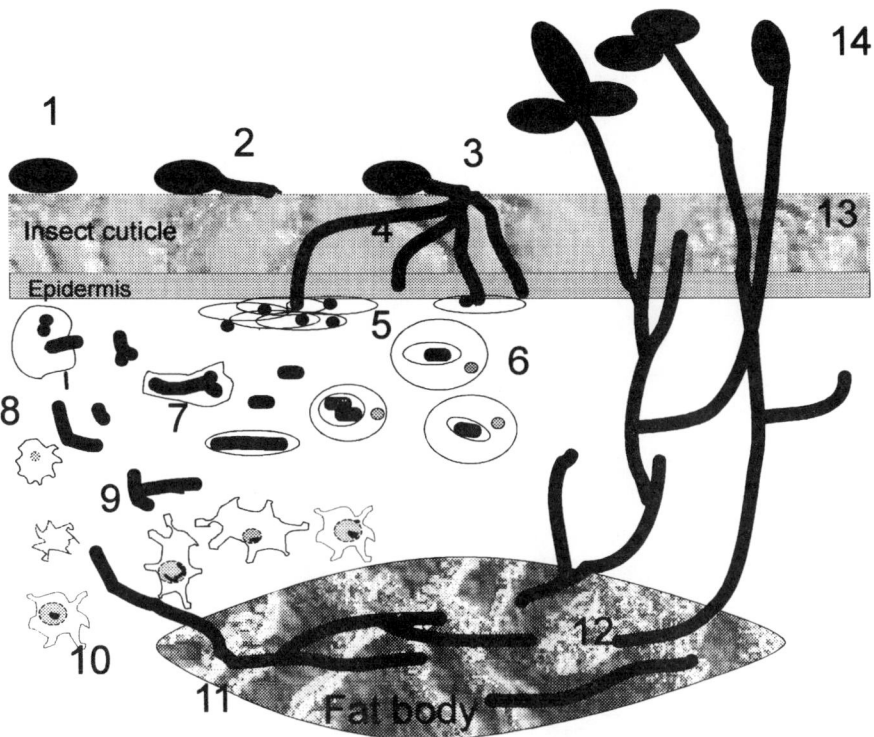

Figure 2 Schematic illustration of the development of entomopathogenic fungi such as *B. bassiana* and *M. anisopliae* in an insect host. 1. Attachment of a conidium to the host cuticle. 2. Germination and appressoria formation by the conidium. 3. Cuticle penetration by infecting hyphae. 4. Lateral growth and penetration of the epidermis. 5. Aggregation of host haemocytes at sites of fungal penetration. 6. Phagocytosis of hyphal bodies by phagocytic host cells (e.g. plasmatocytes in *G. mellonella*). 7. Ingested hyphal bodies are not killed within the endocytic vacuole, but rather grow and multiply. 8. At a later stage, when free circulating hyphal bodies occur, phagocytic activity against fungal cells decreases, and this is accompanied by impaired attachment, spreading and cytoskeleton formation. 9. Propagation of hyphal bodies within the haemolymph does not elicit a cellular immune response. 10. Plasmatocytes undergoing apoptosis exhibit swollen nuclei, clumped chromatin and bleb formation on their surfaces. 11. Infiltration of host tissues by invading hyphae. 12. Vegetative fungal growth causes the death of the infected hosts. 13. Hyphae penetrate insect cuticle. 14. Formation of conidia producing conidiophores.

mechanisms behind the evasion of cellular defence appear more sophisticated, since evidence is available that hyphal bodies possess a surface coat that either mimics host components or is comprised of absorbed haemolymph proteins (Boucias *et al.*, 1995).

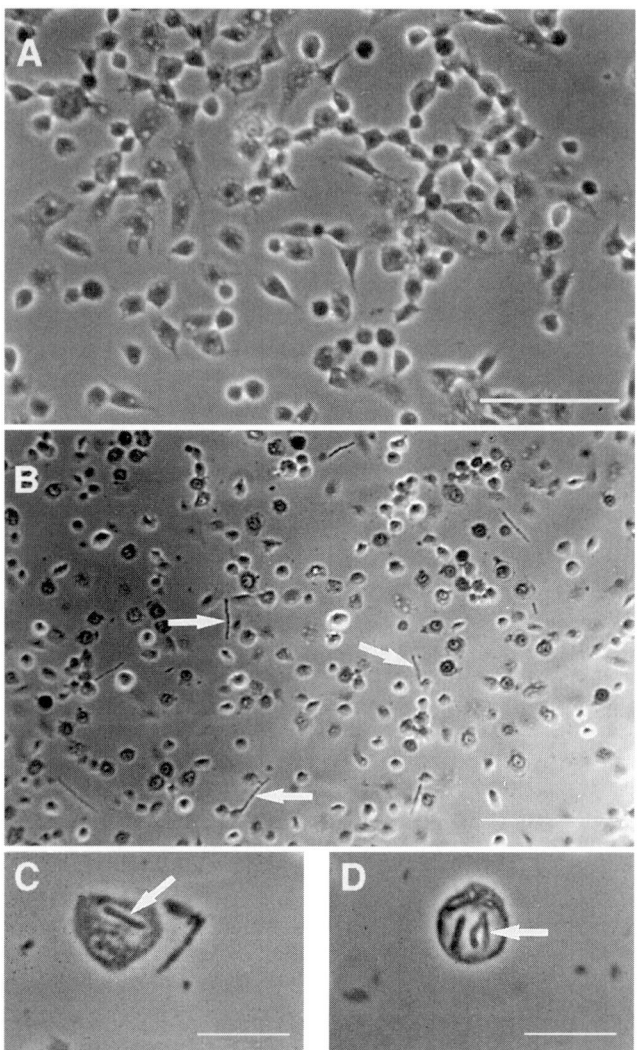

Figure 3 Isolated *G. mellonella* plasmatocytes in monolayers after two hours' incubation in a culture medium. Plasmatocytes were isolated from healthy larvae (A) or from larvae infected with *M. anisopliae* (B) three days after the injection of blastospores (10^4/larva). Plasmatocytes from healthy larvae spread out and exhibited filopodia formation whereas those from infected larvae remained rounded. Haemolymph samples from larvae infected with *M. anisopliae* contained hyphal bodies (arrows) that were not retained by the nylon wool and were neither attached nor ingested by plasmatocytes (A and B, scale bars = 100 μm). Two days after topical application of *M. anisopliae* conidia, a limited number of hyphal bodies phagocytosed by plasmatocytes (arrow) was observed in the haemolymph of infected *G. mellonella* larvae (C). Ingested hyphal bodies were not killed, but multiplied and grew within the plasmatocytes (D). (C and D, scale bars = 10 μm).

Furthermore, the occurrence of free circulating hyphal bodies during infection with *B. bassiana* or *M. anisopliae* is accompanied by suppression of haemocyte reactions such as spreading and filopodia formation (Hung et al., 1993; Mazet et al., 1994; Vilcinskas et al., 1997b). Inhibition of haemocyte spreading and filopodia formation has also been noted as a key determinant in preventing cellular recognition of non-self (Rizki and Rizki, 1986, 1990; Guzo and Stolz, 1987; Tanaka, 1987; Vinson, 1991). Toxins of low molecular mass and exogenous proteases of fungal origin have been shown to inhibit spreading and attachment as well as the formation of cytoskeleton structures or filopodia of isolated *G. mellonella* plasmatocytes *in vitro* (Vilcinskas et al., 1997b; Griesch and Vilcinskas, 1998). The number of circulating haemocytes decreases during the propagation of hyphal bodies (Hung and Boucias, 1992). This observation can be explained by the assumption that circulating haemocytes become adhesive during pathogenesis and adhere to the basal lamina of the haemocoel (Boucias et al., 1995).

Although cellular reactions in insects against fungi are well documented (Vey and Götz, 1986; Götz, 1991), humoral immunity is largely unknown (St Leger and Bidochka, 1996). Recent reports provide evidence that the humoral reactions in response to injected fungal elicitors are similar to those observed after the injection of bacterial provocators. Protein patterns of electrophoretically separated cell-free haemolymph samples from *G. mellonella* or *M. sexta* larvae pre-injected with yeast cells or *B. bassiana* blastospores showed several new or enhanced bands compared to haemolymph samples from untreated larvae (Bidochka et al., 1997; Vilcinskas and Matha, 1997b). Among the proteins which are induced in response to challenge with fungal elicitors, at least lysozyme and cecropins have been identified (Vilcinskas and Matha, 1997a).

In contrast, propagation of hyphal bodies within the haemocoel during natural mycosis neither induces humoral immune response nor impairs in general the capacity of the infected host to synthesize proteins (Boucias et al., 1994; Mazet and Boucias, 1996; Vilcinskas and Matha, 1997b). Studies utilizing sodium dodecylsulphate gel electrophoresis in combination with autoradiography and inhibition zone assays showed that protein synthesis by infected larvae (even when moribund) was not suppressed, only the neosynthesis of certain proteins such as lysozyme and cecropins. These results exclude the possibility that the observed suppression of humoral immune responses in infected host insects is due to generally reduced metabolism caused by fungal infection. They provide instead evidence that parasitic fungi may actively avoid or suppress the humoral immune responses of host insects (Vilcinskas and Matha, 1997a,b).

2.2.4. *Host Death, Saprophytic Development and Conidia Formation*

Vegetative replication and growth of hyphal bodies in the haemocoel is followed by the invasion of host tissues. Infected hosts at this stage of mycosis appear moribund. Entomopathogenic fungi infiltrate all tissues by means of invasive hyphae and subsequently kill the infected host. After the host's death, the cadaver becomes hard and mummified, and sometimes also melanized. The life cycle of the entomopathogenic fungus is completed by the saprophytic phase and formation of reproductive structures of sexual or asexual origin. In appropriate conditions, such as high humidity and adequate temperatures, vegetative hyphae grow out from the host through the cuticle and form conidiophores outside. The whole insect cadaver is usually transformed into fungal biomass (Müller-Kögler, 1965; Ferron, 1981; Tanada and Kaya, 1993; Hayek and St Leger, 1994; Khachatourians, 1996).

3. FUNGAL FACTORS DETERMINING VIRULENCE

3.1. Fungal Toxins and their Effects on Insects

Entomopathogenic fungi are known to produce several classes of molecules which are toxic to their host insects and to other organisms. Fungal toxins can be differentiated into peptidic and non-peptidic compounds (Khachatourians, 1996). This chapter is focused on the products of low molecular mass, which are also known as fungal secondary metabolites because they are not part of primary metabolism.

It has been concluded that the capacity to release secondary metabolites must improve the survival fitness of the producing organism, since the biosynthesis of such complex molecules is programmed by many thousands of DNA bases, and natural selection precludes the expenditure of so much metabolic energy without any functional role (Williams *et al.*, 1989). A large and still increasing number of toxic or otherwise biologically active secondary metabolites has been isolated from culture media or the mycelia of numerous entomopathogenic fungal species. Some of these fungal metabolites were also identified in their host insects during mycosis, and at least some of them exhibit insecticidal activity. The exact function of these so-called fungal toxins is not always clear. In principle they could be released directly to poison and kill the infected host or to suppress its immune system, thus facilitating development of the mycosis. Secondary metabolites produced by entomopathogenic fungi have received considerable interest in recent research. Because of their insecticidal activity, and because of the medical importance of particular fungal toxins such as cyclosporin A, numerous investigations have been made to elucidate their exact role in the molecular interactions between parasitic fungi and their host insects and to explore their potential application in medicine or biological control of pest insects.

We will not attempt to provide an overview of the broad spectrum of identified fungal secondary metabolites, but rather to outline some advances in the understanding of the mode of action of representative members which have recently been achieved. Many reports are focused on cyclic peptides of fungal origin. Their putative role during fungal pathogenesis will be evaluated in the light of recent work.

M. anisopliae and *B. bassiana* represent intensively examined entomopathogenic fungi because they are used world-wide to control insect pests (Roberts, 1981; Tanada and Kaya, 1993; Clarkson and Charnley, 1996; Khachatourians, 1996). *M. anisopliae* produces at least two families of secondary metabolites, destruxins and cytochalasins. Additionally, we will focus on two families of peptidic secondary metabolites which are synthesized by *B. bassiana*: beauverolides and cyclosporins.

3.1.1. *Destruxins*

Destruxins are the most abundant secondary metabolites produced by *M. anisopliae* (see Jegorov *et al.*, 1993). They represent cyclic hexadepsipeptides composed of five amino acids: L-proline, L-isoleucine, N-methyl-L-valine, N-methyl-L-alanine, β-alanine and a variable D-α-hydroxy acid which characterizes the destruxin types A–E (Figure 4).

Insect species treated with destruxins varied considerably in their susceptibility to these toxins (Roberts, 1981). In *G. mellonella*, destruxins caused tetanic paralysis and often death after intrahaemocoelic injection (Roberts, 1966; Dumas *et al.*, 1996a). The presence of destruxins in insects infected with *M. anisopliae* was first described by Suzuki *et al.* (1971). Their potency can be estimated by the fact that injection of 0.0006 μmol of destruxin is sufficient to induce paralysis and to kill insect larvae (Kodaira, 1962). Destruxins inhibit haemocytic activation and nodule formation in insects (Vey *et al.*, 1985; Huxham *et al.*, 1989). Destruxin E was reported to trigger degranulation of isolated granular haemocytes from the crayfish *Pacifastacus leniusculus* (see Cerenius *et al.*, 1990), but this observation was not confirmed by our experiments using insect granular cells treated with the same compound. Muroi *et al.* (1994) reported that destruxin B, produced by *M. anisopliae*, is a specific and reversible inhibitor of vacuolar-type H^+-translocating adenosine triphosphatase, which controls acidification in intracellular vesicles. Phagocytically active haemocytes,

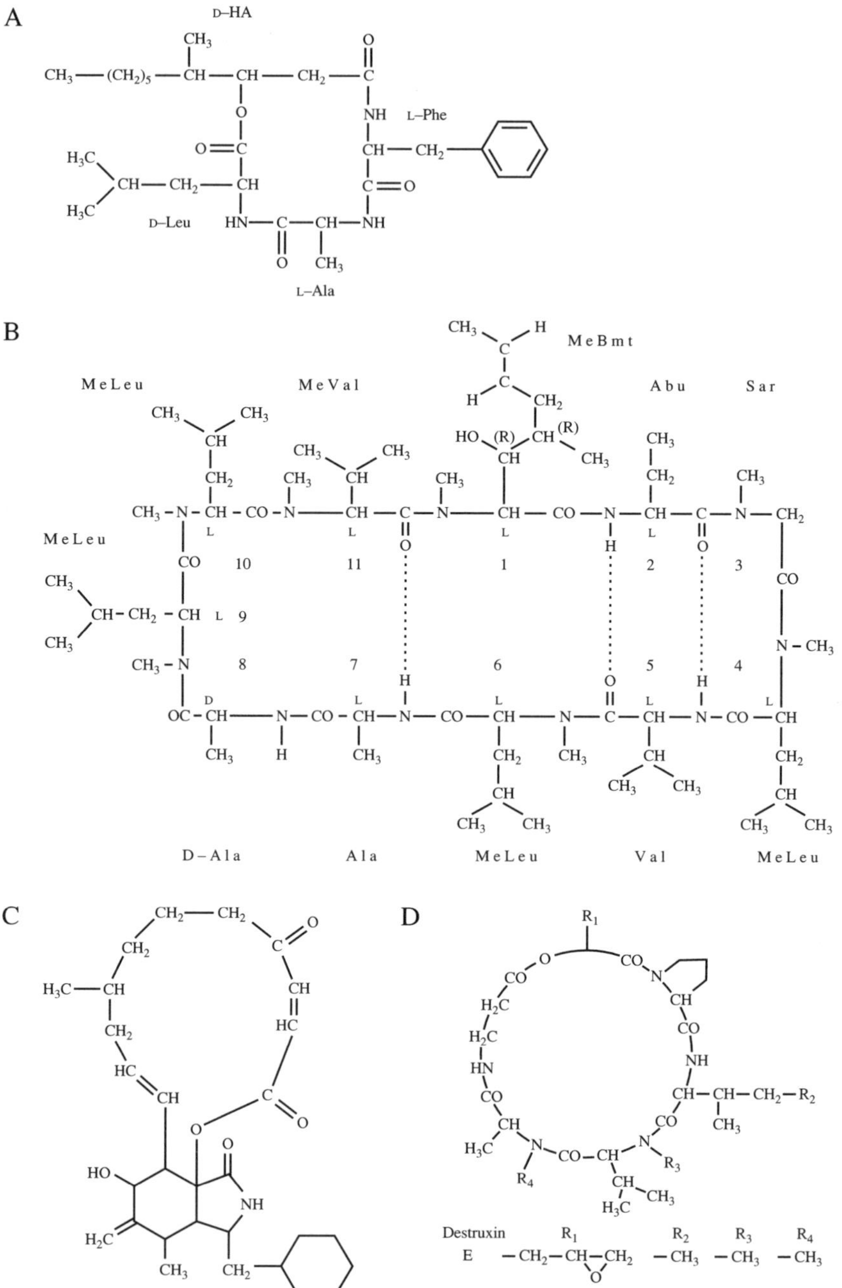

Figure 4 Molecular structure of beauverolide L (A), cyclosporin A (B), cytochalasin B (C) and destruxins A and E (D).

such as the plasmatocytes of *G. mellonella*, employ digestive enzymes which are activated through acidification and released into an endocytic vacuole to kill and digest phagocytosed pathogens. This led us to speculate that *M. anisopliae* may release destruxin B to prevent acidification inside endocytic vacuoles after being phagocytosed. Invading hyphal bodies ingested by *G. mellonella* plasmatocytes were neither killed nor digested within endocytic vacuoles, but grew and multiplied inside (Vilcinskas *et al.*, 1997b). Inhibition of acidification in phagocytic vacuoles by release of destruxins would not only support survival of ingested fungal cells but also permit their distribution within the host by circulating infected haemocytes.

It is possible that the different types of destruxins released during mycosis act synergistically. Recent data suggest that destruxins also directly incapacitate the cellular part of the host immune system. Isolated plasmatocytes from *G. mellonella* larvae infected with *M. anisopliae* exhibited reduced phagocytic activity against fungal cells compared to plasmatocytes obtained from healthy larvae. Such inhibition of phagocytic activity was observed after the incubation of isolated plasmatocytes in monolayers with destruxins A or E in sublethal concentrations (Vilcinskas *et al.*, 1997a).

Further investigations revealed specific effects of these fungal toxins on a subcellular level, especially with regard to the cytoskeleton. Plasmatocytes obtained from *G. mellonella* larvae infected with *M. anisopliae*, as well as those isolated from healthy larvae and treated *in vitro* with destruxins in sublethal concentrations, remained round in shape and exhibited reduced capacity to adhere and spread on glass surfaces. In contrast, plasmatocytes obtained from healthy larvae spread and attached after two hours' incubation in a cell culture. In addition, plasmatocytes from non-infected larvae exhibited filopodia formation accompanied by formation of typical cytoskeleton structures such as bundled actin filaments, so-called stress fibres and a network of microtubulues. Pathogenic effects concerning the intracellular distribution of actin and tubulin were also observed in isolated plasmatocytes from larvae infected with *M. anisopliae* and in plasmatocytes isolated from healthy larvae and subsequently incubated with destruxins *in vitro*. Therefore, fungal secondary metabolites produced during mycosis appear to impair cellular immune responses by interfering with subcellular structures (Vilcinskas *et al.*, 1997b).

G. mellonella plasmatocytes isolated from larvae infected with *M. anisopliae* exhibit symptoms of apoptosis. The observed alterations in morphology and cytoskeleton formation are typical of cells undergoing programmed cell death, e.g. reduced capacity to attach and spread on foreign surfaces and impaired formation of filopodia and particular cytoskeleton structures such as bundled actin filaments and microtubules. The nuclei of plasmatocytes from infected larvae are swollen and contain obviously clumped chromatin. Furthermore, we observed blebs on their cell membrane (Figure 5) in which actin accumulated. Incubation of isolated plasmatocytes with destruxins *in vitro* resulted in similar morphological and cytoskeleton alterations (Vilcinskas *et al.*, 1997b). Cells of the Malpighian tubules or the midgut of *G. mellonella* larvae treated with destruxins also exhibited pycnosis of nuclei, which was documented by electron microscopical studies (Dumas *et al.*, 1996b). All of these morphological alterations are reported to be characteristic of cells undergoing apoptosis (Cohen, 1993; Hale *et al.*, 1996). Interestingly, isolated granular cells of *G. mellonella* recovered from incubation with destruxin after removal of this toxin by medium exchange, whereas isolated plasmatocytes treated in the same manner remained affected and subsequently died (Griesch, 1998). The assumption that destruxin induces programmed cell death only in plasmatocytes was confirmed by fluorescent staining with antibodies against apoptotic cells (A. Vilcinskas, unpublished observation).

The hypothesis that destruxins promote apoptosis in particular target cells is not only supported by the above-mentioned results but further strengthened by reports in the literature. Destruxins exhibit a strong antiviral effect against baculoviruses in insect cell cultures (Quiot *et al.*, 1985). Replication *in vitro* and infectivity *in vivo* of baculoviruses was later shown to be reduced in apoptotic cells. The best antiviral effect of destruxin was achieved in insect cell lines pre-treated with this toxin before virus application (Kopecky *et al.*, 1992). In conclusion, the putative induction of programmed cell death in target cells of destruxin could explain their insusceptibility to viruses. Induced apoptosis of target cells may also be responsible for the effects of destruxins on humoral immune reactions in insects.

Antimicrobial activity in the haemolymph of *G. mellonella*, detected in response to injected fungal cells, was significantly lower when destruxin A was simultaneously injected in a sub-lethal concentration, suggesting inhibitory activity of this toxin on humoral immune reactions (Vilcinskas, 1994). *G. mellonella* plasmatocytes, identified target cells of certain destruxins, represent the major phagocytic cell type and contribute to multicellular defence reactions in this insect species (Ratcliffe, 1993). Plasmatocytes undergoing programmed cell death can probably neither ingest and encapsulate injected fungal cells nor release factors which stimulate synthesis of immune proteins responsible for enhanced antimicrobial activity within the haemolymph. According to our knowledge, these results provide the first evidence that parasitic fungi may inhibit host immune responses by induction of apoptosis in particular immune competent host cells, and that this mechanism is mediated by secondary metabolites.

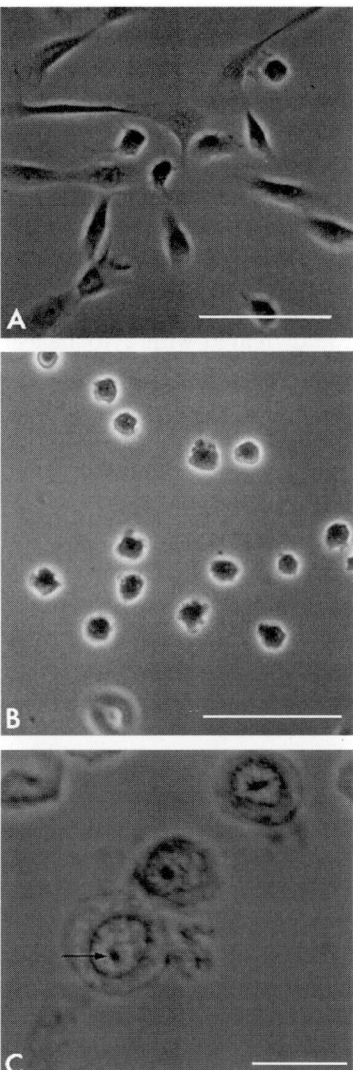

Figure 5 Plasmatocytes isolated from untreated *G. mellonella* larvae spread out and showed filopodia formation after two hours' incubation in monolayers (A), whereas plasmatocytes from larvae infected with *M. anisopliae* were rounded with blebs on their surface (B), swollen nuclei and clumped chromatin (arrow) (C). Such morphological alterations are typical of cells undergoing apoptosis. Scale bars (A and B) = 50 μm; (C) = 10 μm.

3.1.2. Cytochalasins

Cytochalasins are closely related perhydro-indoles with a macrocyclic ring which are produced by members of several fungal families (Figure 4). At least cytochalasin B and D are released by *M. anisopliae*. The best known effect of cytochalasins is their inhibitory activity on formation of filamentous actin in experimental conditions (Schliwa, 1986; Bershadsky and Vasilev, 1988). However, no discernible immediate or long-term effect has been observed after intrahaemocoelic injection of cytochalasin B in concentrations up to 150 µg/g into *G. mellonella* larvae (Roberts, 1981). In contrast, plasmatocytes isolated from this insect and treated with cytochalasins *in vitro* in sub-lethal concentrations exhibited strongly reduced phagocytic activity, attachment and spreading accompanied by impaired cytoskeleton formation. The actin distribution in cytochalasin-treated cells, visualized with fluorescein isothiocyanate-labelled phalloidin, differed remarkably from that observed in plasmatocytes after incubation with destruxins, suggesting different mechanisms behind their effects (Vilcinskas *et al.*, 1997a, b). Cytochalasins are reported specifically to affect actin polymerization by reversibly binding to the barbed ends of the actin filaments (Brown and Spudrich, 1979; Flanagan and Lin, 1980). It is noteworthy that cytochalasin-treated cells of vertebrate and insect origin exhibit similar morphological alterations and distribution of fluorescent-labelled actin. Incubation of cultured human fibroblasts or isolated *G. mellonella* plasmatocytes with cytochalasin resulted in spindle-shaped cells. Neither actin filaments nor stress fibres, but rather irregularly located granular spots, were seen within the cells (Bershadsky and Vasilev, 1988; Vilcinskas *et al.*, 1997b).

Such typical distribution of actin was not observed in plasmatocytes isolated from *G. mellonella* larvae infected with *M. anisopliae*, suggesting that cytochalasins play a minor role during mycosis in comparison to destruxins. This assumption is confirmed by the very low yield of cytochalasin D compared to other fungal toxins in cultures of *M. anisopliae* (see Roberts, 1981). Nevertheless, we can speculate that cytochalasins and destruxins act synergistically when released during mycosis. Impaired cytoskeleton formation in target cells reduces their capacity to attach and spread on foreign surfaces. Phagocytosis and multicellular encapsulation are based on complex movements of host haemocytes which require a functioning cytoskeleton.

3.1.3. Beauverolides

B. bassiana produces the cyclic depsipeptides beauvericin, bassianolide and beauvellide and the pigments bassianin, tenellin and oosporein (Ferron, 1981; Roberts, 1981; Gillespie and Claydon, 1989; Khachatourians, 1991).

Beauvellide was later identified as a mixture of cyclic depsipeptides which have been designated beauverolides (Figure 4). Until now, fifteen different members of this family have been characterized. No direct toxic effect of beauverolides against insects or other arthropods was noted after feeding or intrahaemocoelic injection (Jegorov et al., 1994). Because of the lack of further reports, we shall focus in the following paragraphs on beauverolide L, which was first isolated by Jegorov et al. (1994).

Although beauverolide L is highly abundant in fungal extracts, reports concerning its effects and role during mycosis have only recently been published (Vilcinskas et al., 1997e, 1998). Injection of beauverolide L, either dissolved in diluted ethanol or coated on silica particles, into G. mellonella larvae caused no mortality but induced strong humoral and cellular immune responses. The challenge induced both the formation of melanized nodules and the release of lysozyme and cecropin-like activity within the haemolymph, suggesting stimulatory activity on humoral immune responses. This observation is inconsistent with the inhibition of cellular immune responses in S. exigua and G. mellonella larvae 24 hours after being challenged with B. bassiana reported by Hung and Boucias (1992), Hung et al. (1993) and Vilcinskas et al. (1997a). Beauverolides and most identified fungal cyclic peptides are non-polar molecules. Because of the hydrophilic milieu in the haemolymph of infected insects, hydrophobic toxins are unlikely to be released, but rather to be exposed on the surface of the secreting fungal cells. Beauverolide L coated on particles offered for phagocytosis exhibited direct cytotoxic effects on G. mellonella plasmatocytes, suggesting that this compound may display immune suppressive activity only when in a surface-bound form. Hydrophobicity may explain the observed immune stimulatory effect of beauverolide-coated silica beads (Vilcinskas et al., 1998), because hydrophobic surfaces of bacteria are reported to promote adhesion of haemocytes (Absolom, 1988).

3.1.4. Cyclosporins

Cyclosporins are hydrophobic, cyclic undecapeptides (Figure 4) produced by several members of the fungal genera *Beauveria*, *Verticilium* and *Tolypocladium* (see Jegorov et al., 1990). Cyclosporin A has become one of the most important medical drugs, used to suppress immune rejection after organ transplantation. Therefore the interaction of cyclosporin A with the vertebrate immune system has been intensively examined, but this is beyond the scope of this review.

Surprisingly, it is still not clear why particular entomopathogenic fungi produce this cyclic peptide. Cyclosporin A exhibits insecticidal activity against mosquito larvae (Weiser and Matha, 1988) but not against

G. mellonella larvae (Vilcinskas *et al.*, 1997c). Its insecticidal activity against mosquitoes was confirmed by experiments with *Tolypocladium* conidia inactivated by ultraviolet light and extracted with methanol. The latter were inactive, whilst the crude methanolic extract of inactivated conidia exhibited an insecticidal effect accompanied by characteristic ultrastructural alterations in cells of treated mosquitoes (Matha *et al.*, 1992).

Neither feeding nor intrahaemocoelic injection of cyclosporin A into *G. mellonella* larvae caused obvious toxic or adverse effects (Vilcinskas *et al.*, 1997c). Isolated plasmatocytes of this insect treated with cyclosporin A exhibited moderate, concentration-dependent inhibition of phagocytic activity, whereas attachment and spreading were apparently not affected. Incubation with this cyclic peptide influenced the actin distribution within plasmatocytes. Formation of stress fibres was not noted, but a thin line of actin occurred at the edges of the attached and spread cells (Vilcinskas *et al.*, 1998). Dumas *et al.* (1996b) showed that incubation of *G. mellonella* Malpighian tubules *in vitro* with medium containing cyclosporin A resulted in the typical histopathological findings seen in treated mosquito larvae (Weiser *et al.*, 1989). This obvious effect of dissolved cyclosporin A on isolated plasmatocytes and Malpighian tubules is inconsistent with the pronounced resistance of *G. mellonella* larvae to injected molecules. This contradiction and the insecticidal activity of cyclosporin A against other insects stimulated us to investigate the mechanisms behind the resistance of *G. mellonella* larvae to cyclosporin A. Using a radiolabelled analogue of cyclosporin A, its distribution within the body of injected larvae was monitored. Injected cyclosporin A was rapidly removed from haemolymph and then accumulated and stored in the fat body. Only a limited amount of radioactivity was excreted in faeces. Injected hydrophobic molecules were transported through the haemolymph by at least four binding proteins. The major cyclosporin-binding protein was lipophorin, an abundant insect lipoprotein (Vilcinskas *et al.*, 1997c). Comparable binding of radiolabelled cyclosporin A to proteins in the haemolymph of mosquito larvae was not observed. Therefore it is quite possible that the selective insecticidal activity of fungal secondary metabolites is influenced by the presence or absence of haemolymph binding proteins. This report provided the first evidence for interaction between secondary fungal metabolites and insect haemolymph proteins. Such interactions could influence the specificity and/or virulence of the fungi, if these secondary metabolites play a significant role in the development of mycosis.

3.2. Interactions of Fungal Proteases with the Insect Immune System

The chitinous exoskeleton of insects forms a primary barrier against invading microorganisms and thus represents a part of their defence system.

The ability of parasitic fungi to invade susceptible insect species directly through the cuticle is mediated by their capacity to release digestive enzymes which assist the penetration process. Extracellular fungal proteases have been thought to play a predominant role because more than 70% of the dry weight of the cuticle consists of proteins. The research on fungal infection of insects has been comprehensively reviewed by several authors (Charnley, 1984; Boucias and Pendland, 1991a, b; Charnley and St Leger, 1991; St Leger, 1991; Clarkson and Charnley, 1996; St Leger and Bidochka, 1996).

The capacity of certain fungal species such as *B. bassiana* to produce extracellular proteases has been proposed to determine their virulence towards their insect hosts (Bidochka and Khachatourians, 1990; Gupta, S.C. et al., 1994). Numerous extracellular proteases released by *M. anisopliae* have been purified and characterized (St Leger et al., 1986, 1987, 1989, 1994a, b).

However, interactions of fungal proteases with the immune system of the host insect have so far been neglected, although the effects of fungal development on cellular immunity in infected hosts has been examined intensively (Bidochka and Khatchatourians, 1987; Hung and Boucias, 1992, 1993; Pendland et al., 1993). Suppression of immune responses by invading fungi has been attributed to their low molecular mass secondary metabolites, released during mycosis (Vilcinskas et al., 1997a, b). Evidence for the participation of high molecular mass fungal metabolites in suppression of host cellular immune responses was provided by Mazet et al. (1994), who reported a fungal molecule of molecular mass greater than 10 kDa that inhibited spreading and filopodia formation by host haemocytes.

Compared with the large spectrum of extracellular proteases isolated and characterized from culture filtrate or extracts of entomopathogenic fungi grown *in vitro*, the number of proteases identified as being released within the haemocoel of infected insects is rather limited. A basic elastase-type serine protease released by *B. bassiana* has been detected within the haemolymph of infected *Bombyx mori* larvae, but it exhibited only very low proteolytic activity in the presence of haemolymph because of host protease inhibitors (Shimizu et al., 1993a, b).

Toxins of high molecular mass may represent proteases from entomogenous fungi, because these enzymes have recently been shown to impair *G. mellonella* haemocytes *in vitro* (Griesch and Vilcinskas, 1998). Proteases released by *B. bassiana* or *M. anisopliae* cultivated in medium with casein as a source of nutrients inhibited phagocytic activity of isolated *G. mellonella* plasmatocytes and impaired attachment, spreading and cytoskeleton formation. The same effects have been observed *in vivo* in larvae infected with one of these fungal species (Vilcinskas et al., 1997a, b). These reports do not allow a reliable estimation of the degree to which fungal extracellular proteases contribute to suppression of immune responses in infected hosts, but they lend some credit to our hypothesis that low and high molecular mass metabolites of fungal origin synergistically facilitate the overcoming of the host's defence system.

A thermolysin-like metalloprotease has been shown to be produced by *M. anisopliae* by St Leger *et al.* (1994a). Metalloproteases have been demonstrated to affect phagocytic activity, attachment and spreading of isolated *G. mellonella* plasmatocytes to a higher degree than serine proteases such as trypsin and chymotrypsin. The cytotoxic effects were characterized by cytoskeleton alterations such as impaired formation of stress fibres and adhesion plaques (Griesch and Vilcinskas, 1998). Such effects could be explained by the proteolytic digestion of receptors mediating the binding of plamatocytes to extracellular matrix proteins. As a result, focal contacts and other sites of cell adhesion are partially disrupted and the associated actin filaments detach from the cell membrane.

Apart from the described immuno-suppressive effects of fungal proteases produced *in vitro*, we have provided evidence that microbial proteases exhibit stimulatory activity on humoral immune responses. Injection of thermolysin or other proteases from *M. anisopliae* in sublethal concentrations into *G. mellonella* larvae induced remarkable humoral immune responses consisting of conspicuous release of lysozyme and expression of cecropin-like and protease inhibitory activity. Subsequent experiments revealed that proteolytic digestion of haemolymph proteins resulted in the production of small (<3 kDa) heat-stable molecules which are potent inducers of immune responses in *G. mellonella* (see Griesch, 1998).

4. HUMORAL IMMUNE REACTIONS OF INSECTS AGAINST FUNGI

Humoral immunity is based on molecules produced and released or activated directly to kill invading pathogens and parasites or to provide assistance to mechanisms which mediate resistance of the host organism. It includes all types of antibiotics employed in host defence reactions. Plants and animals have developed a broad spectrum of compounds with bacteriostatic or bactericidal activity, whereas the number of known compounds with antifungal properties appears rather limited. Even in the well-examined vertebrate immune system, humoral responses against parasitic fungi are poorly understood, and only a few antifungal peptides have been recognized. This is probably due to the eukaryotic nature of fungal cells, which share a large degree of ultrastructrual and physiological similarity with the cells of their potential hosts. Bacterial pathogens, as prokaryotic organisms, differ in many aspects from their hosts' cells. The degree of similarity between a host and a pathogen or parasite limits the possibility of developing molecules which are exclusively targeted at invading organisms. Molecules with antimicrobial activity which also affect host cells are inappropriate for incorporation in humoral defence reactions.

Although insect immunity against bacteria has attracted considerable interest in past decades, advances in understanding the humoral responses against fungi and identification of antifungal peptides have emerged only in recent years. This is probably due to the fact that antifungal activity is not so obvious and therefore not so easily detectable as antibacterial effects, which can be demonstrated by well-established methods such as inhibition and lysis zone assays. According to St Leger (1991), two criteria have to be fulfilled to prove that particular compounds contribute to the restriction of mycoses. First, they must be present in a form available to the pathogen at a time, place and concentration necessary to arrest the development of a particular fungus in the host. Second, it must be demonstrated by genetical, physical or chemical manipulation that changes in the concentration of the antifungal factor result in a corresponding change in susceptibility to the pathogen. Considering these criteria, in the following paragraphs we will review identified proteins from insects which exhibit antifungal activity and discuss their putative contribution to humoral defence reactions against fungi.

The first evidence that immunization changes the sensitivity of a particular insect to lethal doses of fungal cells was provided by experiments in which *G. mellonella* larvae were pre-injected with fungal elicitors such as zymosan or heat-inactivated *Saccharomyces cerevisiae* cells. Immunized larvae exhibited increased survival rates after the injection of normally lethal doses of yeast cells. Susceptibility to the entomopathogen *B. bassiana* decreased in immunized (= pre-injected) larvae compared to untreated larvae. Although no protective immunity against entomopathogenic fungi in terms of decreased mortality was detected, immunized *G. mellonella* larvae survived injection of viable blastospores or topical infection with conidia of *B. bassiana* significantly longer (Vilcinskas and Wedde, 1997).

4.1. Inducible Antifungal Proteins

Attempts to identify humoral immune factors in insects have been made by comparative analysis of the protein composition of the haemolymph of pre-injected and untreated insects. *G. mellonella* haemolymph from immunized larvae exhibited weak inhibitory activity against yeast cells, but not against developmental stages of *B. bassiana* or *M. anisopliae* (see Vilcinskas *et al.*, 1997d). Injection of suspended bacterial or fungal cells as well as soluble lipopolysaccharide or zymosan preparations induced the synthesis and release of novel or constitutive proteins within the haemolymph of larvae of

S. exigua (see Boucias *et al.*, 1994), *G. mellonella* (see Vilcinskas and Matha, 1997a,b) and *Manduca sexta* (see Bidochka *et al.*, 1997). Among the induced proteins detected by electrophoretical separation of haemolymph samples, the lysozyme band (around 14 kDa) was always enlarged, independent of the type of elicitor used, suggesting the non-specific release of this protein during humoral immune responses in insects.

However, different protein patterns were induced after challenge with either bacterial or fungal elicitors, suggesting humoral responses that are more or less specific for bacteria or fungi. Indeed, evidence for different pathways leading to expression of genes encoding either antifungal or antibacterial proteins has been reported by Lemaitre *et al.* (1995). They identified a recessive mutation in *Drosophila* which impaired the inducibility of all genes encoding antibacterial peptides during the immune response in this dipteran species, whereas the antifungal peptide drosomycin (Fehlbaum *et al.*, 1994; see below) remained fully inducible in homozygous immune-deficiency mutants.

A common strategy for the detection and identification of inducible antifungal proteins in haemolymph samples from untreated, wounded, injected or infected insects employs living microorganisms which are cultivated in stationary or submerged cultures. Several assays have been established for comparative monitoring *in vitro* of germination, growth and survival of particular fungal cells exposed to haemolymph samples from untreated and treated insects (Vilcinskas, 1994). Using individually modified assays for measurement of fungal development employing light microscope inhibition zone assays or photometric measurement of fungal cell density in submerged cultures, the following antifungal proteins (AFPs) have been identified in insects.

The first AFP isolated from an insect source was reported by Iijima *et al.* (1993). They purified a histidine-rich protein consisting of 67 amino acid residues with a molecular mass of 7 kDa from the haemolymph of third instar larvae of the flesh fly *Sarcophaga peregrina*. The molecule was constitutively present in the haemolymph and inhibited growth *in vitro* of *Candida albicans*. It induced the loss of substance(s) with absorbance at 260 nm from fungal cells, and its antifungal activity was synergistically enhanced in the presence of the antibacterial protein sarcotoxin IA (Iijima *et al.*, 1993).

An inducible AFP from insects was first reported by Fehlbaum *et al.* (1994). The authors identified a small, slightly cationic, 44-residue peptide, processed from a 70-residue precursor molecule in larvae and adults of immunized *Drosophila*, which was therefore designated as drosomycin. Eight cysteine residues were detected in this peptide, which were thought to form four intramolecular disulphide bridges, thus explaining its particular resistance to heat treatment, pH variations and proteolytic digestion.

Interestingly, drosomycin shows a significant similarity in amino acid sequence with the 5 kDa cysteine-rich plant AFPs reported from Brassicaceae. Drosomycin exhibits fungistatic and fungicidal activity *in vitro*; however, the mechanism behind its antifungal activity remains to be elucidated (Fehlbaum et al., 1994).

Further AFPs isolated from insects were reported in 1995. Two AFPs, tenecin-3 and holtricin-3, were purified from haemolymph of *Tenebrio molitor* by Jung et al. (1995) and *Holotrichia diomphalia* by Lee et al. (1995). Both consist of approximately 80 amino acid residues and show similarities with the AFP from *S. peregrina*.

Several inducible proteins from insect sources with recognized antibacterial activity have recently been shown also to exhibit antifungal activity. For example, metchnikowin, an antibacterial protein from *Drosophila*, also inhibited growth of the fungus *Neurospora crassa in vitro* (Levashina et al., 1995). Thanatin, an inducible 21-residue peptide was also shown to exhibit both antibacterial and antifungal activity. This peptide has similarities in amino acid sequence with frog antimicrobial proteins (Fehlbaum et al., 1996). Cecropins, another class of inducible peptides isolated from several insect species, are reported to exhibit lytic activity against both bacterial and fungal cells (Jaynes, 1989; Vilcinskas and Matha, 1997a). Lysozyme was found to inhibit growth *in vitro* of non-pathogenic fungi such as *Saccharomyces cervisiae* and the mycelium forming fungus *Absidia glauca*, but not that of virulent strains of *B. bassiana* and *M. anisopliae* (see Vilcinskas and Matha, 1997a). This observation was surprising because the mode of action of lysozyme is well characterized. It hydrolyses β-1-4 linkages between N-acetylglucosamine and N-acetylmuramic acid residues in bacterial cell wall peptidoglycans. Antifungal activity of lysozyme was therefore difficult to understand. Lysozyme is suggested to remove the murein sacculus of bacteria after the action of other antibacterial proteins rather than actively to kill invading bacteria (Boman and Hultmark, 1987). Interestingly, fragments of peptidoglycan hydrolysis have been shown to elicit humoral immune responses in insects (Kanost et al., 1988). Lysozyme from *G. mellonella* was the first antibacterial factor to be purified and completely characterized from insects (Powning and Davidson, 1976), and is reported to share structural similarity with chicken (C) type lysozyme (Jollès et al., 1979; Kanost et al., 1990). Both types, lysozyme isolated from haemolymph of *G. mellonella* larvae and hen egg white lysozyme, exhibit antifungal activity against *Saccharomyces cerevisiae* (see Vilcinskas and Matha, 1997a). However, detection of inducibility and antifungal activity *in vitro* is insufficient to prove the contribution of a particular protein to humoral immune responses against fungi. In accordance with the previously mentioned criteria, which have to be fulfilled to demonstrate the contribution of a compound to antifungal defence (St Leger, 1991), a correlation was found between induced increases in lysozyme concentration in the haemolymph of *G. mellonella* larvae and changes in their susceptibility to non-pathogenic yeast. Injection of fungal elicitors such as zymosan or heat-inactivated *Saccharomyces cerevisiae* into larvae resulted in strongly enhanced lysozyme activity within the haemolymph and these immunized larvae exhibited increased survival rates after a second injection with a normally lethal dose of living yeast cells (Vilcinskas *et al.*, 1997d).

Further evidence confirming the putative contribution of a particular compound to immunity against fungi should be provided by experiments which elucidate the mechanism behind its antifungal activity. The mode of action of the currently identified antifungal proteins in insects is not at all clear and literature data imply critical questions. For example, how can the same compound inhibit growth of prokaryotic bacteria and eukaryotic fungi without being directly cytotoxic to host cells? How do pathogenic fungi avoid or overcome antifungal proteins? How can it be demonstrated that the antifungal activity of a particular compound detected *in vitro* is not restricted to artificial conditions but also exerted *in vivo*?

Answering these questions will open a new frontier in the research concerning insect antifungal proteins and provide better evaluation of their significance in humoral immune responses against fungi. Advances in elucidation of the mechanism behind the antifungal activity of lysozyme will be discussed in the following paragraph.

The antifungal activity of lysozyme was first detected against the human-pathogenic fungi *Coccidoides immitis* and *Candida albicans* by Collins and Pappagianis (1974) and Kamaya (1970). Later, it was suggested that cationic proteins including lysozyme may synergistically kill fungal cells (Lehrer *et al.*, 1975; Waldorf and Diamond, 1989). The fungistatic activity of lysozyme has been attributed to its capacity to induce the loss of potassium in fungal cells (Collins and Pappagianis, 1974), and this hypothesis has recently been confirmed. The antifungal effect of lysozyme against *S. cerevisiae* can be inhibited in a concentration-dependent manner by potassium or the specific potassium channel blocker tetraethylenammonium chloride (Vilcinskas and Matha, 1997a). In this context it is noteworthy that the lytic activity of lysozyme against bacteria is also dependent on the ionic strength of the medium (Gupta, D.K. *et al.*, 1987). Furthermore, it has been reported that basic proteins interact with phospholipid membranes, thereby influencing their permeability for ions (Kimmelberg and Papahadjopoulos, 1971). Induced efflux of potassium ions appears to be a plausible mode of action for the antifungal activity of lysozyme, because potassium plays a major role in the regulation of the ion concentration in fungal cells. Potassium loss through the fungal cell membrane interferes with a variety of metabolic processes, including glycolysis and respiration. The same mode of action is also attributed to commercially available fungicides such as nystatin, candicidin or amphotericin B. These compounds bind to sterols in the fungal membrane and cause flux of up to 90% of the intracellular potassium content (Gale, 1974; Hamond et al., 1974; Garraway and Evans, 1984).

A mechanism influencing the permeability of fungal cell membranes is putatively responsible for the antifungal properties of many, if not all, of the identified antifungal proteins. We suspect that this mode of action will prove to be the rule, rather than an exception, when the mechanisms of other antifungal proteins are investigated. The currently available data support this suggestion. The AFP isolated from *S. peregrina*, drosomysin, as well as cecropins and lysozyme, are relatively small cationic peptides. Cecropins have been shown to influence permeability of lipid bilayers by channel formation (Christensen et al., 1988). Insect defensins, another group of cationic antibacterial proteins in insects (Hoffmann and Hetru, 1992), are reported to induce voltage-dependent channels in bacterial cells, leading to rapid leakage of potassium (Cocianich et al., 1993).

We can expect that entomopathogenic fungi developed mechanisms that promote avoidance or inactivation of antifungal proteins. *B. bassiana* is reported to suppress the synthesis of antimicrobial proteins during mycosis in *G. mellonella*, but this could be a side effect of general destruction of host tissues (Vilcinskas and Matha, 1997b). Furthermore, it was shown that virulent strains of both fungi, *M. anisopliae* and *B. bassiana*, release proteases that digest lysozyme added to the medium of stationary or submerged cultures. In contrast, growth *in vitro* of a non-virulent form of *B. bassiana*, which exhibited reduced capacity to release proteases, was inhibited by lysozyme in a concentration-dependent manner (Vilcinskas and Matha, 1997a).

If proteases released by fungi during mycosis are also capable of inactivating antifungal peptides within the infected insect, more complex molecular host–pathogen interactions are predictable, because the activity of fungal proteases may, in turn, be influenced by protease inhibitors produced by the insect. In fact, the haemolymph of a number of insect species has been found to contain inhibitor activity against trypsin, chymotrypsin and elastase (Polanowski et al., 1992).

4.2. Inducible Protease Inhibitors

It has been outlined above that proteases released by parasitic fungi during mycosis play an important role in cuticle penetration and may contribute to suppression of cellular and humoral immune responses in the host insect. The virulence of particular entomopathogenic fungi has been reported to be determined by their capacity to release an appropriate quantity and spectrum of proteases (Bidochka and Khachatourians, 1990; Gupta, S.C. et al., 1994).

Therefore we can correspondingly speculate that the susceptibility of a particular insect to pathogens is influenced by its capacity to recruit endogenous protease inhibitors which inactivate microbial proteases. In contrast to AFPs, which directly affect fungal cells, host protease inhibitors may thus indirectly inhibit growth or propagation of invading fungi.

4.2.1. Contribution of Protease Inhibitors to Humoral Antifungal Defence

The putative contribution of particular protease inhibitors to humoral defence reactions of insects against fungi is supported by their detected increase in response to injected elicitors. *S. exigua* larvae challenged with vegetative cells of the entomopathogenic fungus *Nomuraea rileyi* exhibited enhanced inhibitory activity against several proteases (Boucias and Pendland, 1987). Injection of microbial elicitors into *G. mellonella* larvae enhanced antimicrobial activity and inhibitory effects against proteases released by *B. bassiana* or *M. anisopliae*. The time-dependent increase of lysozyme and cecropin-like activity in the haemolymph of challenged larvae correlated with an increase of haemolymph inhibitory activity against several protease types and protease mixtures tested (Wedde *et al.*, 1998).

The detected increase of protease inhibitors in the haemolymph during the humoral immune response alone does not itself demonstrate inducibility, because this term is restricted to molecules that are synthesized *de novo*. Enhanced protease inhibition could be due to stored molecules released by haemocytes in the course of an immune response. For example, the protein serpin, a member of a superfamily composed of a large number of serine protease inhibitors, is localized within the granules of granular cells in *M. sexta* (see Kanost *et al.*, 1995). This cell type is known to discharge the content of its intracellar granules during immune responses (Ratcliffe *et al.*, 1991; Ratcliffe, 1993).

Recently, evidence has been provided that neosynthesis of protease inhibitors is in fact induced during the humoral immune response in *G. mellonella*. Their synthesis *de novo* was demonstrated by inhibition of both transcription and ribosomal translation of protease inhibitors using actinomycin D and cycloheximide, respectively (Griesch, 1998). The relative increase of lysozyme and cecropin-like activity within the haemolymph correlated with an increase of inhibitory activity against several proteases, suggesting simultaneous induction of molecules responsible for antimicrobial and protease inhibitory activity (Wedde *et al.*, 1998).

Protease inhibitors purified from haemolymph of at least three insect species have been reported to inactivate exogenous fungal proteases. Three inhibitors acting on extracellular proteases of *M. anisopliae* which are toxic to *G. mellonella* were partially purified and characterized from the haemolymph of this insect species (Kucera, 1982, 1984). Inhibitory activity against fungal proteases was also discovered in the haemolymph of the silkworm *Bombyx mori* by Eguchi (1982). A group of protease inhibitors from the integument and haemolymph of *B. mori* inhibited proteases of *Aspergillus melleus* and *B. bassiana*. One of these inhibitors was shown to inhibit the germination of conidia and germ tube development of *B. bassiana*. This inhibitor of fungal proteases was later purified and characterized (Yoshida *et al.*, 1990; Eguchi *et al.*, 1993, 1994). A site-specific serine protease inhibitor extracted from haemolymph of resistant *Anticarsia gemmatalis* larvae inhibited germination *in vitro* of *N. rileyi* conidia and subsequent germ tube formation (Boucias and Pendland, 1987). Preparations containing purified heat and acid stable protease inhibitors from *G. mellonella* haemolymph inhibited significantly, but not completely, the activity of exogenous proteases released by *B. bassiana* and delayed growth of the fungus in medium containing proteins as a source of nutrients. This inhibitory activity on *B. bassiana* proteases and growth was enhanced in haemolymph samples obtained from pre-injected larvae. Correspondingly, pre-challenged *G. mellonella* larvae exhibited prolonged survival after subsequent injection of *B. bassiana* blastospores or topical application of its conidia, suggesting delayed development of mycosis in immunized larvae (Vilcinskas and Wedde, 1997).

However, the reports mentioned in the last paragraph share the same weakness, previously mentioned in the section on AFPs (Section 4.1). Detection *in vitro* of inhibition of extracellular fungal proteases and of fungal growth by protease inhibitors purified from a particular insect does not necessarily indicate that the inhibitor efficiently regulates the same proteases *in vivo*. But stringent proof of this will be difficult due to the complexity of the molecular interactions between parasitic fungi and their insect hosts. We suggest that inducibility of protease inhibitors should be considered to be a strong indication of an active role in insect immune responses regulating the proteolytic activity released by damaged cells, wounded tissues, or invading pathogens.

Advances in understanding the functions of inducible protease inhibitors in insects recently followed the detection of an inducible metalloprotease inhibitor (MPI) in the haemolymph of *G. mellonella* by Wedde *et al.* (1998). This type of specific protease inhibitor had not previously been reported from an invertebrate source (Kanost and Jiang, 1996). In contrast to all other protease inhibitors previously identified in insects, which represent constitutive molecules, the MPI was detectable only in immunized *G. mellonella* larvae or those infection with *M. anisopliae*. Topical application of *M. anisopliae* conidia induced neither enhanced lysozyme secretion nor detectable cecropin-like activity, but it increased inhibitory activity against the bacterial metalloprotease thermolysin. The

MPI was also shown to mediate insusceptibility to injected thermolysin, which is normally extremely toxic to *G. mellonella* larvae. Isolation and characterization of the MPI revealed a new type of inhibitor molecule (Wedde *et al.*, 1998). The assignment of a contribution of the MPI to humoral immune reactions is based on its inducibility during humoral immune responses. Furthermore, many groups of entomopathogens are reported to release metalloproteases, whereas their production by insects has so far not been documented. Attempts to detect metalloproteases in the haemolymph of *G. mellonella* yielded negative results (Griesch, 1998). This also supports our hypothesis that the MPI is targeted at microbial metalloproteases.

The entomopathogen *M. anisopliae* is reported to produce a thermolysin-like metalloprotease that is supposed to substitute for Prl, the first identified and characterized exogenous protease of this fungus, if this protease is inhibited by corresponding inhibitors within the infected host (St Leger *et al.*, 1994a). It is also reported that enhancin, a metalloprotease associated with particular entomopathogenic viruses, mediates their virulence and triggers fusion of the infectious virion with the cell membrane of host cells (Lepore *et al.*, 1996). Interestingly, we observed that both the entomopathogenic fungus *M. anisopliae* and a nuclear polyhedrovirus infecting *G. mellonella* induced the release of an MPI within the haemolymph during natural pathogenesis. Other humoral immune responses, such as induction of lysozyme or cecropin-like activity, were not detected during the same experiments (A. Vilcinskas *et al.*, unpublished observations). This observation raises the question of the recognition and inactivation of non-regulated proteases within the insect haemolymph, in cases when proteolytic activity is released by damaged cells, wounded tissues, or invading pathogens.

4.2.2. *Recognition and Inactivation of Non-regulated Proteases*

We will briefly summarize our recent attempts to elucidate the mechanism of the regulation of released proteases in the haemocoel of *G. mellonella*. Exploitation of host proteins as a source of nutrients for invading pathogens is obviously mediated by associated proteolytic microbial enzymes. Metalloproteases released by human pathogens have been reported strongly to affect host cells and even to inactivate endogenous inhibitors, thus allowing unrestricted activity of metabolic processes in the host which are normally under the control of a proteolytic cascade (Maeda, 1996). Because of the significant impact of non-regulated proteases on the homeostatic balance within an organism, we expected there to be an efficient mechanism in insects to recognize and inactivate dangerous non-regulated proteases.

The existence of such a mechanism was proven by experiments in which G. *mellonella* larvae were injected with either proteases or proteolytically digested haemolymph molecules. Injection of the metalloprotease thermolysin resulted in strong humoral immune responses accompanied by the release of inducible proteins such as lysozyme and protease inhibitors, e.g. an MPI. We assumed that non-regulated proteolytic activity is recognized by particular products of haemolymph protein hydrolysis. To prove this hypothesis, cell-free haemolymph from G. *mellonella* larvae was incubated *in vitro* with thermolysin. After removal of the metalloprotease by ultrafiltration, small molecule fractions (<3 kDa) were injected into recipient larvae. The antimicrobial and protease inhibitory activity within their haemolymph was measured 24 hours after injection. The same was done with haemolymph samples from immunized larvae containing the induced MPI. The results clearly indicated that treatment of haemolymph with thermolysin led to the formation of highly heat stable molecules with a molecular mass below 3 kDa which are very potent elicitors of humoral immune responses (Griesch, 1998).

These results provided a new aspect concerning recognition of 'foreignness'. If the products of non-regulated proteolytic degradation within the haemolymph of insects are really potent elicitors of immune responses, then one can postulate a new pathway of non-self recognition. This hypothesis explains, for example, why no or only limited proteolytic digestion of host haemolymph proteins is detectable during early development of pathogenic fungi in living hosts. Thrifty release of proteases by the fungus may help to avoid strong activation of host immune responses.

Identification of the molecules produced by proteolytic digestion of haemolymph proteins and studies on the mechanisms that lead to the induction of immune responses will be a major part of our future work.

4.3. Detoxification Proteins

As previously outlined, fungal parasites of insects are known to produce toxins which participate in the infection process. An increasing number of low molecular mass secondary metabolites is being reported to exhibit either insecticidal or immune suppressive activity (Roberts, 1981; Gillespie and Claydon, 1989; Khachatourians, 1991). The capacity to produce such toxins seems to determine the virulence of entomogenous fungi such as *M. anisopliae* (see Clarkson and Charnley, 1996). Correspondingly, the susceptibility or resistance of particular insect hosts against pathogenic fungi may also depend on their ability to inactivate fungal toxins. The literature on mechanisms responsible for detoxification of fungal toxins in insects is rather limited (Dowd, 1992). Recent advances in the research concerning destruxins and cyclosporin have provided new insights into the detoxification process of secondary metabolites released by *M. anisopliae* or *B. bassiana*.

Injection of destruxins in sublethal concentrations into *G. mellonella* larvae results in reversible tetanic paralysis (Roberts, 1981). The recovery of treated insects suggests a detoxification process. In fact, literature data indicate that these fungal cyclic peptides are detoxified in treated insects such as *G. mellonella* or *Locusta migratoria* via enzymatic hydrolysis. This leads to the formation of polar, linear metabolites which are not toxic. The major process in detoxification of destruxin E is the formation of destruxin–glutathione conjugates mediated by glutathione-S-transferase and subsequent biological degradation into cysteinyl–destruxin conjugates. Destruxin A is metabolized by enzymatic hydrolysis of the lactone function whereas in destruxin E hydrolysis occurs at the epoxide function (Jegorov et al., 1992; Loutelier et al., 1994, 1995).

Attempts to identify decomposition products of cyclosporin A after injection into *G. mellonella* larvae, which might suggest detoxification of this toxin, yielded negative results. In contrast to destruxins, cyclosporin A is inactivated by binding proteins in the haemolymph of this insect which mediate its transport to the fat body, where it is stored in an inactive form (Vilcinskas et al., 1997c).

Cyclosporin A is reported to interfere with a special kind of detoxification mechanism with particular importance in medical research. This secondary metabolite inhibits P-glycoprotein, a cell membrane-associated transporter molecule responsible for the so-called multidrug resistance of certain cancer cells (Sonneveld et al., 1992). P-glycoprotein functions as an adenosine triphosphate-dependent efflux pump which removes drugs from the cell before they have a chance to exert their cytotoxic effects. Over-expression of this pump in multidrug resistant cancer cells limits the effectiveness of many clinically important drugs. P-glycoprotein-mediated resistance of cancer cells against cytotoxic therapeutic agents can be modulated by a variety of compounds including cyclosporin A (Gottesman and Pastran, 1993; Bellamy, 1996). Immunohistochemical detection of P-glycoprotein in the human blood–brain barrier suggests that this pump restricts access of circulating drugs and xenobiotics to the central nervous system. A P-glycoprotein-related pump was also detected in the blood–brain barrier of a herbivorous insect which feeds on plants containing neurotoxic allelochemicals. It has been reported that such a pump probably accounts for the resistance of the tobacco hornworm *Manduca sexta* to certain neurotoxic alkaloids (nicotine, morphine, atropine) present in its plant diet (Murray et al., 1994). Since the host range of fungi producing cyclosporin A reportedly includes numerous herbivorous insects, a sophisticated role for this secondary metabolite in fungal pathogenesis can be postulated. If cyclosporin A inhibits P-glycoprotein-related pumps in herbivorous insects, its release by a parasitic fungus during infection could suppress resistance of the feeding host to toxic compounds in the food. Therefore cyclosporin A could mediate poisoning of infected hosts, thus promoting fungal development. P-glycoprotein probably contributes to the resistance of the integument against penetration by certain xenobiotics and thereby also to the resistance of particular insects against pesticides (Lanning et al., 1996). The putative significance of such a multidrug transporter for resistance against xenobiotics is supported by its detection in several invertebrates and vertebrates. Two genes encoding for P-glycoproteins were identified in *Drosophila melanogaster*. They shared high sequence similarity with the mammalian P-glycoprotein gene (Wu et al., 1991). Since parallels between multi-insecticide resistance in insects and multidrug resistance in mammalian tumour cells have been recognized, further research is required to explore the potential of fungal toxins such as cyclosporin A in medicine and insect pest control (Sonneveld et al., 1992; Murray et al., 1994).

5. COMPARISON WITH OTHER PATHOGENS AND PARASITES OF INSECTS

The vast majority of known entomogenous fungi is capable of infecting a large number of insect species, whereas the host range of entomopathogenic viruses, bacteria and protozoa is rather restricted. Among these groups of entomopathogens, only parasitic fungi can infect a susceptible insect host directly through the cuticle, whereas entomopathogenic viruses, bacteria and protozoa have to invade their host via the alimentary canal. Only parasitic nematodes can also penetrate the insect cuticle, e.g., at thin sites within the tracheal system (Bedding and Molyneux, 1982), and insect parasitoids may introduce their eggs into the insect body by transcuticular injection. At least two genera of nematodes, *Steinernema* and *Heterorhabditis*, are associated with symbiotic bacteria in the intestine of the free-living infective stage of these nematodes. Such nematodes have been reported to impair the humoral immune reactions of the infected host, thereby also protecting its symbiotic bacteria (Götz et al., 1981). Recently, it was confirmed that the virulence of the nematodes is synergistically enhanced by the symbiosis with bacteria (Ehlers et al., 1997). However, all pathogens or parasites which gain access to the haemocoel are exposed to the humoral and cellular defences of the host insect. Thus the ability to avoid or suppress immune responses is a prerequisite for survival and development of any insect pathogen.

It has been shown in this review that information about the suppression of insect humoral immune responses by fungi is relatively new, whereas other groups of entomopathogens have already been reported actively to counteract the immune mechanisms. For example, *Bacillus thuringiensis*, a bacterial entomopathogen which is used world-wide to control insect pests, has been shown to release inhibitors which interfere with the humoral defence system, thereby contributing to its virulence (Edlund *et al.*, 1976).

The effects of entomogenous bacteria and the above-mentioned nematodes on humoral immunity of infected hosts are probably due to released proteases capable of digesting antibacterial proteins (Dalhammar and Steiner, 1984). However, proteolytic digestion of haemolymph proteins by fungi in living insects has not yet been proven, although exogenous fungal proteases produced *in vitro* can digest immune proteins such as lysozyme (Vilcinskas, 1994). *G. mellonella* larvae infected with *B. bassiana* exhibit a reduced capacity to release immune proteins in response to injected microbial elicitors (Vilcinskas and Matha, 1997b). This suppression of humoral responses seems to be a side effect of impaired cellular immunity. Induction of apoptosis in phagocytic insect haemocytes has been reported for fungi (see Section 3.1.1) and a similarly sophisticated mechanism of immunosuppression has also been noted for *Microplis demolitor*, a parasitic wasp which attacks the larval stages of the lepidopteran *Pseudoplusia includens*. Female wasps inject their eggs, together with calyx fluid consisting of suspended proteins and polydnaviruses, into their hosts. The polydnavirus occurs only associated with the parasitic wasp and has been shown specifically to induce apoptosis in granular cells of the host. Granular cells contribute to multicellular encapsulation and would normally attack the developing eggs of the wasp. Thus induction of apoptosis in immune competent host cells promotes development of both *M. demolitor* eggs and associated viruses (Strand and Pech, 1995). Apoptotic granular cells of *P. includens* exhibited morphological alterations, such as blebs on the cell membrane and impaired spreading and cytoskeleton formation, similar to those observed in *G. mellonella* plasmatocytes which underwent programmed cell death during *M. anisopliae* infection. Suppression of the insect immune system, including humoral responses, by parasitic Hymenoptera has been extensively studied (Vinson, 1993). The current knowledge indicates parallels in the strategies of parasitic fungi and Hymenoptera to overcome host immune responses.

We have indicated in this review that inducible protease inhibitors probably contribute to humoral immune responses in insects and that natural fungal infection induces the release of a specific MPI in *G. mellonella* larvae (whereas other humoral reactions, such as induction of lysozyme, have not been detected in this species or in other insects). The MPI is also induced during infection of *G. mellonella* with nuclearpolyhedrovirus (A. Vilcinskas *et al.*, unpublished observations). Futhermore, an increase of protease inhibitory activity within the haemolymph of this insect was observed in response to infection with parasitic protozoa such as *Nosema* or *Vairimorpha* by Kucera and Weiser (1985). From all these observations we conclude that protease inhibitors are non-specifically induced in insects in reponse to infection. The degree of specificity of insect immune responses is at present obscure. Recent observations on *Drosophila* indicated that genes encoding antibacterial and antifungal proteins are independently regulated (Lemaitre *et al.*, 1995). Studies on the regulation of other antiparasitic immune responses in insect vectors of eukaryotic parasites such as protozoa or metazoa are at present rather limited, but may open the gate to the discovery of several mechanims of non-self recognition with different pathways for signal transduction leading to the expression of effector molecules.

REFERENCES

Absolom, D.R. (1988). The role of hydrophobicity in infection: bacterial adhesion and phagocytic ingestion. *Canadian Journal of Microbiology* **34**, 287–298.

Bedding, R.A. and Molyneux, A.S. (1982). Penetration of insect cuticle by infective juveniles of *Heterorhabditis* spp. (Heterorhabditidae: Nematoda). *Nematologica* **28**, 345–359.

Bellamy, W.T. (1996). P-glycoproteins and multidrug resistance. *Annual Reviews in Pharmacology and Toxicology* **36**, 161–183.

Bershadsky, D. and Vasilev, J.M. (1988). *Cytoskeleton*. New York and London: Plenum Press.

Bidochka, M.J. and Khachatourians, G.G. (1987). Hemocytic defense response to the entomopathogenic fungus *Beauveria bassiana* in the migratory grasshopper *Melanoplus sanguinipes*. *Journal of Invertebrate Pathology* **59**, 165–173.

Bidochka, M.J. and Khachatourians, G.G. (1988). Regulation of extracellular protease in the entomopathogenic fungus *Beauveria bassiana*. *Experimental Mycology* **12**, 161–168.

Bidochka, M.J. and Khachatourians, G.G. (1990). Identification of *Beauveria bassiana* extracellular protease as a virulence factor in pathogenicity toward the migratory grasshopper, *Melanoplus sanguinipes*. *Journal of Invertebrate Pathology* **56**, 362–370.

Bidochka, M.J., St Leger, R.J. and Roberts, D.W. (1997). Induction of novel proteins in *Manduca sexta* and *Blaberus giganteus* as a response to fungal challenge. *Journal of Invertebrate Pathology* **70**, 184–189.

Boman, D.G. and Hultmark, D. (1987). Cell-free immunity in insects. *Annual Reviews in Microbiology* **41**, 103–126.

Boucias, D.G. and Pendland, J. (1987). Detection of protease inhibitors in the hemolymph of resistant *Anticarsia gemmatalis* which are inhibitory to the entomopathogenic fungus, *Nomuraea rileyi*. *Experientia* **43**, 336–339.

Boucias, D. G. and Pendland, J. (1991a). Attachment of mycopathogens to cuticle: the initial event of mycosis in arthropod hosts. In: *The Fungal Spore and Disease*

Initiation in Plants and Animals (G. T. Cole and H.C. Hoch, eds), pp. 101–128. New York: Plenum Press.

Boucias, D.G. and Pendland, J. (1991b). The fungal cell wall and its involvement in the pathogenic process in insect hosts. In: *Fungal Cell Wall and Immune Response* (J.P. Latge and D.G. Boucias, eds), pp. 302–316. *NATO ASI Series* **H53** (23).

Boucias, D.G., Hung, S.-Y., Mazet, I. and Azbell, J. (1994). Effect of the fungal pathogen, *Beauveria bassiana*, on the lysozyme activity in *Spodoptera exigua* larvae. *Journal of Insect Physiology* **40**, 385–391.

Boucias, D.G., Mazet, I., Pendland, J. and Hung, S.-Y. (1995). Comparative analysis of the *in vivo* and *in vitro* metabolites produced by the entomopathogen *Beauveria bassiana*. *Canadian Journal of Botany* **73**, 1092–1099.

Brey, P.T., Lee, W.-J., Yamakawa, M., Koizumi, Y., Perrot, S., Francois, M. and Ashida, M. (1993). Role of the integument in insect immunity: epicuticular abrasion and induction of cecropin synthesis in cuticular epithelial cells. *Proceedings of the National Academy of Sciences of the USA* **90**, 6275–6279.

Brown, S. and Spudrich, J.A. (1979). Cytochalasin inhibits the rate of elongation of actin filaments. *Journal of Cell Biology* **83**, 657–662.

Butt, T.M., Hajek, A.E. and Humber, R.A. (1996). Gypsy moth immune defenses in response to hyphal bodies and natural protoplasts of entomophtoralean fungi. *Journal of Invertebrate Pathology* **68**, 278–285.

Cerenius, L., Thörnqvist, P.-O., Vey, A., Johannson, M.W. and Söderhäll, K. (1990). The effect of the fungal toxin destruxin E on isolated crayfish haemocytes. *Journal of Insect Physiology* **36**, 785–789.

Charnley, A.K. (1984). Physiological aspects of destructive pathogenesis in insects by fungi: a speculative overview. In: *Invertebrate–Microbial Interactions* (J.M. Anderson, D.M. Rayner and D.W. Walton, eds), pp. 229–270. London: Cambridge University Press.

Charnley, A.K. and St Leger, R.J. (1991). The role of cuticle-degrading enzymes in fungal pathogenesis in insects. In: *The Fungal Spore and Disease Initiation in Plants and Animals* (G.T. Cole and H.C. Hoch, eds), pp. 181–190. New York: Plenum Press.

Christensen, B., Fink, J., Merrifield, R.B. and Mauzerall, D. (1988). Channel-forming properties of cecropins and related model compounds incorporated into planar lipid membranes. *Proceedings of the National Academy of Sciences of the USA* **85**, 5072–5076.

Clarkson, J.M. and Charnley, A.K. (1996). New insights into the mechanisms of fungal pathogenesis in insects. *Trends in Microbiology* **4**, 197–203.

Cociancich, S., Ghazi, A., Hetru, C., Hoffmann, J.A. and Letellier, L. (1993). Insect defensin, an inducible antibacterial peptide, forms voltage dependent channels in *Micrococcus luteus*. *European Journal of Biochemistry* **268**, 19239–19245.

Cociancich, S., Bulet, P., Hetru, C. and Hoffmann, J.A. (1994). The inducible antibacterial peptides of insects. *Parasitology Today* **10**, 132–139.

Cohen, J. J. (1993). Apoptosis. *Immunology Today* **14**, 123–130.

Collins, M. S. and Pappagianis, D. (1974). Inhibition by lysozyme of the growth of the spherule phase of *Coccidoides immitis in vitro*. *Infection and Immunity* **10**, 613–623.

Dalhammar, G. and Steiner, H. (1984). Characterization of inhibitor A, a protease from *Bacillus thuringiensis* which degrades attacins and cecropins, two classes of antibacterial proteins in insects. *European Journal of Biochemistry* **139**, 247–252.

David, W.A. (1967). The physiology of the insect integument in relation to the invasion of pathogens. In: *Insects and Physiology* (J.W. Beament and J.E. Treherne, eds), pp. 17–35. Edinburgh: Oliver and Boyd.

Dowd, P. (1992). Detoxification of mycotoxins by insects. In: *Molecular Mechanisms of Insecticide Resistance: Diversity among Insects* (C.A. Mullin and J.G. Scott, eds), pp. 264–275. Washington DC: Chemical Society Press.

Dumas, C., Matha, V., Quiot, J.-M. and Vey, A. (1996a). Effects of destruxines, cyclic depsipeptide mycotoxins, on calcium balance and phosphorylation of intracellular proteins in lepidopteran cell lines. *Comparative Biochemistry and Physiology* **114C**, 213–219.

Dumas, C., Ravallec, M., Matha, V. and Vey, A. (1996b). Comparative study of the cytological aspects of the mode of action of destruxins and other peptidic fungal metabolites on target epithelial cells. *Journal of Invertebrate Pathology* **67**, 137–146.

Edlund, T., Siden, I. and Boman, H.G. (1976). Evidence for two inhibitors from *Bacillus thuringiensis* interfering with the host humoral defense system of saturniid pupae. *Infection and Immunity* **14**, 934–941.

Eguchi, M. (1982). Inhibition of fungal protease by hemolymph protease inhibitors of the silkworm *Bombyx mori*. *Applied Entomology and Zoology* **17**, 589–590.

Eguchi, M., Itoh, M., Chou, L.Y. and Nishino, D. (1993). Purification and characterization of a fungal protease specific protein inhibitor (FPI-F) in the silkworm haemolymph. *Comparative Biochemistry and Physiology* **104B**, 537–543.

Eguchi, M., Itoh, M., Nishino, K., Shibata, H., Tanaka, T., Kamei-Hayashi, K. and Hara, S. (1994). Amino acid sequence of an inhibitor from the silkworm (*Bombyx mori*) hemolymph against fungal protease. *Journal of Biochemistry* **115**, 881–884.

Ehlers, R.-U., Wulff, A. and Peters, A. (1997). Pathogenicity of axenic *Steinernema feltiae*, *Xenorhabdus bovienii*, and the bacto-helminthic complex to larvae of *Tipula oleracea* (Diptera) and *Galleria mellonella* (Lepidoptera). *Journal of Invertebrate Pathology* **69**, 212–217.

Fehlbaum, P., Bulet, P., Michaut, L., Lagueux, M., Broeckaert, W., Hetru, C. and Hoffmann, J.A. (1994). Septic injury of *Drosophila* induces the synthesis of a potent antifungal peptide with sequence homology to plant antifungal peptides. *Journal of Biological Chemistry* **269**, 33159–33163.

Fehlbaum, P., Bulet, P., Chernysh, S., Briand, J.-P., Roussel, J.-P., Letelier, L., Hetru, C. and Hoffmann, J.A. (1996). Structure–activity analysis of thanatin, a 21-residue inducible insect defense peptide with sequence homology to frog skin antimicrobial peptides. *Proceedings of the National Academy of Sciences of the USA* **93**, 1221–1225.

Ferron, P. (1981). Fungal control. In: *Microbial Control of Pest and Plant Diseases* (H.D. Burghes, ed.), pp. 465–482. London: Academic Press.

Flanagan, M.D. and Lin, S. (1980). Cytochalasins block actin filament elongation by binding to high affinity sites associated with F-actin. *Journal of Biological Chemistry* **255**, 835–838.

Gale, E.F. (1974). The release of potassium ions from *Candida albicans* in the presence of polyene antibiotics. *Journal of General Microbiology* **80**, 451–456.

Garraway, M.O. and Evans, R.C. (1984). *Fungal Nutrition and Physiology*. New York etc.: John Wiley and Sons.

Gillespie, A.T. and Claydon, N. (1989). The use of entomogenous fungi for pest control and the role of toxins in pathogenesis. *Pesticide Science* **27**, 203–215.

Gillespie, J.P. and Khachatourians, G.G. (1992). Characterization of the *Melanoplus sanguinipes* hemolymph after infection with *Beauveria bassiana*. *Comparative Biochemistry and Physiology* **103B**, 455–463.

Gillespie, J.P., Kanost, M.R. and Trenczek, T. (1997). Biological mediators of insect immunity. *Annual Reviews in Entomology* **42**, 611–643.
Goettel, M.S., St Leger, R.J., Rizzo, N.W., Staples, R.C. and Roberts, D. W. (1989). Ultrastructural localization of a cuticle-degrading protease produced by the entomopathogenic fungus *Metarhizium anisopliae* during penetration of host (*Manduca sexta*) cuticle. *Journal of General Microbiology* **135**, 2233–2239.
Gottesmann, M.M. and Pastran, I. (1993). Biochemistry of multidrug resistance mediated by the multidrug transporter. *Annual Reviews in Biochemistry* **62**, 385–427.
Götz, P. (1986). Encapsulation in arthropods. In: *Immunity in Invertebrates* (M. Brehelin, ed.), pp. 153–170. Berlin and Heidelberg: Springer Verlag.
Götz, P. (1988). Immunologie. Immunreaktionen bei Wirbellosen, insbesondere Insekten. *Verhandlungen der Deutschen Gesellschaft für Zoologie* **81**, 113–129.
Götz, P. (1991). Invertebrate immune response to fungal cell wall components. In: *Fungal Cell Wall and Immune Response* (J.P. Latge and D.G. Boucias, eds), pp. 317–329. *NATO ASI Series* **H53** (23).
Götz, P. and Boman, H.G. (1985). Insect immunity. In: *Comprehensive Insect Physiology, Biochemistry and Pharmacology* (G.A. Kerkut and L.J. Gilbert, eds), vol. 3, pp. 453–485. Oxford and New York: Pergamon Press.
Götz, P., Boman, A. and Boman, H.G. (1981). Interaction between insect immunity and an insect-pathogenic nematode with symbiotic bacteria. *Proceedings of the Royal Society* **212**, 333–350.
Griesch, J (1998). *Wirkung, Erkennung und Inaktivierung von Proteasen in der Hämolymphe der Großen Wachsmotte* Galleria mellonella *— ein neuer Mechanismus der Fremderkennung im Immunsystem von Insekten.* PhD Thesis, Institute of Zoology, Free University of Berlin.
Griesch, J. and Vilcinskas, A. (1998). Proteases released by entomopathogenic fungi impair phagocytic activity, attachment and spreading of plasmatocytes isolated from haemolymph of the greater wax moth, *Galleria mellonella*. *Biocontrol Science and Technology* **8**, 517–531.
Gross, J., Müller, C., Vilcinskas, A. and Hilker, M. (1998). Antimicrobial activity of exocrine gland secretions, hemolymph and larval regurgitate of the mustard leaf beetle, *Phaedon cochlearia*. *Journal of Invertebrate Pathology* **72**, 296–303.
Gunnarsson, S.G.S. (1988). Infection of *Schistocerca gregaria* by the fungus, *Metarhizium anisopliae*: cellular reactions in the integument studied by scanning electron and light microscopy. *Journal of Invertebrate Pathology* **52**, 9–17.
Gunnarsson, S.G.S. and Lackie, A.M. (1985). Haemocytic aggregation in *Schistocerca gregaria* and *Periplaneta americana* as a response to injected substances of microbial origin. *Journal of Invertebrate Pathology* **46**, 312–319.
Gupta, D.K., von Figura, K. and Hasilik, A. (1987). Conditions for a reliable application of the lysoplate method in the determination of lysozyme. *Clinica Chimica Acta* **165**, 73.
Gupta, S.C., Leathers, T.D., El-Sayed, G.N. and Ignoffo, C.M. (1994). Relationship among enzyme activities and virulence parameters in *Beauveria bassiana* and *Trichoplusia ni*. *Journal of Invertebrate Pathology* **64**, 13–17.
Guzo, D. and Stoltz, D.B. (1987). Observations on cellular immunity and parasitism in the tussock moth. *Journal of Insect Physiology* **33**, 19–31.
Hale, A.J., Smith, C.A., Sutherland, L.C., Stoneman, V.E., Longthorne, V.L., Culhane, A.C. and Williams, G.T. (1996). Apoptosis: molecular regulation of cell death. *European Journal of Biochemistry* **236**, 1–26.
Hammond, S.M., Lambert, P.A. and Kliger, B. (1974). The mode of action of

polyene antibiotics: induced potassium leakage in *Candida albicans*. *Journal of General Microbiology* **81**, 325–330.
Hayek, A. E. and St Leger, R. J. (1994). Interactions between fungal pathogens and insect hosts. *Annual Review of Entomology* **39**, 293–322.
Hoffmann, J.A. (1995). Innate immunity of insects. *Current Opinion in Immunology* **7**, 4–10.
Hoffmann, J.A. and Hetru, C. (1992). Insect defensins: inducible antibacterial peptides. *Immunology Today* **13**, 411–415.
Hoffmann, J.A., Reichhart, J.-M. and Hetru, C. (1996). Innate immunity in higher insects. *Current Opinion in Immunology* **8**, 8–13.
Hung, S.H. and Boucias, D. G. (1992). Influence of *Beauveria bassiana* on the cellular defense response of the beet army worm, *Spodoptera exigua*. *Journal of Invertebrate Pathology* **60**, 152–158.
Hung, S.H., Boucias, D.G. and Vey, A. (1993). Effect of *Beauveria bassina* and *Candida albicans* on cellular defense response of *Spodoptera exigua*. *Journal of Invertebrate Pathology* **61**, 179–187.
Huxham, I.M., Lackie, A.M. and McCorcindale, N.J. (1989). Inhibitory effects of cyclodepsipeptides, destuxines, from the fungus *Metarhizium anisopliae* on cellular immunity in insects. *Journal of Insect Physiology* **35**, 97–105.
Iijima, R., Kurata, S. and Natori, S. (1993). Purification, characterization, and cDNA cloning of an antifungal protein from the hemolymph of *Sarcophaga peregrina* (flesh fly). *Journal of Biological Chemistry* **268**, 12055–12061.
Janeway, C.A. (1994). The role of microbial pattern recognition in self:non-self discrimination in innate and adaptive immunity. In: *Phylogenetic Perspectives in Immunity: The Insect Host Defense* (J.A. Hoffmann, C.A. Janeway and S. Natori, eds), pp. 115–122. Boca Raton, Florida: CRC Press.
Jaynes, J. (1989). Peptides to the rescue. *New Scientist* **11**, 42–44.
Jegorov, A., Matha, V. and Weiser, J. (1990). Production of cyclosporins by entomopathogenic fungi. *Microbios Letters* **45**, 65–69.
Jegorov, A., Matha, V. and Hradec, J. (1992). Detoxification of destruxins in *Galleria mellonella*. *Comparative Biochemistry and Physiology* **103C**, 227–229.
Jegorov, A., Sedmera, P. and Matha, V. (1993). Biosynthesis of destruxins. *Phytochemistry* **33**, 1403–1405.
Jegorov, A., Sedmera, P., Matha, V., Simek, P., Zahradnickova, H., Landa, Z. and Eyal, J. (1994). Beauverolides L and La from *Beauveria bassiana* and *Paecilomyces fumosoroseus*. *Phytochemistry* **37**, 1301–1303.
Jolles, J., Schoentgen, F., Crozier, L. and Jolles, P. (1979). Insect lysozymes from three species of Lepidoptera: their structural relatedness to the C (chicken) type of lysozyme. *Journal of Molecular Evolution* **14**, 267–271.
Jung, Y.H., Park, B.Y., Lee, D.K., Hahn, Y.S. and Chung, J. (1995). Biochemical and molecular characterization of an antifungal protein from *Tenebrio molitor*. *Molecular and Cell Biology* **5**, 289–292.
Kamaya, T. (1970). Lytic action of lysozyme on *Candida albicans*. *Mycopathologia et Mycologia Applicata* **31**, 320–330.
Kanost, M.R. and Jiang, H. (1996). Proteinase inhibitors in invertebrate immunity. In: *New Directions in Invertebrate Immunology* (K. Söderhäll, K.S. Iwanaga and G.R. Vasta, eds), pp. 155–173. Fairhaven, New York: SOS Publications.
Kanost, M.R., Dai, M. and Dunn, P.E. (1988). Peptidoglycan fragments elicit antibacterial protein synthesis in larvae of *Manduca sexta*. *Archives of Insect Biochemistry and Physiology* **8**, 147–164.
Kanost, M.R., Kawooya, J.K., Law, J.H., Ryan, R.O., Van Heusden, M.C. and

Ziegler, R. (1990). Insect haemolymph proteins. *Advances in Insect Physiology* **22**, 299–380.
Kanost, M., Prasad, S., Huang, Y. and Willot, E. (1995). Regulation of serpin gene-1 in *Manduca sexta*. *Insect Biochemistry and Molecular Biology* **25**, 285–291.
Khachatourians, G.G. (1991). Physiology and genetics of entomopathogenic fungi. In: *Handbook of Applied Mycology*, vol. 2: *Humans, Animals and Insects* (D.K. Arora, L. Ajello and K.G. Mukerji, eds), pp. 613–663. New York: Marcel Dekker.
Khachatourians, G.G. (1996). Biochemistry and molecular biology of entomopathogenic fungi. In: *The Mycota*, vol. 6: *Human and Animal Relationships* (D.H. Howard and J.D. Miller, eds), pp. 331–363. Berlin–Heidelberg: Springer Verlag.
Kimmelberg, H.K. and Papahadjopoulos, D. (1971). Interactions of basic proteins with phospholipid membranes. *Journal of Biological Chemistry* **246**, 1142–1148.
Kodaira, Y. (1962). Studies on the toxic substances to insects, destruxin A and B, produced by *Oospora destructor*. *Agricultural and Biological Chemistry* **26**, 36–42.
Koidsumi, K. (1957). Antifungal action of cuticular lipids in insects. *Journal of Insect Physiology* **1**, 40–51.
Kopecky, J., Matha, A. and Jegorov, A. (1992). The inhibitory effect of destruxin A on the replication of arboviruses in *Aedes albopictus* C6/36 cell line. *Comparative Biochemistry and Physiology* **103C**, 23–25.
Kucera, M. (1982). Protease inhibitor of *Galleria mellonella* acting on the toxic protease from *Metarhizium anisopliae*. *Journal of Invertebrate Pathology* **38**, 33–38.
Kucera, M. (1984). Partial purification and properties of *Galleria mellonella* proteolytic inhibitors acting on *Metarhizium anisopliae* toxic protease. *Journal of Invertebrate Pathology* **43**, 190–196.
Kucera, M. and Weiser, J. (1985). Different course of proteolytic inhibitory activity and proteolytic activity in *Galleria mellonella* larvae infected by *Nosema algerae* and *Varimorpha heterosporum*. *Journal of Invertebrate Pathology* **45**, 41–46.
Lanning, C.L., Ayad, H.M. and Abou-Donia, M.B. (1996). P-glycoprotein involvement in cuticular penetration of [^{14}C]thiocarp in resistant tobacco budworms. *Toxicology Letters* **85**, 127–133.
Lee, S.Y., Moon, H.J., Kawabata, S.I., Kurata, S., Natori, S. and Lee, B.L. (1995). A sepecin homologue of *Holotrichia diomphalia*: purification, sequencing and determination of disulfide pairs. *Biological and Pharmaceutical Bulletin* **18**, 457–459.
Lehrer, R.I., Landra, K.M. and Hake, R.B. (1975). Nonoxidative fungicidal mechanisms of mammalian granulocytes: demonstration of components with candidacidal activity in human rabbit and guinea pig leucocytes. *Infection and Immunity* **11**, 1226–1234.
Lemaitre, B., Kromer-Metzger, E., Michaut, L., Nicolas, E., Meister, M., Georgel, P., Reichart, J.-M. and Hoffmann, J.A. (1995). A recessive mutation, immune deficiency (*imd*), defines two distinct pathways in the *Drosophila* host defense. *Proceedings of the National Academy of Sciences of the USA* **92**, 9465–9469.
Lemaitre, B., Nicolas, E., Michaut, L., Reichhart, J.-M. and Hoffmann, J.A. (1996). The dorsoventral regulatory gene cassette spätzle/Toll/cactus controls the potent antifungal response in *Drosophila* adults. *Cell* **86**, 973–983.
Leopold, J. and Samsinakova, A. (1970). Quantitative estimation of chitinase and several other enzymes in the fungus *Beauveria bassiana*. *Journal of Invertebrate Pathology* **15**, 34–42.
Lepore, L.S., Roelvink, P.R. and Granados, R.R. (1996). Enhancin, the granulosis

virus protein that facilitates nucleopolyhedrovirus (NPV) infections, is a metalloprotease. *Journal of Invertebrate Pathology* **68**, 131–140.

Levashina, E.A., Ohresser, S., Bulet, P., Reichhart, J.-M., Hetru, J.A. and Hoffmannn, J.A. (1995). Metchnikowin, a novel immune inducible proline-rich peptide from *Drosophila* with antibacterial and antifungal properties. *European Journal of Biochemistry* **233**, 694–700.

Loutelier, C., Lange, C., Cassier, P., Vey, A. and Cherton, J.-C. (1994). Non-extractive metabolism study of E and A destruxins in the locust, *Locusta migratoria* L. III. Direct high-performance liquid chromatographic analysis and parallel fast atom bombardment mass spectrometric monitoring. *Journal of Chromatography B* **656**, 281–292.

Loutelier, C., Marcual, A., Cassier, P., Cherton, J.-C. and Lange, C. (1995). Desorption of ions from locust tissues. III. Study of a metabolite of A-destruxin using fast-atom bombardment linked-scan mass spectrometry. *Rapid Communications in Mass Spectrometry* **9**, 408–412.

Maeda, H. (1996). Role of microbial proteases in pathogenesis. *Microbiology and Immunology* **40**, 685–699.

Matha, V., Grubhoffer, L. and Söderhäll, K. (1990a). New glucose and zymosan specific lectin from *Galleria mellonella* hemocyte lysate. In: *Lectins—Biology, Biochemistry, Clinical Biochemistry*, vol. 7, pp. 91–94. St. Louis, Missouri: Sigma Chemical Company.

Matha, V., Grubhoffer, L., Weyda, F. and Hermanova, L. (1990b). Detection of β-1,3-glucan-specific lectin on the surface of plasmatocytes, immunocompetent cells of greater wax moth, *Galleria mellonella*. *Cytobios* **64**, 35–42.

Matha, V., Jegorov, A., Weiser, J. and Pillai, J.S. (1992). The mosquitocidal activity of conidia of *Tolypocladium tundrense* and *Tolypocladium terricola*. *Cytobios* **69**, 163–170.

Mazet, I. and Boucias, D.G. (1996). Effects of the fungal pathogen, *Beauveria bassiana*, on protein synthesis of infected *Spodoptera exigua* larvae. *Journal of Insect Physiology* **42**, 91–99.

Mazet, I., Hung, S.Y. and Boucias, D.G. (1994). Detection of toxic metabolites in the hemolymph of *Beauveria bassiana* infected *Spodoptera exigua* larvae. *Experentia* **50**, 142–147.

Müller-Kögler, E. (1965). *Pilzkrankheiten bei Insekten*. Hamburg: Verlag Paul Parey.

Muroi, M. Shiragami, N. and Takasuki, A. (1994). Destruxin B, a specific and readily reversible inhibitor of vacuolar-type H^+-translocating ATPase. *Biochemical and Biophysical Research Communications* **205**, 1358–1365.

Murray, C.L., Quaglia, M., Arnason, J.T. and Morris, C.E. (1994). A putative nicotine pump at the metabolic blood–brain barrier of the tobacco hornworm. *Journal of Neurobiology* **25**, 23–34.

Ochiai, M. and Ashida, M. (1988). Purification of a β-1,3-glucan recognition protein in the prophenoloxidase activating system from hemolymph of the silkworm, *Bombyx mori*. *Journal of Biological Chemistry* **263**, 12056–12062.

Pendland, J. and Boucias, D.G. (1996). Phagocytosis of lectin-opsonized fungal cells and endocytosis of the ligand by insect *Spodoptera exigua* granular hemocytes: an ultrastructural and immunocytochemical study. *Cell and Tissue Research* **285**, 57–67.

Pendland, J., Hung, S.-Y. and Boucias, D.G. (1993). Evasion of host defense by *in vivo*-produced protoplast-like cells of the insect mycopathogen *Beauveria bassiana*. *Journal of Bacteriology* **175**, 5962–5969.

Pendland, J., Lopez-Lastra, D. and Boucias, D.G. (1994). Laminarin-binding sites on cell walls of the entomopathogen *Nomurea rileyi* associated with growth and adherence to host tissues. *Mycologia* **86**, 327–335.

Persson, M., Häll, L. and Söderhäll, K. (1984). Comparison of peptidase activities in some fungi pathogenic to arthropods. *Journal of Invertebrate Pathology* **44**, 342–348.

Polanowski, A., Wilusz, T., Blum, M., Escoubas, P., Schmidt, J. and Travis, J. (1992). Serine proteinase inhibitor profiles in the hemolymph of a wide range of insect species. *Comparative Biochemistry and Physiology* **102B**, 757–760.

Powning, R.F. and Davidson, W.J. (1976). Studies on insect bacteriolytic enzymes—II. Some physical and enzymic properties of lysozyme from hemolymph of *Galleria mellonella*. *Comparative Biochemistry and Physiology* **55B**, 221–228.

Quiot, J.-M., Vey, A. and Vago, C. (1985). Effect of mycotoxins on invertebrate cell lines *in vitro*. In: *Advances in Cell Culture* (K. Maramorosh, ed.), pp. 199–212. London: Academic Press.

Ratault, C. and Vey, A. (1977). Production d'estérase et de N-acétyl-β-D glucosaminidase dans l'intégument du coléoptère *Oryctes rhinoceros* par le champignon entomophthoène. *Entomophaga* **22**, 289–294.

Ratcliffe, N.A. (1993). Cellular defense responses of insects: unresolved problems. In: *Parasites and Pathogens of Insects* (S.N. Thompson and D.A. Frederici, eds), vol. 1, pp 267–304. London: Academic Press.

Ratcliffe, N.A., Brookman, J.L. and Rowley, A.F. (1991). Activation of prophenoloxidase cascade and initiation of nodule formation in locusts by bacterial lipopolysaccharides. *Developmental and Comparative Immunology* **15**, 33–39.

Richman, A. and Kafatos, F. (1995). Immunity to eukaryotic parasites in vector insects. *Current Opinion in Immunology* **8**, 14–19.

Rizki, T.M. and Rizki, R.M. (1986). Surface changes on hemocytes during encapsulation in *Drosophila melanogaster* Meigen. In: *Hemocytic and Humoral Immunity in Arthropods* (A.P. Gupta, ed.), pp. 157–189. New York: John Wiley and Sons.

Rizki, R.M. and Rizki, T.M. (1990). Parasitoid virus-like particles destroy *Drosophila* cellular immunity. *Proceedings of the National Academy of Sciences of the USA* **87**, 8388–8392.

Roberts, D.W. (1966). Toxins from entomogenous fungus *Metarhizium anisopliae* II. Symptoms and detection in moribund hosts. *Journal of Invertebrate Pathology* **8**, 222–227.

Roberts, D.W. (1981). Toxins of entomopathogenic fungi. In: *Microbial Control of Pests* (H.D. Burghes, ed.), pp. 441–464. London: Academic Press.

Roberts, D.W. and Hajek, A.E. (1992). Entomopathogenic fungi as bioinsecticides. In: *Frontiers in Industrial Mycology* (G.F. Leatham, ed.), pp. 144–159. New York: Chapman and Hall.

Roberts, D.W. and Humber, R.A. (1981). Entomogenous fungi. In: *The Biology of Conidial Fungi* (G.T. Cole and B. Kendrick, eds), vol. 2, pp. 201–236. New York: Academic Press.

Schliwa, M. (1986). *The Cytoskeleton — an Introductory Survey*. Vienna and New York: Springer Verlag.

Shimizu, S., Tsuchitani, Y. and Matsumoto, T. (1993a). Serology and substrate specificity of extracellular proteases from four species of entomopathogenic hyphomycetes. *Journal of Invertebrate Pathology* **61**, 192–195.

Shimizu, S., Tsuchitani, Y. and Matsumoto, T. (1993b). Production of an extracellular protease by *Beauveria bassiana* in the haemolymph of *Bombyx mori*. *Letters in Applied Micobiology* **16**, 291–294.

Smith, R.J. and Grula, E.A. (1982). Toxic components on the larval surface of the

corn earworm (*Heliothis zea*) and their effects on germination and growth of *Beauveria bassiana*. *Journal of Invertebrate Pathology* **39**, 15–22.
Söderhäll, K., Rögener, W., Söderhäll, I., Newton, R.P. and Ratcliffe, N.A. (1988). The properties and purification of a *Blaberus craniifer* plasma protein which enhances the activation of haemocyte prophenoloxidase by a β-1,3-glucan. *Insect Biochemistry* **18**, 323–330.
Söderhäll, K., Cerenius, L. and Johansson, M. (1996). The prophenoloxidase activating system in invertebrates. In: *New Directions in Invertebrate Immunology* (K. Söderhäll, S. Iwanaga and G.R. Vasta, eds), pp. 229–254. Fair Haven, New York: SOS Publications.
Sonneveld, P., Durie, B.G., Lokhorst, H.M., Marie, J.-P., Solbu, G., Suciu, S., Zittoun, R., Löwenberg, B. and Nooter, K (1992). Modulation of multiple drug-resistant myeloma by cyclosporin. *Lancet* **340**, 255–259.
St Leger, R.J. (1991). Integument as a barrier to microbial infections. In: *Physiology of the Insect Epidermis* (K. Binnington and I. Retnakan, eds), pp. 284–306. Melbourne: CSIRO Publications.
St Leger, R.J. and Bidochka, M.J. (1996). Insect–fungal interactions. In: *New Directions in Invertebrate Immunology* (K. Söderhäll, K.S. Iwanaga and G.R. Vasta, eds), pp. 443–479. Fairhaven, New York: SOS Publications.
St Leger, R.J., Charnley, A.K. and Cooper, R.M. (1986). Cuticle-degrading enzymes of entomopathogenic fungi: mechanisms of interaction between pathogen enzymes and insect cuticle. *Journal of Invertebrate Pathology* **47**, 295–302.
St Leger, R.J., Charnley, A.K. and Cooper, R.M. (1987). Characterization of cuticle-degrading proteases produced by the entomopathogen *Metarhizium anisopliae*. *Archives of Biochemistry and Biophysics* **253**, 221–232.
St Leger, R.J., Cooper, R.M. and Charnley, A.K. (1988). The effect of melanization of *Manduca sexta* cuticle on growth and infection by *Metarhizium anisopliae*. *Journal of Invertebrate Pathology* **52**, 459–470.
St Leger, R.J., Butt, T.M., Staples, R.C. and Roberts, D.W. (1989). Synthesis of proteins including a cuticle-degrading protease during differentiation of the fungus *Metarhizium anisopliae*. *Experimental Mycology* **13**, 253–262.
St Leger, R.J., Goettel, M., Roberts, D.W. and Staples, R.C. (1991). Prepenetration events during infection of host cuticle by *Metarhizium anisopliae*. *Journal of Invertebrate Pathology* **58**, 168–179.
St Leger, R.J., Frank, D.C., Roberts, D.W. and Staples, R.C. (1992). Molecular cloning and regulatory analysis of the cuticle-degrading protease structural gene from the entomopathogenic fungus, *Metarhizium anisopliae*. *European Journal of Biochemistry* **204**, 991–1001.
St Leger, R.J., Bidochka, M.J. and Roberts, D.W. (1994a). Isoforms of the cuticle-degrading Pr1 proteinase and production of a metalloproteinase by *Metarhizium anisopliae*. *Archives of Biochemistry and Biophysics* **313**, 1–7.
St Leger, R.J., Bidochka, M.J. and Roberts, D.W. (1994b). Characterization of a novel carboxypeptidase produced by the entomopathogenic fungus *Metarhizium anisopliae*. *Archives of Biochemistry and Biophysics* **314**, 392–398.
Strand, M.R. and Pech, L.L. (1995). *Microplis demolitor* polydnavirus induces apoptosis of a specific haemocyte morphotype in *Pseudoplusia includens*. *Journal of General Virology* **76**, 283–291.
Sugumaran, M. and Kanost, M.R. (1993). Regulation of insect hemolymph phenoloxidase. In: *Parasites and Pathogens of Insects* (N.E. Beckage, S.N. Thompson and B.A. Frederici, eds), vol. 1, pp. 317–342. San Diego: Academic Press.
Suzuki, A., Kawakami, K. and Tamura, S. (1971). Detection of destruxins in

silkworm larvae infected with *Metarhizium anisopliae*. *Insect Biochemistry and Molecular Biology* **23**, 43–46.

Tanada, Y. and Kaya, H.K. (1993). *Insect Pathology*. London: Academic Press.

Tanaka, T. (1987). Morphological changes in haemocytes of the host, *Pseudaletia separata*, parasitized by *Microplitis mediator* or *Apanteles kariyai*. *Developmental and Comparative Immunology* **11**, 57–67.

Trenzcek, T. (1992). Immunität bei Insekten. *Biologie in unserer Zeit* **22**, 212–217.

Trenczek, T. and Faye, I. (1988). Synthesis of immune proteins in primary cultures of fat body from *Hyalophora cecropia*. *Insect Biochemistry* **18**, 299–312.

Vey, A. and Götz, P. (1986). Antifungal cellular defence mechanism in insects. In: *Hemocytic and Humoral Immunity in Arthropods* (A.P. Gupta, ed.), pp. 89–115. New York: John Wiley and Sons.

Vey, A., Quiot, J.M., Vago, C. and Fargues, J. (1985). Effet immunodépresseur de toxines fongiques. Inhibition de la réaction d'encapsulement multicellulaire par les destruxines. *Comptes Rendues de l'Académie des Sciences (Paris), Série III* **300**, 647–651.

Vilcinskas, A. (1994). *Biochemische und immunologische Untersuchungen zur humoralen Abwehr von Pilzinfektionen bei Insekten am Beispiel der Großen Wachsmotte Galleria mellonella (Lepidoptera)*. PhD thesis, Institute of Zoologie, Free University of Berlin.

Vilcinskas, A. and Matha, V. (1997a). Antimycotic activity of lysozyme and its contribution to humoral antifungal defence reactions of *Galleria mellonella*. *Animal Biology* **6**, 19–29.

Vilcinskas, A. and Matha, V. (1997b). Effect of the entomopathogen *Beauveria bassiana* on the humoral immune response of *Galleria mellonella* larvae. *European Journal of Entomology* **94**, 461–472.

Vilcinskas, A. and Wedde, M. (1997). Inhibition of *Beauveria bassiana* proteases and fungal development by inducible protease inhibitors in the haemolymph of *Galleria mellonella* larvae. *Biocontrol Science and Technology* **7**, 591–601.

Vilcinskas, A., Matha, V. and Götz, P. (1997a). Inhibition of phagocytic activity of plasmatocytes isolated from *Galleria mellonella* by entomogenous fungi and their secondary metabolites. *Journal of Insect Physiology* **43**, 475–483.

Vilcinskas, A. Matha, V. and Götz, P. (1997b). Effects of the entomopathogenic fungus *Metarhizium anisopliae* and its secondary metabolites on morphology and cytoskeleton of plasmatocytes isolated from the greater wax moth *Galleria mellonella*. *Journal of Insect Physiology* **43**, 1149–1159.

Vilcinskas, A., Kopacek, P., Jegorov, A., Vey, A. and Matha, V. (1997c). Detection of lipophorin as the major cyclosporin-binding protein in the hemolymph of the greater wax moth, *Galleria mellonella*. *Comparative Biochemistry and Physiology* **117C**, 41–45.

Vilcinskas, A., Wedde, M., Griesch, J. and Götz, P. (1997d). Humoral immune reactions of insects against fungi. In: *Entomopathogenic Fungi — Fundamental and Applied Aspects* (Z. Landa, ed.), pp. 19–32. Czeske Budejovice: Czech Academy of Sciences.

Vilcinskas, A., Podsiadlowski, L. and Matha, V. (1997e). The role of fungal secondary metabolites in entomopathogenic processes. In: *Entomopathogenic Fungi — Fundamental and Applied Aspects* (Z. Landa, ed.), pp. 33–43. Czeske Budejovice: Czech Academy of Sciences.

Vilcinskas, A., Jegorov, A., Landa, Z. and Matha, V. (1998). Effects of beauverolide L and cyclosporin A on humoral and cellular immune response of the greater wax moth, *Galleria mellonella*. *Comparative Biochemistry and Physiology*, in press.

Vinson, S.B. (1991). Suppression of the insect immune system by parasitic

Hymenoptera. In: *Insect Immunity* (J.P.N. Pathak, ed.), pp. 171–187. Dordrecht, Boston and London: Kluwer Academic Publishers.

Walldorf, A.R. and Diamond, R.D. (1989). Aspergillosis and mucormycosis. In: *Immunology of Fungal Diseases* (R.D. Cox, ed.). Boca Raton, Florida: CRC Press.

Wedde, M. (1997). *Isolierung, Charakterisierung und Funktion eines induzierbaren Metalloprotease-Inhibitors der Großen Wachsmotte* Galleria mellonella. PhD thesis, Institute of Zoology, Free University of Berlin.

Wedde, M., Weise, C., Kopacek, P., Franke, P. and Vilcinskas, A. (1998). Purification and characterization of an inducible metalloprotease inhibitor from the hemolymph of greater wax moth larvae, *Galleria mellonella*. *European Journal of Biochemistry* **255**, 535–543.

Weiser, J. and Matha, V. (1988). The insecticidal activity of cyclosporines on mosquito larvae. *Journal of Invertebrate Pathology* **51**, 92–93.

Weiser, J., Matha, V., Zizka, Z. and Jegorov, A. (1989). Pathology of cyclosporin A in mosquito larvae. *Cytobios* **59**, 143–150.

Williams, D.H., Stone, M.J., Hauck, P.R. and Rahman, S.K. (1989). Why are secondary metabolites (natural products) biosynthesized? *Journal of Natural Products* **55**, 1189–1208.

Wu, C.T., Budding, M., Griffin, M.S. and Croop, J.M. (1991). Isolation and characterization of *Drosophila* multidrug resistance gene homologs. *Molecular and Cellular Biology* **11**, 3940–3948.

Yoshida, S., Yamashita, M., Yonehara, S. and Eguchi, M. (1990). Properties of fungal protease inhibitors from the integument and hemolymph of the silkworm and effect of an inhibitor on the fungal growth. *Comparative Biochemistry and Physiology* **95B**, 559–564.

Zacharuk, R.Y. (1970). Fine structure of the fungus *Metarhizium anisopliae* infecting three species of larval Elateridae. III. Penetration of the host integument. *Journal of Invertebrate Pathology* **15**, 372–396.

Index

Page numbers in *italic* indicate illustrations and tables.

adhesion and host cell interaction, *Neospora caninum* 73–7, *87*, *88*
African trypanosomes 117–19
AIDS, cryptosporidiosis 136
Aleurain-like proteinases 244
American trypanosomiasis 119–20, 121
amoebapain 132–3
amoebiasis 131–2, 133
Ancylostoma caninum, proteinases *204*, 212, 214
    cathepsin B occluding loop 238
    cDNA, cysteine proteinase genes 213
Ancylostoma duodenale, proteinases 212
Anisakis, proteinases 201, 208–9
Anisakis simplex, proteinases *203*
anti-schistosome vaccine, schistosome calpain as target 177–8
antifungal proteins identified in insects 292–3
antipain 116–17
antiprotozoan drugs, protease inhibitors as 110
Apicomplexa
    adhesion and invasion of host cells 68–9, 69–73
    secretory organelles in adhesion, invasion and intracellular development 71–3
    see also Neospora caninum
apoptosis, induced, entomopathogenic fungi 283–4, *285*, 302
archaeon putative cysteine proteinases, relationship, papain superfamily 226, *227*
Ascaridoidea, proteinases 201, *202–3*, 207–9

Ascaris suum, proteinases *202*, 207, 238
asparaginyl endopeptidases 187
    Fasciola hepatica 193–4
    schistosome legumain 186–8
aspartic protease inhibitors as antimalarials 130, 131
aspartic proteinases 107, 108
    cathepsin D *168*, 184–6
    cathepsin D-like, trichostrongylid spp. laboratory models 221
    Plasmodium falciparum 127–8
    Taenia 198
aspartyl proteinases
    Caenorhabditis elegans 224
    Dirofilaria immitis 211
    Toxoplasma circumcincta 220

B1 gene, PCR diagnosis, *T. gondii* 66
Beauveria bassiana 270, 271, 281
    pathogenesis 276–7, 279
beauverolides *282*, 286–7
blastospores 275
bovine neosporosis 68
    abortion 48, 50, 58
    congenital, clinical signs 58–9
bradyzoites
    Neospora caninum 51–2, 53–4, 54, *86*
    Neospora and *Toxoplasma* 53–4
    Toxoplasma gondii 51, 52, 53–4
branch A, papain superfamily 231, *232*, *235*, 237–40
    cathepsin B-like proteinases 237–9
    cathepsin C-like proteinases 239–40
    see also cathepsin B; cathepsin C; cathepsin X

branch B papain superfamily 231, *232*, *236*, 240–4
  cathepsin L-like proteinases 107–8, 241–2
    Haemonchus 216
    Heterodera glycines 224
    Schistosoma japonicum 167
    schistosome cercariae 169–70
    Toxocara 207
  cruzipain-like proteinases 240–1
Brugia pahangi, proteinases *203*
Brugia, proteinases 211

Caenorhabditis elegans 206, 224–5
  cathepsin B
    genes 238
    occluding loop 238, 239
  cathepsin B-like proteinases 237
  cruzipain-like proteinases 241
calpain *168*
  functions 174
  large subunit 176, *177*
  phylogenic analysis, evolutionary divergence 225, 226
  schistosomes 174, 176–8
cathepsin B 107, *168*
  Fasciola hepatica 192–3
  occluding loop 238–9
  schistosomes 178–80
cathepsin B-like proteinases 107–8, 237–9
  Caenorhabditis elegans 24
  Haemonchus contortus 219, 237, 238
  Schistosoma mansoni 166
  schistosome cercariae 169–70
cathepsin C 182–4, 231
  Fasciola hepatica 193
cathepsin C-like proteinases 239–40
  Giardia 239–40
  Haemonchus 215
cathepsin D *168*, 184–6
  amino acid sequences *185*
cathepsin D-like aspartic proteinase, trichostrongylid laboratory models 221
cathepsin H proteinase 244
cathepsin L 107, 166, 190–2
  L1 *168*, 180–1
  L2 *168*, 181–2
cathepsin L-like proteinases 107–8, 241–2
  Ancylostoma caninum 213
  Haemonchus 216
  Heterodera glycines 224

  Schistosoma japonicum 167
  schistosome cercariae 169–70
  Toxocara 207
cathepsin X 231
cathepsin X-like proteinases 240
cats, *Neospora caninum* experimental infection 55
cattle
  Neospora caninum experimental infection 85–7
  neosporosis 68
    bovine abortion 48, 50, 58
    congenital, clinical signs 58–9
cecropins 279, 293, 295
cell fusion, evidence for, *Trypanosoma brucei* 16, 17
cell surface-associated proteins, *Neospora caninum* tachyzoites, identification and characterization 80–3
cellular defence system, insect 269–70
cercarial elastase, schistosomes *168*, 170–3
  functions 170, 171–2
  genes encoding 170–1, 171–2
  immunological studies 173
cestodes, proteinases of 196–200
Chagas disease 119–20
  treatment target, cruzain as 121
chloroquine resistance, malaria 123
chromosomal recombination, *Trypanosoma brucei*, genetic exchange 14
chymostatin 124, 130
Clan CA, family C1 papain superfamily, classification *233–4*
Clans, cysteine proteinases 225
clindamycin, neosporosis 58
Clonorchis sinensis, proteinases 196
CNS, neosporosis 54, 65, 68
collagenase activity, *Entamaeba histolytica* 135
complement
  activation, *Entamaeba histolytica* 134
  schistosomes, resistance to 169, 173
congenital
  neosporosis, subclinical 58
  toxoplasmosis 138
conidia, entomopathogenic fungi 271–2
  attachment 273
  formation 280
  host cuticle penetration 273–4
conoid, cell invasion, apicomplexan parasites 71
Crithidia fasciculata, experimental genetic exchange 20

cruzain (cruzipain) 120–2
  function 121
  gene encoding 120–1
cruzipain-like proteinases 240–1
Cryptosporidium, proteases *109*
  Cryptosporidium parvum 136–7
cutaneous leishmaniasis 110–11
cyclosporins 287–8
  cyclosporin A *282*, 287, 301–2
cysteine protease inhibitors
  African trypanosomiasis, potential treatment 118
  as antimalarials 131
  antipain and leupeptin 116–17, 130
  blocking *Cryptosporidium parvum* sporozoite invasion 137
  Entomoeba histolytica, potential for treatment 134–5
  fluoromethyl ketone cysteine protease inhibitors 121–2, 130, 136
cysteine proteases 107
  Entomoeba histolytica 132–5
  Giardia lamblia 136
  Leishmania 114–17
  M_r 35–40,000 cysteine protease 125
  M_r 68,000 cysteine protease 125
  Plasmodium falciparum 125, 127, 128, 129
  Trichomonas vaginalis 137–8
  Trypanosoma brucei 117–18
  Trypanosoma cruzi 120–2
cysteine proteinases 245–6
  Caenorhabditis elegans 224–5
  Dirofilaria immitis 210
  Haemonchus 216, 218
  Heterodera glycines 223–4, 242
  hypothetical evolution and diversification 243
  legumains *see* legumains
  Plasmodium 236
  Schistosoma mansoni 166
  schistosomes 174, 176–8
  Trichomonas circumcincta 219, 220
  Trichomonas foetus 234–5
  Trichomonas vaginalis 234–5
  Taenia 198
  Toxocara 207–8
  trichostrongylid laboratory models 221
  Trichuris 222
  see also cathepsin B; cathepsin C; cathepsin L; cruzain (cruzipain)

cysteine proteinases of papain superfamily, phylogenetic relationships *see* papain superfamily, phylogenetic relationships
cytochalasins 286
cytokines, *Neospora caninum* infections 84
dense granules
  Neospora and *Toxoplasma* 53, 73
  proteins, *Neospora caninum* 80–1
destruxins 281, 283-*5*
  injection of sublethal concentration into *Galleria mellonella* 300
  molecular structure of A and E *282*
  promotion of apoptosis 283–4, *285*
Dictyocaulus viviparus 204, 214, 215
Dictyostelium discoideum 240, 241
digenean trematodes 164–5, 245
  see also Fasciola hepatica; *Schistosoma japonicum*; *Schistosoma mansoni*; schistosomes
dipeptidylpeptidase *see* cathepsin C
Diplostomum pseudospathaceum 196
Dirofilaria immitis 203
  aspartyl proteinases 211
  cysteine proteinases 210, 211
  metalloproteinases 210
disease control, implications, genetic exchange in trypanosomatids 31–2
DNA content, *Trypanosoma brucei* hybrids 8, 9, *10*
dogs
  Neospora caninum histopathological effects 68
  neosporosis 57–8
drosomycin 292, 292–3, 295
drug resistance, *Trypanosoma brucei* experimental crosses 18

E-64, cysteine protease inhibitor 130
Echinococcus granulosus, proteinases *197*, 198–9
EL2 gene 172
ELISA, *Neospora caninum* 50, 60–2, 78–9
endopeptidases, definition 106–7
enhancin 298
Entamoeba, cysteine proteinase genes 235–6
Entamoeba dispar 132
  non-pathogenicity 133

Entamoeba histolytica 131–2
   pathogenicity 134
   proteases *109*, 135
      cysteine 132–5
   virulence factors 133, 134, 134–5
entomogenous fungi, proteinases 289–90
entomopathogenic fungi 270–1
   comparison with other pathogens and parasites of insects 301–3
   development in insect host *277*
   fungal factors determining virulence 280–90
      fungal toxins 280–8
      proteases, interactions with insect immune systems 288–90
   pathogenesis 271–80
      adhesion and germination, fungal spores 271–3
      host cuticle penetration 273–4, 288–9
      interactions with insect immune system 275–9
'epidemic' population structure, *T. brucei* 26
epidemiological
   and ecological factors, genetic exchange in *Trypanosoma brucei* 26
   implications, genetic exchange in trypanosomatids 31–2
ER60 proteinase *168*, 174, 188–9
ERFNIN motif, cysteine proteinases 230–1, 234–7
erythrocyte invasion and rupture, *Plasmodium* proteases mediating 123–6
eukaryotic serine proteases 107
exopeptidases 107, 182, 184
   see also cathepsin C
exoskeleton of insects, fungal ability to invade 273–4, 288–9

falcipain 127, 128, 129
   inhibition 130
families, cysteine proteinases 225
Fasciola gigantica, proteinases 194
Fasciola hepatica, proteinases 190–4
   asparaginyl endopeptidases 193–4
   cathepsin B 192–3, 239
   cathepsin B-like genes 238
   cathepsin L 190–2
   cathepsin L-like proteinases 241
   dipeptidylpeptidase (DPP) and leucine aminopeptidase (LAP) 193
   other proteinases 193–4

fasciolosis, liver pathology 191
Filarioidea, proteinases *203*, 209–11
fluoromethyl ketone cysteine protease inhibitors 121–2, 130, 136
fungal
   proteases, interactions with insect immune system 288–90
   toxins 280-*5*, 289
      beauverolides *282*, 286–7
      cyclosporins *282*, 287–8
      cytochalasins 286
      destruxins 281, *282*, 283-*5*
      detoxification of 299–301
fungi, entomopathogenic 270–80
   pathogenesis 271–80
      adhesion and germination of spores 271–3
      host death, saprophytic development and conidia formation 280
      insect immune system, interactions with 275–9, 288–90
      molecular interactions with host *272*
      penetration of cuticle 273–4, 288–9
   see also insect immune system

Galleria mellonella 270
   plasmocytes, healthy larvae and *Metarhizium anisopliae* infected larvae *278*
   protease inhibitor neosynthesis, humoral antifungal response 296
   susceptibility to destruxins 281
gene duplication, *Trypanosoma brucei* 32–3
genetic exchange
   evidence from population genetics analysis, *Trypanosoma vivax*, *Trypanosoma evansi*, *Trypanosoma equiperdum* 28–9
   population genetics analysis, *T. congolense* 27–8
genetic exchange, Trypanosomatidae 1–35
   laboratory experiments 3–20
Giardia
   cathepsin B occluding loop 239
   cathepsin C-like proteinases 239–40
   proteases *109*
Giardia lamblia 135–6
   cysteine protease genes 136
giardiasis 135–6
Globodera pallida 206, 223, 223–4

glycocalyx release, schistosomal
        metamorphosis 169, 171, 173
granular cells, insect immune system 269–70

haematocytes, insect immune system 269–70
haemoglobin degradation
    plasmodial proteases mediating 126–30
    schistosomes
        catabolism *175*
        exoproteinases 184
        proteases 174, 178–9, 183
Haemonchus
    cysteine proteinases, potential vaccine
        candidates 218
    galactose-containing glycoprotein
        complex 217
    H11 217
Haemonchus contortus 214, 215–19
    geographical strain divergence 219
    immunization against 216–17
    proteinases *204*, 212–13, 214
        cathepsin B occluding loop 238
        cathepsin B-like 219, 237, 238
        cathepsin C-like 239
Haplometra cylindracea, proteinases 195
Heligmosomoides polygyrus, proteinases *206*
Heligmospiroides polygyrus, laboratory model
        systems 220–1
helminths, parasitic
    cestodes 196–200
    nematodes 200–25
    trematodes 164–5, 190–4, 245
        schistosomes see *Schistosomatium
            douthitti*; *Schistosoma japonicum*;
            Schistosoma mansoni; schistosomes
        see also *Fasciola hepatica*
Heterodera glycines *206*, 223–4, 242
histolysin 132
holtricin-3 293
hookworm infection 211
human infectivity, trypanosomatids 31–2
humoral immune reactions of insects against
        fungi 269, 290–301
    detoxification proteins 299–301
    inducible antifungal proteins 291–5
    inducible protease inhibitors 295–9
        recognition and de-activation of non-
            regulated proteases 298–9
        inhibition 302
hybridization and human infectivity,
        trypanosomatids 31–2

hyphal bodies 275
    phagocytosis of 276

IFN-gamma, *Neospora caninum* infection 84
IL-10, acute murine toxoplasmosis 84
IL-12, *Neospora caninum* infection 84
immune protein synthesis, insect 269
immunosuppression, neosporosis,
        reactivation 58
indirect fluorescence antibody technique
        (IFAT) *Neospora caninum* 59–60
induced apoptosis, entomopathogenic fungi
        283–4, *285*, 302
infectivity, human, *Trypanosoma brucei*,
        experimental crosses 17–18
insect defensins 295
insect immune system 268–70
    cellular immunity 269–70
        destruxins, effects 283
        evasion 275–7, 279
        suppression by invading fungi 289
    humoral immunity 269
        against fungi 290–301
        destruxins, effects 284
        evasion 279
    interactions with parasitic
        fungal proteases 288–90
    fungi 275–9
iscom ELISA, *Neospora caninum* 61, 78–9
ITS 1, PCR target sequence, *Neospora* 66

kala-azar 110
kallikrein-like proteinase, schistosomes *168*,
        189–90
kinetoplast DNA, *Trypanosoma brucei*,
        genetic exchange 11–14

legumains
    Fasciola hepatica 193–4
    schistosomes *168*, 186–8
Leishmania
    genetic exchange, evidence for 30–1
    hybrids, natural occurrence 30
    infectivity 113, 116
    ploidy 22
    proteases *109*, 110–17
        cysteine 114–17
        metalloprotease 111–14
Leishmania mexicana, cysteine protease genes
        114–15
leishmaniasis 2, 110–11

leishmanolysin 111–14
  activity, control of 112–13
  function 113–14
  genes encoding 112
leucine aminopeptidase (LAP)
  Fasciola hepatica 193
  schistosomes 167
leupeptin 116–17, 130
lymphopains 241
lysozyme 274, 279, 292, 293–4, 295
  potassium loss in fungal cells 294–5

malaria 122–3
malaria parasites 122–31
  proteases identified, erythrocyte invasion
    and rupture 125–6
  proteases mediating haemoglobin
    degradation 126–30
  proteases as potential chemotherapy
    targets 130–1
mating compatibility, *Trypanosoma brucei*
  experimental crosses 4–6
maxicircles, *Trypanosoma brucei* 11, *12*, 13
meiosis, *Trypanosoma brucei* 16, 17, 25–6
melanization, insect immune system 270, 274
Meloidogyne incognita, proteinases *206*, 223
Mendelian inheritance, *Trypanosoma brucei*
  genetic exchange 8
metalloproteases 107, 108
  Leishmania 111–14
  Trypanosoma cruzi 122
metalloproteinase inhibitor, *Galleria*
  mellonella 297–8
metalloproteinases
  Ancylostoma caninum 212, 213
  Ancylostoma duodenale 212
  Brugia 211
  Dictyocaulus 215
  Dirofilaria immitis 210
  Haemonchus 215
  Metarhizium anisopliae 290, 298
  Necator americanus 212
  Onchocerca 209
  Schistosomatium douthitti 170
  schistosomes 167
  Strongyloides stercoralis (Ss40) 201
  Taenia 198
  Teladorsagia circumcincta 219–20
  Trichinella 222
  trichostrongylid laboratory models 221
  Trichostrongylus 220

Trichuris 222
Metarhizium anisopliae 270, 271, 281
  conidia 272
  pathogenesis 276–7, 279
  secondary metabolites *see* cytochalasins;
    destruxins
Metastrongylidae 214
metchnikowin 293
micronemes, apicomplexans 71–2
  Neospora 52–3
minicircles, *Trypanosoma brucei* 11, *12*, 13
molecular karyotype analysis, *Trypanosoma*
  brucei genetic exchange 14
monoclonal antibodies (mAbs), against
  Neospora caninum tachyzoites 77–8
monogenean ectoparasites 245
mouse models, *Neospora caninum* infection
  55–6
M_r 35–40,000 cysteine protease 125
M_r 37,000 protease 125–6
M_r 68,000 cysteine protease 125
M_r 75,000 serine protease 125
M_r 76,000 serine protease 125
mucosal leishmaniasis 111
multicellular encapsulation, insect immunity
  270
  fungal cells, survival of 276
murine models, neosporosis 55–6, 83–4, 85

Nc5-PCR 67
Nc-p33 81, 82–3
Nc-p36 81, 82
  polyclonal antibodies against 65
Nc-p43 81, 81–2, 82
  polyclonal antibodies against 65
NCDG1 80
NCDG2 80–1
Necator americanus, proteinases *203*, 211–12
nematodes
  free-living, proteinases *206*
  parasitic, proteinases 200–25
    of plants *206*, 223–5
Neospora caninum
  adhesion and host cell interactions 68–9,
    73–7, *87*, *88*
  historical background and phylogenic
    status 49–51
  host-parasite interactions 67–8
  in vitro
    cultivation 56–7
    isolation 62–4

INDEX

life cycle 51–2
morphology and ultrastructure 52–4
tachyzoites 51, 52, 68, *86*
  interactions with host cells 73–7
  mAbs 77–9
  polyclonal antisera 80–3
  proteins, intracellular and cell-surface associated 77–83
  tissue cysts 53–4
    histology and immunohistochemistry 64–5
Neospora and *Toxoplasma*
  phylogeneticla relations 50–1
  secretory organelles 52–3
neosporosis 48, 52, 57–67
  acute, histopathological lesions 64–5
  in cattle 48, 50, 58–9, 68
  clinical signs 57–9
  diagnostic tools 59–67
    direct detection techniques 62–7
    indirect detection 59–62
  in dogs 68
    treatment 58
  immune reaction to 68
  immunology 83–7
  mouse models 55–6, 83–4, 85
  natural infection 54, 59
  serological diagnosis 50
  transplacental transmission 50, 52, 54–5
neutral thiol proteinase (NTP), *P. westermani* 194, 195
Nippostrongylus brasiliensis
  laboratory model systems 220–1
  proteases *205*

Onchocerca cervicalis, proteinases *203*
Onchocerca lienalis, proteinases *203*
Onchocerca, proteinases 209–10
Onchocerca volvulus, proteinases *203*, 240
onchocercal dermatitis 209–10
oral infection, neosporosis 54
Ostertagia ostertagi 205, 219, 220
  cathepsin B occluding loop 238
  cathepsin B-like genes 237

p30 (*SAG1*) gene, PCR diagnosis, *Toxoplasma gondii* 66
papain superfamily, phylogenic relationships 226–30
  branch A 231, *232, 235*, 237–40
  branch B *232, 236*, 240–4

clan CA, family C1, classification 226–7, *233–4*
ERFNIN motif 230–1, 234–7
evolution 227, 234
  hypothetical diversification of cysteine proteinases *243*
  table *228–30*
  uprooted neighbour-joining phylogenic tree *232, 235, 236*
Paragonimus westermani
  immune tolerance 195
  proteinases 194–5, 241
parasitophorous vacuole membrane (PVM), *Neospora* and *Toxoplasma* 53
pattern recognition receptors, insect immune system 269
pepstatin 130
peptide protease inhibitors, antimalarial effects 130
Phaedon cochleariae 273
phagocytosis, insect cellular immune system 270, 276
phenotypic traits, inheritance, *Trypanosoma brucei* 17–19
phylogenetic
  methods, population genetics studies, trypanosomatids 24–5
  tree
    papain superfamily *232, 235, 236*
    Trypanosomatidae 35
plasmatocytes, insect immune system 269, *278*
plasmepsin I and II 127–8
Plasmodium
  cysteine proteinases 236
  food vacuole proteases 127–30
  micronemal proteins 71–2
  proteases *109*
    mediating erythrocyte invasion and rupture 123–6
    mediating haemoglobin degradation 126–30
    as potential chemotherapeutic targets 130–1
  rhoptry proteins 72
Plasmodium falciparum 122–3, 127–8, 129
  haemoglobin degradation 127
ploidy, *Trypanosoma brucei*, genetic exchange 9–11, 22
polyclonal antibodies, against Nc-p43 and Nc-p36 65

polyclonal antisera, identification and characterization, intracellular and cell surface-associated *Neospora caninum* tachyzoite proteins 80–3
polymerase chain reaction (PCR)
  Neospora and *Toxoplasma* 66–7
  Neospora-specific 50, 67
  Toxoplasma gondii 66
population genetics and evolutionary studies, trypanosomatids 20–31
  evidence for genetic exchange 25–31
  methods 21–5
propeptide 246
protease, M_r 37,000 125–6
protease inhibitors
  as antiprotozoan drugs 110
  contribution to insect humoral antifungal defence 296–8
proteases, classification 106–8
  see also aspartic proteases; cysteine proteases; metalloproteases; serine proteases
proteasome 108
protein disulfide isomerase (PDI) family, ER60 188–9
Proteocephalus, proteinases *197*, 199
proteolytic enzymes, evolution of 163–4
protoplast 275
protozoan parasites
  Cryptosporidium 109, 136–7
  Giardia lamblia 135–6
  infectious diseases caused 106
  protease inhibitors as antiprotozoan drugs 110
  proteases 105–39
    survey of identified 108–10
  Trichomonas 109, 137–8, 234–5
  see also African trypanosomes; *Entamoeba histolytica*; malaria parasites; *Toxoplasma gondii*; *Trypanosoma cruzi*
pyrimethamine, neosporosis 58

recombinant antigen ELISA, *Neospora caninum* 61–2
recombination and linkage methods, trypanosomatids, population genetics studies 23–4
Rhabditoidea, proteinases 201, *202*
rhoptries 53, 72

salivarian trypanosomes 34
Schistocephalus, proteinases *199*
Schistocephalus solidus, proteinases *197*
Schistosoma japonicum
  cathepsin B-like proteinase 238
  cathepsin C 183–4
  cathepsin C-like proteinase 239
  cathepsin D 184, *185*, 186
  cathepsin L1 180–1
  cathepsin L 166
  cathepsin L-like proteinases 167, 241, 242
  cruzipain-like proteinase 241
  Sj32 186
  SjCL2 181–2
Schistosoma mansoni
  amino acid sequences *185*
  calpain 176, *177*
  cathepsin B-like proteinases 166, 238
  cathepsin C, cDNA 183
  cathepsin C-like proteinases 239
  cathepsin D 184, 186
  cathepsin L2, SmCL2 181–2
  cathepsin L-like activity 167
  cathepsin L-like proteinases 241–2
  cruzipain-like proteinases 241
  leucine aminopeptidase activity 167
  Sm31 178–80
  Sm32 186–7
  Sm480 167
  SmCL1 181
  SmCL2 181–2
  SmSP1 189
Schistosomatium douthitti 170
schistosomes
  adults and schistosomules, proteinases 174–90
    calpain 174, 176–8
    catabolism of haemoglobin *175*
    cathepsin C 182–4
    cathepsin D 184–6
    cathepsin L1 180–1
    cathepsin L2 181–2
    cathepsin-B 178–80
    cercarial elastase 170–3
    ER60 188–9
    kallikrein-like proteinase 189–90
  cercariae, proteinases 169–73
  eggs and miracidia, proteinases 165–7
  legumain *168*, 186–8
  proteinases, gene and/or cDNA reported in literature *168*

proteolytic pathway *175*
vaccination studies 178
schistosomiasis, liver pathogenesis 166
schistosomular surface m28 172–3
SDS-PAGE, *Neospora caninum* host-parasite interactions 81
secretory organelles, apicomplexan parasites 52–3, 71–3
segregation tests and ploidy, trypanosomatids, population genetics studies 22–3
serine proteases 107
  malarial parasites 125, 126
  M_r 75,000 serine protease 125
  M_r 76,000 serine protease 125
  Trypanosoma brucei 118–19
serine proteinase inhibitor, insect humoral antifungal response 297
serine proteinases
  Anisakis 208–9
  Dictyocaulus viviparus 215
  Dirofilaria immitis 210
  Necator americanus 212
  Onchocerca 209
  schistosomes 167, 169
  Teladorsagia circumcincta 219, 220
  Toxocara canis 207
  Toxocara 207
  Trichinella 222
  trichostrongylid laboratory models 221
  Trichostrongylus 220
  Trichuris 222
serpin 296
sheep, *Neospora caninum* experimental infection 85
sleeping sickness 31, 32
sparganosis 200
Spirometra, proteinases *197*, 199–200
ssrRNA gene, PCR detection, *Neospora caninum* 66–7
Strongyloidea *203–4*, 211–14
Strongyloides ratti 202, 237
Strongyloides simiae, proteinases *202*
Strongyloides spp., proteinases 201
Strongyloides stercoralis 202, 237
Strongylus vulgaris, proteinases *204*, 213–14
sulfadiazine, neosporosis 58
surface glycoproteins, variant, *Trypanosoma brucei* experimental crosses 18–19
Syngamus trachea, proteinases *204*, 214

tachyzoites
  Neospora caninum 51, 52, 54, *86*
  cell destruction 68
  in vitro cultivation 56–7
  physical interaction with host cells 73–7
Neospora and *Toxoplasma* 52–3
Taenia, proteinases 196, *197*, 198
Teladorsagia circumcincta 205, 219–20
tenectin-3 293
thanatin 293
thermolysin 299
thermolysin-like metalloproteinase 290, 298
tissue cysts, *Neospora caninum* 51–2, 54, 55–6
Toxocara canis 207
  proteinases *202*, 207–8, 244
toxocariasis 207
Toxoplasma gondii 138
  adhesion and invasion of host cells *in vitro* 69–70
  dense granules 73
  evasion of endocytic pathway 70
  histology and immunohistochemistry 64–5
  life cycle, infective stages 52
  micronemal proteins 72
  morphology and ultrastructure 52–4
  PCR diagnosis 66
  proteases *109*
  rhoptry proteins 72–3
  secretory organelles 52–3
  tachyzoite–host cell interactions, adhesion 76–7
  tissue cysts 53–4
    histology and immunohistochemistry 64–5
  transmission electron microscopy 62, *63*
toxoplasmosis 52
  immunology, acute murine 84
transmission electron microscopy (TEM), *Neospora* and *Toxoplasma* 62, *63*
transplacental transmission, neosporosis 50, 52, 54–5
trematodes, digenean 164–5
  see also Chlonorchis; *Fasciola hepatica*; *Paragonimus*; schistosomes
triatomine bug 19, 20
Trichinella spiralis 206, 221, 222–3
Trichomonas, proteases *109*
Trichomonas vaginalis 137–8, 234–5
trichostrongylid laboratory models, proteinases 220–1

Trichostrongyloidea *204–6*, 214–21
Trichostrongylus colubriformis *205*, 220
Trichostrongylus vitrinus, proteinases *205*, 220
Trichuris muris *206*, 222
Trichuris suis *206*, 222
Trichuroidea 221–3
triploid progeny, Trypanosoma brucei genetic exchange 9, 11
Tritrichomonas foetus 234–5
trypanopain 117–18
Trypanosoma brucei
    African trypanosomiasis 117
    leishmanolysin homologue, genes encoding 119
    proteases *109*
        cysteine (trypanopain) 117–18
        serine 118–19
Trypanosoma brucei, genetic exchange 2
    experimental crosses 3–19
        experimental design 3–4
        mating compatibility 4–6
        published crosses 5
    experimental versus population genetics analysis 32–4
        biological considerations 33–4
        molecular considerations 32–3
    hybridization 31
    location and timing in tsetse fly 6–8
    mechanisms
        changes in ploidy 9–11
        kDNA, inheritance 11–14
        Mendelian inheritance of markers 8
        models of genetic exchange 14–17
        molecular karyotype analysis 14
    phenotypic traits, inheritance 17–19
    ploidy 9–11, 22
    population genetics analysis 25–7
Trypanosoma congolense, genetic exchange 27–8
Trypanosoma cruzi 119–22
    cysteine protease *109*, 120–2
    ploidy 22

population genetics analysis, evidence for genetic exchange 29–30
Trypanosoma equiperdum 28–9
Trypanosoma evansi 28–9
Trypanosoma (Megatrypanum) conorhini 19
Trypanosoma rangeli 19–20
Trypanosoma vivax 28–9
trypanosomatids, genetic exchange 35
    biological context 31–5
    epidemiology and disease control, implications 31–2
    phylogenetic perspective 34-5
    experimental versus population genetics analysis 32–4
    phylogenetic perspective 34-5
    population genetics and evolutionary studies 20–31
        evidence for genetic exchange 25–31
        phylogenetic methods 24–5
        recombination and linkage studies 23–4
        segregation tests and ploidy 22–3
    see also Trypanosoma brucei, genetic exchange
trypanosomes
    African 117–19
        other proteases 118–19
    salivarian, evolution 34
trypanosomiasis 2
    African 117
    American 119–20, 121
tsetse fly
    mixed trypanosomatid infection 33–4
    Trypanosoma brucei, genetic exchange, location and timing 6–8

uprooted neighbour-joining phylogenetic tree, papain superfamily *232*, *235*, *236*
Urechis caupo 240

vaccination studies, schistosome calpain 178
vinyl sulfone inhibitors, falcipain 130
visceral leishmaniasis 110

Contents of Volumes in This Series

Volume 31

Parasitic Infections in Women and their Consequences
    L. Brabin and B. J. Brabin
The Pathophysiology of Malaria
    N. J. White and M. Ho
The Interaction of *Leishmania* Species with Macrophages
    J. Alexander and D. G. Russell
The Effects of Trypanosomatids on Insects
    G. A. Schaub
Echinococcus multilocularis Infection: Immunology and Immunodiagnosis
    B. Gottstein
Nematodes as Biological Control Agents: Part II
    I. Popiel and W. M. Hominick

Volume 32

Blastocystis in Humans and Animals: Morphology, Biology, and Epizootiology
    P. F. L. Boreham and D. J. Stenzel
Giardia and Giardiasis
    R. C. A. Thompson, J. A. Reynoldson and A. H. W. Mendis
Immunology of Leishmaniasis
    F. Y. Liew and C. A. O'Donnell
Transport of Nutrients and Ions across Membranes of Trypanosomatid Parasites
    D. Zilberstein
The Biology of Fish Coccidia
    A. J. Davies and S. J. Ball
The Sexuality of Parasitic Crustaceans
    A. Raibaut and J. P. Trilles

Volume 33

The Treatment of Human African Trypanosomiasis
    J. Pépin and F. Milord
Plasmodium Species Infecting *Thamnomys rutilans*: a Zoological Study
    I. Landau and A. Chabaud
Metacercarial Excystment of Trematodes
    B. Fried
The Minor Groups of Parasitic Platyhelminthes
    K. Rohde
Sarcoptes scabiei and Scabies
    I. Burgess

Volume 34

Molecular Studies for Insect Vectors of Malaria
    J. M. CRAMPTON
The Ribosomal RNA Genes of *Plasmodium*
    A. P. WATERS
Molecular Mimicry
    R. HALL
Relationships Between Chemotherapy and Immunity in Schistosomiasis
    P. J. BRINDLEY
Regulatory Peptides in Helminth Parasites
    D. W. HALTON, C. SHAW, A. G. MAULE AND D. SMART
Bait Methods for Tsetse Fly Control
    C. H. GREEN

Volume 35

Chemotherapy of Nematode Infections of Veterinary Importance, with Special Reference to Drug Resistance
    GEORGE A. CONDER AND WILLIAM C. CAMPBELL
Parasites as Indicators of Water Quality and the Potential Use of Helminth Transmission in Marine Pollution Studies
    K. MACKENZIE, H. H. WILLIAMS, B. WILLIAMS, A. H. MCVICAR AND R. SIDDALL
Variation in *Echinococcus*: Towards a Taxonomic Revision of the Genus
    R. C. A. THOMPSON, A. J. LYMBERY AND C. C. CONSTANTINE
How Schistosomes Profit From the Stress Responses They Elicit in Their Hosts
    MARIJKE DE JONG-BRINK
Myiasis of Humans and Domestic Animals
    MARTIN HALL AND RICHARD WALL
Parasitism and Parasitoidism in Tarsonemia (Acari: Heterostigmata) and Evolutionary Considerations
    MAREK KALISZEWSKI, FRANÇOISE ATHIAS-BINCHE AND EVERT E. LINDQUIST

Volume 36

Rare, New and Emerging Helminth Zoonoses
    J. D. SMYTH
Population Genetics of Parasitic Protozoa and Other Microorganisms
    M. TIBAYRENC
The Biology of Fish Haemogregarines
    A. J. DAVIES
The Taxonomy and Biology of Philophthalmid Eyeflukes
    P. M. NOLLEN AND I. KANEV
Human Lice and Their Management
    I. F. BURGESS
Ticks and Lyme Disease
    C. E. BENNETT

Volume 37

Nitric Oxide and Parasitic Disease
   I. A. CLARK AND K. A. ROCKETT
Molecular Approaches to the Diagnosis of Onchocerciasis
   J. E. BRADLEY AND T. R. UNNASCH
The Evolution of Life History Strategies in Parasitic Animals
   R. POULIN
The Helminth Fauna of Australasian Marsupials: Origin and Evolutionary Biology
   I. BEVERIDGE AND D. M. SPRATT
Malarial Parasites of Lizards: Diversity and Ecology
   J. J. SCHALL

Volume 38

Intracellular Survival of Protozoan Parasites with Special Reference to *Leishmania* spp., *Toxoplasma gondii* and *Trypanosoma cruzi*
   J. MAUËL
Regulation of Infectivity of *Plasmodium* to the Mosquito Vector
   R. E. SINDEN, G. A. BUTCHER, O. BILLKER AND S. L. FLECK
Mouse–Parasite Interactions: from Gene to Population
   C. MOULIA, N. LE BRUN AND F. RENAUD
Detection, Screening and Community Epidemiology of Taeniid Cestode Zoonoses: Cystic Echinococcosis, Alveolar Echinococcosis and Neurocysticercosis
   P. S. CRAIG, M. T. ROGAN AND J. C. ALLAN
Human Strongyloidiasis
   D. I. GROVE
The Biology of the Intestinal Trematode *Echinostoma caproni*
   B. FRIED AND J. E. HUFFMAN

Volume 39

Clinical Trials of Malaria Vaccines: Progress and Prospects
   C. A. FACER AND M. TANNER
Phylogeny of the Tissue Cyst-forming Coccidia
   A. M. TENTER AND A. M. JOHNSON
Biochemistry of the Coccidia
   G. H. COOMBS, H. DENTON, S. M. A. BROWN AND K.–W. THONG
Genetic Transformation of Parasitic Protozoa
   J. M. KELLY
The Radiation-attenuated Vaccine against Schistosomes in Animal Models: Paradigm for a Human Vaccine?
   P. S. COULSON

Volume 40

Part 1 *Cryptosporidium parvum* **and related genera**

Natural History and Biology of *Cryptosporidium parvum*
  S. TZIPORI AND J. K. GRIFFITHS
Human Cryptosporidiosis: Epidemiology, Transmission, Clinical Disease, Treatment, and Diagnosis
  J. K. GRIFFITHS
Innate and Cell-mediated Immune Responses to *Cryptosporidium parvum*
  C. M. THEODOS
Antibody-based Immunotherapy of Cryptosporidiosis
  J. H. CRABB
Cryptosporidium: Molecular Basis of Host–Parasite Interaction
  H. WARD AND A. M. CEVALLOS
Cryptosporidiosis: Laboratory Investigations and Chemotherapy
  S. TZIPORI
Genetic Heterogeneity and PCR Detection of *Cryptosporidium parvum*
  G. WINDMER
Water-borne Cryptosporidiosis: Detection Methods and Treatment Options
  C. R. FRICKER AND J. H. CRABB

Part 2 *Enterocytozoon bieneusi* **and Other Microsporidia**

Biology of Microsporidian Species Infecting Mammals
  E. S. DIDIER, K. F. SNOWDEN AND J. A. SHADDUCK
Clinical Syndromes Associated with Microsporidiosis
  D. P. KOTLER AND J. M. ORENSTEIN
Microsporidiosis: Molecular and Diagnostic Aspects
  L. M. WEISS AND C. R. VOSSBRINCK

Part 3 *Cyclospora cayetanensis* **and related species**

Cyclospora cayetanensis
  Y. R. ORTEGA, C. R. STERLING AND R. H. GILMAN

Volume 41

Drug Resistance in Malaria Parasites of Animals and Man
  W. PETERS
Molecular Pathobiology and Antigenic Variation of *Pneumocystis carinii*
  Y. NAKAMURA AND M. WADA
Ascariasis in China
  PENG WEIDONG, ZHOU XIANMIN AND D. W. T. CROMPTON
The Generation and Expression of Immunity to *Trichinella spiralis* in Laboratory Rodents
  R. G. BELL
Population Biology of Parasitic Nematodes: Applications of Genetic Markers
  T. J. C. ANDERSON, M. S. BLOUIN AND R. M. BEECH
Schistosomiasis in Cattle
  J. DE BONT AND J. VERCRUYSSE

Volume 42

The Southern Cone Initiative Against Chagas Disease
  C. J. SCHOFIELD AND J. C. P. DIAS
Phytomonas and Other Trypanosomatid Parasites of Plants and Fruit
  E. P. CAMARGO
Paragonimiasis and the Genus *Paragonimus*
  D. BLAIR, Z.-B. XU AND T. AGATSUMA
Immunology and Biochemistry of *Hymenolepis diminuta*
  J. ANREASSEN, E. M. BENNET-JENKINS AND C. BRYANT
Control Strategies for Human Intestinal Nematode Infections
  M. ALBONICO, D. W. T. CROMPTON AND L. SAVIOLI
DNA Vaccines: Technology and Applications as Anti-parasite and Anti-microbial Agents
  J. B. ALARCON, G. W. WAINE AND D. P. MCMANUS

Volume 43

Genetic Exchange in the *Trypanosomatidae*
  W. GIBSON AND J. STEVENS
The Host–Parasite Relationship in Neosporosis
  A. HEMPHILL
Proteases of Protozoan Parasites
  P. J. ROSENTHAL
Proteinases and Associated Genes of Parasitic Helminths
  J. TORT, P. J. BRINDLEY, D. KNOX, K. H. WOLFE AND J. P. DALTON
Parasitic Fungi and their Interactions with the Insect Immune System
  A. VILCINSKAS AND P. GÖTZ